Springer Optimization and Its Applications

VOLUME 118

Managing Editor
Panos M. Pardalos (University of Florida)

Editor–Combinatorial Optimization
Ding-Zhu Du (University of Texas at Dallas)

Advisory Board
J. Birge (University of Chicago)
C.A. Floudas (Texas A & M University)
F. Giannessi (University of Pisa)
H.D. Sherali (Virginia Polytechnic and State University)
T. Terlaky (Lehigh University)
Y. Ye (Stanford University)

Aims and Scope
Optimization has been expanding in all directions at an astonishing rate during the last few decades. New algorithmic and theoretical techniques have been developed, the diffusion into other disciplines has proceeded at a rapid pace, and our knowledge of all aspects of the field has grown even more profound. At the same time, one of the most striking trends in optimization is the constantly increasing emphasis on the interdisciplinary nature of the field. Optimization has been a basic tool in all areas of applied mathematics, engineering, medicine, economics, and other sciences.

The series *Springer Optimization and Its Applications* publishes undergraduate and graduate textbooks, monographs and state-of-the-art expository work that focus on algorithms for solving optimization problems and also study applications involving such problems. Some of the topics covered include nonlinear optimization (convex and nonconvex), network flow problems, stochastic optimization, optimal control, discrete optimization, multi-objective programming, description of software packages, approximation techniques and heuristic approaches.

More information about this series at http://www.springer.com/series/7393

Lina Mallozzi • Egidio D'Amato
Panos M. Pardalos

Editors

Spatial Interaction Models

Facility Location Using Game Theory

 Springer

Editors

Lina Mallozzi
Department of Mathematics
  and Applications
University of Naples Federico II
Napoli, Italy

Egidio D'Amato
Department of Industrial and Information
  Engineering
Second University of Naples
Aversa, Italy

Panos M. Pardalos
Department of Industrial and Systems
  Engineering
University of Florida
Gainesville, FL, USA

ISSN 1931-6828   ISSN 1931-6836   (electronic)
Springer Optimization and Its Applications
ISBN 978-3-319-52653-9   ISBN 978-3-319-52654-6   (eBook)
DOI 10.1007/978-3-319-52654-6

Library of Congress Control Number: 2017934484

Mathematics Subject Classification (2010): 90B80, 90B85, 91A10, 91A12

Printed on acid-free paper

This Springer imprint is published by Springer Nature
The registered company is Springer International Publishing AG
The registered company address is: Gewerbestrasse 11, 6330 Cham, Switzerland

Preface

This book aims to provide a comprehensive overview of facility location models that can be investigated in a game theoretical environment. Facility location theory develops the idea of locating one or more facilities optimizing suitable criteria such as minimizing transportation cost or capturing the largest market share, and a huge number of papers have been devoted to this research.

In this volume, we focus on situations where the location decision is faced by several decision makers, leading to a game theoretical framework in a noncooperative way, as well as in a cooperative one. Some chapters are surveys of models and methods regarding this part of the facility location using game theory; other chapters illustrate applications in different contexts such as economics, engineering, and physics. This makes the book useful for a broad audience of researchers working on theory, applications, and computational aspects of location problems.

We would like to express our thanks to all the contributors of chapters and also the valuable assistance of Springer for the publication of this book.

Napoli, Italy
Aversa, Italy
Gainesville, FL, USA

Lina Mallozzi
Egidio D'Amato
Panos M. Pardalos

Contents

Bilevel Models on the Competitive Facility Location Problem 1
Necati Aras and Hande Küçükaydın

**Partial Cooperation in Location Choice: Salop's Model
with Three Firms** 21
Subhadip Chakrabarti and Robert P. Gilles

A Class of Location Games with Type Dependent Facilities 39
Imma Curiel

**Location Methods and Nash Equilibria for Experimental
Design in Astrophysics and Aerospace Engineering** 53
Elia Daniele, Pierluigi De Paolis, Gian Luca Greco,
and Alessandro d'Argenio

Leader-Follower Models in Facility Location 73
Tammy Drezner and Zvi Drezner

Asymmetries in Competitive Location Models on the Line 105
H.A. Eiselt and Vladimir Marianov

Huff-Like Stackelberg Location Problems on the Plane 129
José Fernández, Juana L. Redondo, Pilar M. Ortigosa,
and Boglárka G.-Tóth

A Game Theoretic Approach to an Emergency Units Location Problem .. 171
Vito Fragnelli, Stefano Gagliardo, and Fabio Gastaldi

**An Equilibrium-Econometric Analysis of Rental Housing
Markets with Indivisibilities** 193
Mamoru Kaneko and Tamon Ito

Large Spatial Competition 225
Matías Núñez and Marco Scarsini

Facility Location Situations and Related Games in Cooperation 247
Osman Palanci and S. Zeynep Alparslan Gök

**Sequential Entry in Hotelling Model with Location Costs:
A Three-Firm Case** 261
Stefano Patrí and Armando Sacco

Nash Equilibria in Network Facility Location Under Delivered Prices 273
Blas Pelegrín, Pascual Fernández, and Maria D. García

Sharing Costs in Some Distinguished Location Problems 293
Justo Puerto

Contributors

S. Zeynep Alparslan Gök Faculty of Arts and Sciences, Department of Mathematics, Suleyman Demirel University, Isparta, Turkey

Necati Aras Boğaziçi University, Istanbul, Turkey

Subhadip Chakrabarti Economics Department, Queen's Management School, Queen's University of Belfast, Belfast, UK

Imma Curiel Anton de Kom University of Suriname, Paramaribo, Suriname

Elia Daniele Fraunhofer IWES Institute for Wind Energy and Energy System Technology, Oldenburg, Germany

Alessandro d'Argenio Italian Air Force - Flight Test Centre, Pomezia (RM), Italy

Pierluigi De Paolis Italian Air Force - Flight Test Centre, Pomezia (RM), Italy

Tammy Drezner California State University, Fullerton, CA, USA

Zvi Drezner California State University, Fullerton, CA, USA

H.A. Eiselt Faculty of Business Administration, University of New Brunswick, Fredericton, NB, Canada

José Fernández Department of Statistics and Operations Research, University of Murcia, Espinardo, Murcia, Spain

Pascual Fernández Department of Statistics and Operations Research, University of Murcia, Murcia, Spain

Vito Fragnelli Università del Piemonte Orientale, Alessandria, Italy

Stefano Gagliardo Università degli Studi di Genova, Genova, Italy

Maria D. García San Antonio Catholic University of Murcia, Murcia, Spain

Fabio Gastaldi Università del Piemonte Orientale, Alessandria, Italy

Robert P. Gilles Economics Department, Queen's Management School, Queen's University of Belfast, Belfast, UK

Gian Luca Greco Italian Air Force - Flight Test Centre, Pomezia (RM), Italy

Tamon Ito Saganoseki Hospital, Oita, Japan

Mamoru Kaneko Waseda University, Tokyo, Japan

Hande Küçükaydın MEF University, Istanbul, Turkey

Vladimir Marianov Department of Electrical Engineering, Pontificia Universidad Católica de Chile, Santiago, Chile

Matías Núñez LAMSADE, Université Paris Dauphine, Paris, France

Pilar M. Ortigosa Department of Informatics, University of Almería, Almería, Spain

Osman Palanci Faculty of Arts and Sciences, Department of Mathematics, Suleyman Demirel University, Isparta, Turkey

Stefano Patrí Department of Methods and Models for Economics, Territory and Finance, Sapienza University, Rome, Italy

Blas Pelegrín University of Murcia, Murcia, Spain

Justo Puerto Instituto de Investigación Matemática de la Universidad de Sevilla (IMUS), Edificio Celestino Mutis, Universidad de Sevilla, Sevilla, Spain

Juana L. Redondo Department of Informatics, University of Almería, Almería, Spain

Armando Sacco Department of Methods and Models for Economics, Territory and Finance, Sapienza University, Rome, Italy

Marco Scarsini Dipartimento di Economia e Finanza, LUISS, Rome, Italy

Boglárka G.-Tóth Department of Computational Optimization, University of Szeged, H-6720 Szeged, Arpád tér 2, Hungary

Bilevel Models on the Competitive Facility Location Problem

Necati Aras and Hande Küçükaydın

1 Introduction

Facility location problems (FLPs) that arise as real-life applications in both public
and private sector try to determine the optimal location for facilities such as
warehouses, plants, distribution centers, shopping malls, hospitals, and post offices.
They can have different objectives such as maximization of the profit obtained from
customers and minimization of the costs incurred by locating facilities and serving
customers. The basic FLPs are given in Daskin [14] as p-median, set covering,
maximal covering, fixed-charge, and hub location problems. A p-median model tries
to minimize the demand weighted total or average distance between the customers
and their nearest facility by locating p facilities. A set covering model, on the other
hand, attempts to minimize the number of facilities to be opened necessary to cover
all demand points. In contrast to a set covering model, a maximal covering model
assumes that it may not be possible to cover all the demand points by facilities.
Hence, it locates a fixed number of facilities to cover most of the demand. All
the models mentioned so far are uncapacitated facility location models and neglect
the transportation costs between the customers and facilities as well as fixed costs
of opening facilities. In contrast to these models, fixed-charge models take into
account a limited capacity for facilities and transportation costs to serve customers
using functions of distances and also fixed costs for locating new facilities. Hence,
the total cost which is comprised of fixed cost and transportation cost needs to be
minimized in order to determine the optimal number and locations of facilities, and

N. Aras (✉)
Boğaziçi University, Istanbul, Turkey
e-mail: arasn@boun.edu.tr

H. Küçükaydın
MEF University, Istanbul, Turkey
e-mail: hande.kucukaydin@mef.edu.tr

© Springer International Publishing AG 2017
L. Mallozzi et al. (eds.), *Spatial Interaction Models*, Springer Optimization
and Its Applications 118, DOI 10.1007/978-3-319-52654-6_1

the allocation of demand points to the opened facilities. Unlike the other models, a hub location model includes both the location of facilities referred to as hubs, and the design of the network by determining the hub node that is going to be assigned to every non-hub node.

Many factors influence the location decision for a new facility in a market, but one of the most important factors is related to the existing facilities which belong to competitors offering the same or similar commodities or services. When there is no competitor in the market, the new facility will be the only supplier of the commodity or the service leading to a monopoly in the market. However, if there are already existing facilities in the market, then the new facility will have to compete for customers with the aim of maximizing the market share or the profit (see Drezner [17]). Even the new facilities that are a monopoly at the market entry may face competition later when other competitors enter the market. In this chapter, we consider FLPs in a competitive environment, namely Competitive Facility Location (CFL) problems. These problems are spatial interaction models where a firm or franchise wishes to locate new facilities in a market with already existing or prospective competitors. In some problems, the firm or franchise may have existing facilities with known locations and attractiveness levels in the market, while in others the firm may be a new entrant with no existing facility.

Several CFL models are proposed in the literature so far, see for instance the survey papers by Eiselt et al.[22], Eiselt and Laporte[21], Plastria[38], Drezner[18], Eiselt et al.[23], and the references therein. These survey papers group the studies existing in the literature according to different model components. In fact, the most important factors that can be used to differentiate studies on the CFL problem are twofold: existence of the follower's reaction as well as the timing of action and reaction of the players in the game. CFL models with static competition ignore the reaction of a firm to the opening new facilities or redesigning existing facilities by other competitors. CFL models with foresight, on the other hand, take this reaction into account, and hence are more difficult in general compared to static models. The timing of reaction divides the CFL problems with foresight into two major classes, namely simultaneous-entry CFL problems and sequential-entry CFL models. In simultaneous-entry CFL problems, where a Nash game is involved, firms or franchises simultaneously make their decisions on the facility locations and other design components, if any. In contrast, there exists a precedence of decision making among the competing firms in sequential-entry CFL problems. These problems are generally recognized as a Stackelberg type of game between two firms (see von Stackelberg [49]). These games consist of a new entrant firm or a firm with existing facilities that establishes new facilities in the market, where an existing or a future competitor is present to react to the action of the first firm. This action-reaction situation brings us to the so-called two-level or bilevel programming (BP) problems, which constitutes the backbone of the chapter. BP problems include two independent players, namely the leader and the follower, who act sequentially with the objective of optimizing their own objective functions. The leader selects a strategy to optimize its objective function with the foresight or anticipation that the follower reacts to the chosen strategy with the aim of optimizing its own objective

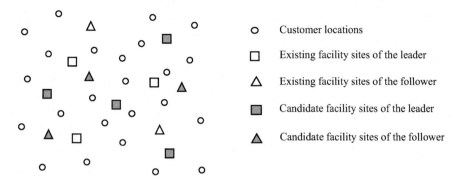

Fig. 1 An example of a CFL problem in discrete space

function. Furthermore, these objective functions are usually in conflict with each other as indicated by Moore and Bard [36]. The purpose of the chapter is to provide a review on the recent studies that include bilevel models within the context of CFL problems.

An example of a general BP problem is depicted in Fig. 1. In this example, both the leader and the follower own existing facilities in the market with known locations and attractiveness levels, and customers are aggregated at demand points as shown by small circles. First the leader, then the follower can install new facilities at candidate facility sites.

Since papers including BP models are reviewed in this chapter, the reader may benefit from visualizing the overall structure of a BP model. By letting $\mathbf{u}$ and $\mathbf{v}$ denote the decision variables of the leader and the follower, respectively, we can define a BP model as follows:

$$\max_{\mathbf{u}} \quad F(\mathbf{u}, \mathbf{v})$$

$$\text{s.t.} \quad G(\mathbf{u}, \mathbf{v}) \leq 0$$

$$\max_{\mathbf{v}} \quad f(\mathbf{u}, \mathbf{v})$$

$$\text{s.t.} \quad g(\mathbf{u}, \mathbf{v}) \leq 0.$$

Here, $G(\mathbf{u}, \mathbf{v})$ and $g(\mathbf{u}, \mathbf{v})$ are the constraints of the leader and follower, respectively, while $F(\mathbf{u}, \mathbf{v})$ and $f(\mathbf{u}, \mathbf{v})$ are their objective functions. Integrality restrictions may exist on the decision variables $\mathbf{u}$ and $\mathbf{v}$. We have to emphasize that such a BP model is formulated from the viewpoint of the leader. It can be easily seen that the so-called upper-level problem (ULP) of the leader incorporates the optimization of the lower-level problem (LLP) of the follower as constraints.

All CFL models attempt to infer the market share captured by each facility. Nevertheless, the captured market share can only be computed when we know the patronizing behavior of customers, i.e. the rules to allocate the customers to facilities. In most of the CFL models, the customers choose the facilities solely based on their proximity to the facility locations or on their preferences towards the facilities. On the other hand, there also exist CFL models that account for the attractiveness level, i.e. the design or the quality of facilities. The first CFL problem presented by Hotelling [27] considers a model involving two identical ice-cream vendors along a beach strip, namely in a linear market, where the customers are attracted to the closest vendor. This kind of facility location games is then developed by Hakimi [25] who also established the first fundamental complexity results. He further introduced two terms into the location science terminology, namely the $(r|p)$-*centroid problem* and the $(r|X_p)$-*medianoid problem* for the leader's and follower's problem, respectively. The $(r|X_p)$-medianoid, or the medianoid problem in short, refers to the follower's problem which tries to optimally locate r new facilities when the locations X_p of leader's p facilities are known. On the other hand, the $(r|p)$-centroid problem solves the leader's problem by opening p new facilities with the anticipation that the follower will locate r facilities in return by solving the $(r|X_p)$-medianoid problem. The CFL literature actually started with identical facilities, but expanded later with models that consider unequally attractive facilities.

Hence, the CFL models can be further divided into two classes by the customer choice rule: deterministic utility models and random utility models. In both classes, customers visit the facilities according to a utility function which is composed of the facility characteristics and the distance between the customers and the facility sites. In other words, if a facility has $n - 1$ attributes such as $a_1, a_2, \ldots, a_{n-1}$, then the utility T of this facility is defined as $T = T(a_1, a_2, \ldots, a_n)$, where a_n represents the distance between the customer and the facility as the facility's nth attribute. The major distinction between the deterministic and random utility models is that in deterministic utility models customers choose the facility that provides the highest utility to them, which implies the all-or-nothing property. In other words, a customer visits a facility with a probability of either zero or one. Drezner [17] points out that the utility function T is usually an additive function such that $T = \sum_{i=1}^{n} \beta_i z_i(a_i)$ where β_i is the weight of the ith attribute and z_i is a function of a_i. If the utility function in a deterministic model is additive, then most of the time the notion of *break-even distance* is used. This is the distance at which utilities of new and existing facilities are equal. In such a case, customers patronize a facility if and only if the facility is opened within the break-even distance. On the contrary, the utility function in a random utility model varies among customers and each customer draws his/her utility from a random distribution of this utility function. Hence, the probability that a customer visits a facility varies between zero and one. Random utility models are basically discrete choice models for short-term travel decisions. A discrete choice model can be employed when an individual decision maker faces a finite number of alternatives, where each alternative is characterized by a set of attributes. In case of a CFL problem, the decision makers are the customers aggregated at demand points, alternatives are the facilities, and attributes are various facility characteristics. The last component that describes a discrete choice model consists of the decision rules according to which the decision makers make a choice among the alternatives (see Ben-Akiva and Bierlaire [4]). The most widely used random utility model in CFL literature is the gravity-based model which was first proposed by Reilly [44] and then employed by Huff [28, 29]. It assumes that the probability that a customer visits a facility is proportional to the attractiveness level of the facility and inversely proportional to a function of the distance between the customer and the facility. The attractiveness level, which can also be named as the design or the quality of the facility, is composed of various facility attributes such as the size or the floor area of the facility, the variety of the products sold, the prices offered by the facility, the existence of a parking lot, and the proximity to public transportation.

We first provide a classification of the studies that are relevant within the scope of this chapter, namely works that develop a BP model for the CFL problem. The second and third columns of Table 1 categorize the studies according to leader's action and follower's reaction. The main objective of any FLP is to find the optimal location for new facilities. Thus, in every CFL study except Küçükaydın et al. [32], both players wish to determine the optimal location ("L") of their new facilities. However, a relatively few number of papers also consider the design ("D") of new or existing facilities as a decision variable. Thus, it is possible that the ULPs and LLPs include only the location decisions ("L") or only the design decisions ("D") or both ("L+D"). In Küçükaydın et al. [32] the follower's reaction only consists of design decisions of the existing facilities. Furthermore, Fischer [24] determines the optimal facility location as well as product prices offered by the facilities for both players which can be recognized as a ("L+D") decision.

It is possible to categorize the CFL problems (as is the case with all FLPs) with regard to the set of candidate sites for opening facilities, which can be a discrete set, a continuous set, or a network. Thus, the next column of Table 1 differentiates the studies according to this criterion. Discrete CFL problems consider predetermined facility sites as candidate locations for the new facilities, whereas in continuous CFL problems it is possible to locate a facility anywhere in the plane. Finally, in network-based problems, the candidate sites are the nodes and edges of the network. All but one study existing in the table are identified as either a discrete ("Disc") or a continuous problem ("Cont"). The only study that involves a CFL problem defined on a network is due to Tóth and Kovács [48]. Thus, it is classified as a network problem ("Netw").

The customer choice rules that are found in the papers reviewed are as follows: "Distance (Dist)", "Preference (Pref)", "Price (Price)", "Gravity-based (Gravity)", and "Proportional (Prop)" in the fifth and sixth columns of Table 1. The first three of these rules can be thought of as a deterministic customer allocation, whereas the last two can be seen as a probabilistic allocation. The papers classified in the deterministic choice rule group allocate the customers either to the closest facility, or to the most preferred facility, or to the facility offering the lowest price. In the case preference is used, each customer ranks the facilities according to a "linear" order. The papers in this group clearly applies a deterministic utility model and preserve the all-or-nothing property assuming that the facilities are either identical or different in terms of their attributes. The papers employing the probabilistic customer allocation employ random utility models, where each customer splits

Table 1 Characteristics of the reviewed papers

	Leader's action	Follower's reaction	Location type	Allocation		No. of facilities		Solution method
				Deterministic	Probabilistic	Leader	Follower	
Alekseeva et al. [1]	L	L	Disc	Dist		M	M	E
Arrondo et al. [2]	L+D	L+D	Cont		Gravity	S	S	H
Ashtiani et al. [3]	L	L	Disc		Gravity	M	M	–
Beresnev [5]	L	L	Disc	Pref		M	M	H
Beresnev [6]	L	L	Disc	Pref		M	M	H
Beresnev [7]	L	L	Disc	Pref		M	M	E
Beresnev and Mel'nikov [8]	L	L	Disc	Pref		M	M	H
Bhadury et al. [9]	L	L	Cont	Dist		M	M	H
Biesinger et al. [10]	L	L	Disc		Gravity	M	M	H
Biesinger et al. [11]	L	L	Disc	Dist		M	M	H
Biesinger et al. [12]	L	L	Disc	Dist	Gravity/Prop	M	M	H
Davydov et al. [15]	L	L	Cont	Dist		M	M	H
Davydov et al. [16]	L	L	Disc	Dist		M	M	H
Drezner and Drezner [19]	L	L	Cont		Gravity	S	S	H
Drezner et al. [20]	L+D	L+D	Disc		Prop	N/M	N/M	H
Fischer [24]	L+D	L+D	Disc	Price		M	M	H
Hendrix [26]	L+D	L+D	Cont		Gravity	S	S	E
Kochetov et al. [30]	L+D	L+D	Disc		Gravity	M	M	H
Kononov et al. [31]	L	L	Disc	Pref		M	M	E
Küçükaydın et al. [32]	L+D	D	Disc		Gravity	M	N	E
Küçükaydın et al. [33]	L+D	L+D	Disc		Gravity	M	M	E/H

Mel'nikov [34]	L	L	Disc	Pref		M	M	H
MirHassani et al. [35]	L	L	Disc	Dist		M	M	H
Panin et al. [37]	L	L	Disc	Price		M	M	H
Plastria and Vanhaverbeke [39]	L	L	Disc	Dist		M	S	E
Rahmani and Yousefikhoshbakht [40]	L	L	Disc	Pref		M	M	–
Ramezanian and Ashtiani [41]	L	L	Disc		Gravity	M	M	E
Redondo et al. [42]	L+D	L+D	Cont		Gravity	S	S	H
Redondo et al. [43]	L+D	L+D	Cont		Gravity	S	S	H
Sáiz et al. [45]	L	L	Disc		Gravity	S	S	E
Shiode et al. [46]	L	L	Cont	Dist		S	S	E
Shiode et al. [47]	L	L	Cont	Dist		S	S	E
Tóth and Kovács [48]	L	L	Netw		Gravity	S	S	E

his/her demand into multiple facilities. All papers with probabilistic allocation utilize the gravity-based model with the following exceptions. The paper by Drezner et al. [20] is classified as "Proportional" since in that work customers may be attracted to multiple facilities if they remain within the region of influence of a facility and their purchasing power is distributed equally among all capturing facilities. Furthermore, the market share captured by a firm from a customer is proportional to the number of its facilities attracting that customer and inversely proportional to the total number of attracting facilities belonging to both the firm and its competitors. Besides Drezner et al. [20], Biesinger et al. [12] also employ a proportional allocation rule in addition to deterministic as well as gravity-based rules. In this work, each customer is attracted to the nearest facility of the leader and follower at the same time and his/her demand is proportionally split between the two facilities according to their attractiveness levels, which only depend on the distance from the customer.

The seventh and eighth columns of Table 1 give information on the number of facilities opened by the leader and the follower: they either open a single facility ("S"), or multiple facilities ("M"), or do not locate any new facility ("N") because their decisions are only related to the redesign of existing facilities. The last column labels the reviewed papers with respect to the solution approach adopted. Two classes of studies are identified: those that are solved by a heuristic method ("H") to find a good feasible solution for the leader, and those that are solved by exact methods ("E") to determine an optimal solution of the leader.

Most of the survey papers on CFL problem group the reviewed works according to various components such as candidate locations, the number of facilities to be opened, the existence of reaction, and the objective function. In this chapter, we select two major ingredients of the CFL problems that include a BP model: leader's decision(s) and the allocation of customers to the facilities. In our opinion, these two issues are important because of two reasons. First, they have a hinge on how realistic the models are. Second, they specify the difficulty level of the developed BP model. Therefore, we partition the papers given in Table 1 into four sections. Sections 2 and 3 include the papers where the leader decides only on the location of new facilities. The difference among the papers in these two sections is that Sect. 2 contains papers with deterministic customer allocation, whereas Sect. 3 consists of the studies with probabilistic allocation. The papers with location and design decisions of the leader are discussed in detail in Sects. 4 and 5. These two sections are differentiated from each other again by the type of the allocation of customers to facilities.

2 Leader's Decision: Location, Allocation: Deterministic

Alekseeva et al. [1] consider an $(r|p)$-centroid problem in the discrete space. Since the facilities in the market are assumed to be identical, the customers visit only the closest open facility, where ties are broken in favor of the leader. The authors propose an iterative exact method which makes use of a single-level reformulation including polynomial number of variables and exponentially many constraints. At each iteration of the method, a family of the constraints representing the follower strategies are extracted and the best strategy of the leader against this family is found by a local search heuristic, which provides an upper bound on the objective function value of the leader. The algorithm stops when the upper bound is equal to a lower bound. To this end, the family of constraints is enlarged at each iteration.

Two different settings are developed in Beresnev [5] for a CFL problem in discrete space with a deterministic customer choice rule. The two settings differ from each other in terms of the LLP. In the first setting, both players try to maximize their profit gained from customers, whereas in the second setting fixed cost of opening facilities is subtracted from the income captured by follower's facilities which represents the number of customers. Moreover, the facilities opened by the follower are not allowed to be detrimental, i.e., the obtained income should exceed the sum of the fixed costs. The leader and the follower share the same set of candidate facility sites, but the follower cannot open its new facility at a site occupied by the leader. Each customer shows his/her preferences towards the new facilities by considering a linear order. Then they select the first open facility according to this order. The author proposes an algorithm which constructs an auxiliary pseudo-Boolean function, called estimation function, whose minimum value is sought to obtain an upper bound on the objective function value of the bilevel models.

Beresnev [6] considers a sequential game between two competing firms that share the same set of candidate facility sites. Each customer is served by a single open facility and is captured by a leader's facility if it is the most desirable one among all open facilities of the leader. However, it is assumed that the customers who do not patronize any of the leader's facilities can be captured by any of the follower's facilities as long as it is more desirable than all of the leader's facilities. A bilevel integer linear programming model is introduced to represent the problem. Since the presence of multiple optima in the LLP poses uncertainty, the author introduces two rules, namely *the rule of cooperative behavior* and *the rule of non-cooperative behavior*. When the rule of cooperative behavior is imposed, the follower chooses the optimal solution to its problem which provides the best outcome for the leader. On the contrary, when the rule of non-cooperative behavior is assumed, then the follower selects the optimal solution which decreases the objective function value of the leader the most. In order to obtain approximate cooperative and non-cooperative solutions, an algorithm that consists of two stages is proposed. At the first stage, a solution of the leader is fixed and the optimal objective function value of the follower is estimated. Then at the second stage, an auxiliary problem is solved, which provides the desired feasible solution of the bilevel problem. In order to obtain these feasible solutions, an algorithm for estimating an upper bound on the objective function value of the bilevel model at any feasible solution is developed. The author represents the problem to find the optimal cooperative and non-cooperative solutions as the maximization of pseudo-Boolean function which is solved by a local search algorithm.

Beresnev [7] is concerned with the same bilevel integer linear programming problem given by Beresnev [6]. In contrast to the study of Beresnev [6], the author suggests a branch-and-bound (B&B) method to find the optimal non-cooperative solution. The problem at hand is first converted into a problem of maximizing a pseudo-Boolean function, where the number of variables coincides with the number of candidate facility sites of the leader. Since this function is implicitly defined, first the LLP and then the auxiliary problem should be solved to optimality. The author adopts the same method used in [6] for estimating an upper bound. In [6], the upper bound is generated for the case where the profits obtained from customers are the same for all facilities. However, in this study the algorithm produces an upper bound in the general form and provides at the same time a feasible solution which yields a lower bound. In order to obtain a good incumbent solution at the root node of the B&B tree, a standard local search algorithm is used which is devised in [8] dealing with a similar problem in [5] and [6].

A centroid problem in the plane is dealt with in Bhadury et al. [9], where the follower locates additional facilities as a reaction. Both players wish to capture most of the demand that is represented by varying the weights at discrete points. The customers choose the closest facility and in case of a tie, leader's facility is preferred. The authors present two heuristic methods to solve the medianoid problem, namely a greedy heuristic and a minimum differentiation heuristic which turns out to be more robust. The latter algorithm is based on an observation of Hotelling: when the prices are fixed and equal, the facilities tend to be located at a central point in the market under duopoly. The idea behind this is basically to locate a new facility at an arbitrarily small distance away from an existing facility. After the medianoid problem is solved with these two heuristics, they make use of them again to solve the ULP. When a solution of the leader is fixed, the follower's problem is solved. Given this solution of the follower, the leader acts like the follower and its problem is solved using the same two heuristic methods. This alternating procedure is repeated until a stopping criterion is met.

Biesinger et al. [11] formulate a bilevel mixed-integer linear model for a discrete $(r|p)$-centroid problem on a weighted complete bipartite graph. Only one facility can be opened at a predetermined candidate site, where each customer is assigned to the closest facility. If there are two closest facilities to a customer, he/she chooses the one which belongs to the leader. A hybrid genetic algorithm is proposed to solve the problem. In order to evaluate a solution of the leader, three methods are considered, which leads to a multi-level evaluation scheme. The first one includes an exact evaluation by solving the follower's problem by means of a commercial solver. The second method solves the LP relaxation of the follower's problem which provides a lower bound on the objective function value of the leader. Finally, the last method employs a greedy algorithm on the follower's problem to get an upper bound on the objective function value of the leader. The best result is obtained by the second method. The proposed genetic algorithm builds up a solution archive for identifying solutions that have been already generated. The generated solutions are stored in a special data structure and the duplicates are converted into new solutions. The algorithm is then locally improved using a tabu search procedure. The best computational results are obtained by the algorithm using the tabu search with a reduced neighborhood. Davydov et al. [16] are concerned with the same discrete $(r|p)$-centroid problem considered in [11]. This time the problem is solved by two heuristic methods, namely a local search with variable neighborhood search and a stochastic tabu search method. Both procedures employ a neighborhood swap over leader's variables. To accelerate the local search, the neighborhood is partitioned into three parts and the two most promising parts are searched thoroughly by finding the ascent direction quickly.

An $(r|p)$-centroid problem in the continuous plane is addressed in Davydov et al. [15] for identical facilities, where the customers select the closest facility. The follower's problem is reformulated as an integer linear programming problem that describes a maximal covering problem introduced by Church and ReVelle [13] and solved exactly by a B&B method. In order to solve the centroid problem, the authors develop a local search heuristic based on variable neighborhood search. In order to find the best neighboring solution according to the swap neighborhood, the $(r|X_{p-1} + 1)$-centroid problem is considered as a subproblem in which it is assumed that the leader has already opened $p - 1$ facilities and wishes to locate one more additional facility. Moreover, it is shown that the $(r|X_{p-1} + 1)$-centroid problem is polynomially solvable for fixed r.

In the CFL problem analyzed in Kononov et al. [31] customers choose a facility to visit on the basis of their preference over a facility. By assuming that the preference h_{ij} of customer j is different for each facility site i, the authors avoid ties between two or more facilities, which eliminates the need to consider optimistic and pessimistic strategies. As a matter of fact, it is not the absolute preference values but their relative values or ranking, which determines the assignment of each customer to a facility. Hence, a deterministic allocation model is developed. Both the leader and the follower open multiple facilities among a set of candidate sites to maximize the profit giving rise to a discrete CFL model. The problem is solved approximately by successively generating upper bounds on the optimal value of the leader's objective function and lower bounds on the optimal value of the follower's objective function. Mel'nikov [34] tackles essentially the same problem considered in [31] except a customer may have the same preference for two different facility sites. A randomized local search algorithm is proposed in the paper for the solution of the problem. The same problem is also studied in MirHassani et al. [35] where customers patronize the facility closest to them. In other words, instead of the preference matrix h_{ij}, distance matrix d_{ij} is prepared among the customer locations and candidate facility sites. As the solution approach, the authors utilize a modified quantum binary particle swarm optimization (QBPSO) method that uses an improvement procedure within the QBPSO. The benefit of the improvement is increasing the speed of convergence and preventing the algorithm from being trapped in local optimal solutions. The CFL problem analyzed in Panin et al. [37] differs from the others in this group in one aspect: the allocation of customers to facilities. Rather than taking into account the distance between the customer location and facility site or a predetermined preference parameter as is the case in [31, 34, 35], customers choose the facility to be visited on the basis of the selling price of the product set by the players differently for each customer. Associated with customer j, there exists a maximal price w_j that this customer is willing to pay for the product. Hence, the firms cannot set a price larger than w_j. Furthermore, there is a cost c_{ij} which represents the cost of serving customer j from an open facility at site i, which constitutes a lower bound on the price. Finally, the number of facilities to be opened by the leader and follower firms is fixed as in the well-known p-median problem as opposed to the other aforementioned papers which treat the number of facilities opened an endogenous variable, i.e., determined by the solution of the model.

The CFL problem analyzed in Plastria and Vanhaverbeke [39] is based on the maximal covering model. The aim of the leader is to maximize the demand covered by its newly opened facilities by taking into account the market entry of a competitor with a single facility. The allocation of customers is modeled using the so-called patronizing sets associated with customer j, namely customer at location j can be served by a facility that is opened within a certain distance away from the customer. Candidate facility sites within this region belong to the patronizing set S_j of customer j. An important assumption regarding the allocation is that if both the leader and follower open a facility in the patronizing set of a customer, then this customer is served from a leader's facility. Although this is a Stackelberg game that can be formulated as a bilevel model, the authors propose a single-level mixed-integer programming model, which is solved using ILOG CPLEX 9.0.

In Rahmani and Yousefikhoshbakht [40] the authors consider a closed-loop supply chain network where customers not only have demand for new products, they also want to return used products. Both players can open one of three kinds of facility at a candidate site: forward, backward, and hybrid processing facility. Moreover, multiple facilities of a given type can be established by both parties. A bilevel mixed-integer nonlinear programming model is developed, which is then transformed into bilevel mixed-integer linear model. No attempt has been made to solve the resulting model.

Shiode et al. [46] deal with the following variant of the CFL problem. Each firm opens a single facility in the continuous plane. Customers choose the nearest facility to them. The feature differentiating the problem from those studied in other papers is that customer demand is dependent on whether the facility belongs to the leader or follower firm. Furthermore, the distance between customers and facilities is measured by the rectilinear or rectangular distance. It is shown that in the case of linear market, where all customers are located on a line, the optimal location for leader's facility coincides with a demand point. In the case of planar market, where customer locations are scattered in the plane, leader's facility is optimally located on one of the grid points that are obtained by drawing horizontal and vertical lines through the customer locations.

Shiode et al. [47] develop a trilevel model rather than a bilevel one because there are three competitors which try to find the best locations for their single facilities. The demand of customers are assumed to be continuously distributed along the line segment and each customer visits the nearest facility. In fact, it turns out that there are infinitely many customers existing along the line segment, but the continuously distributed demand in an interval has a finite value.

3 Leader's Decision: Location, Allocation: Probabilistic

Ashtiani et al. [3] suggest a discrete CFL problem, where the leader wishes to determine the optimal location for p new facilities to maximize its market share. The follower, on the other hand, wants to maximize its market share by locating r new facilities after leader's action, but r is uncertain to the leader. It is assumed that the follower can locate either $1, 2, \ldots, r$ new facilities at candidate sites where each number corresponds to a different scenario. The only information that the leader obtains is the probability of occurrence for each scenario. Since Huff's gravity-based rule is employed, the existing and new facilities have various attractiveness levels. However, the attractiveness level of a new facility is not considered as a decision variable in the model; they are rather predetermined for each candidate facility site. Since the exact number of new facilities of the follower is not known to the leader, robust optimization is employed for the solution of the problem, where the objective function of the leader maximizes the expected value of the market share under various scenarios and minimizes the difference between the optimal solution of a scenario and the expected value. Although the authors claim that the leader's problem is solved to optimality, no solution methodology is given.

Six different bilevel models in discrete space including only location decisions are considered by Biesinger et al. [12]. The various models are based on three customer choice rules: binary, proportional, and partially binary. In a binary model, customers visit only the closest facility, whereas the proportional model considers Huff's gravity-based rule. In partially binary case, the demand is distributed among the closest facilities of the leader and the follower, where the attractiveness depends on the distance between the customer and the facility. Although the attractiveness levels of the facilities are taken into account for the proportional and partially binary cases, they are set equal to one so that the utility of a facility for a customer relies solely on the distance. Each of these customer choice rules is combined with both essential and unessential demand which gives rise to six different bilevel models. In case of essential demand, the entire demand of customers is satisfied, whereas a certain proportion of customers' demand is fulfilled based on the distance to the serving facility for the unessential demand case, i.e., the demand decreases with the increased distance. The bilevel model applying the binary customer choice rule with essential demand coincides with the model by Biesinger et al. [11] given in Sect. 2. The LLP is formulated as a linear mixed-integer programming problem in all scenarios which can be solved by the same three methods suggested by Biesinger et al. [11]. Again an evolutionary algorithm is further improved by a complete solution archive and is turned into a hybrid procedure by employing a tabu search variant which uses the solution archive as the tabu list.

Biesinger et al. [10] suggest the same bilevel model using the proportional rule with essential demand considered in [12], where the attractiveness levels of the facilities are equal to one. The follower's problem is first converted into a mixed-integer problem following a linear transformation. An evolutionary algorithm with a complete solution archive and an embedded tabu search method to optimize leader's locations is employed.

Drezner and Drezner [19] propose three models which take changing market conditions over a planning horizon into consideration. Only the second proposed model, namely the Stackelberg equilibrium model, is discussed here. Both players want to maximize the market share by locating a single facility in the continuous plane. Huff's gravity-based rule is employed to estimate the market share. However, only the location of the facilities is sought. It is assumed that the follower opens its single facility at some time point t_1 over the planning horizon t. It is further assumed that the buying power of customers can change in time, but the facility attractiveness not. In order to solve the problem, the authors suggest three heuristic procedures, namely a brute force approach, a pseudo-mathematical programming approach, and a gradient search approach, but no computational results are reported.

Ramezanian and Ashtiani [41] deal with the version of the CFL problem in which the leader wants to open p new facilities among predefined candidate locations by anticipating that the follower will react by installing r new facilities. Both players are assumed to have already existing facilities. By making use of the Huff's gravity-based model, the authors try to find the optimal facility locations of both parties by an exhaustive enumeration method, where the objective function of the leader and follower is to maximize their own market share. As a consequence, only a small problem instance is solved consisting of 16 demand points and 5 existing facilities three of which belong to the leader while the remaining two are owned by the follower. Parameters $p = r = 2$ implying that two new facilities are opened by the competitors.

Virtually the same problem described above is investigated in Sáiz et al. [45] with $p = r = 1$. The distinctive feature of this paper is the solution approach, which is a branch-and-bound algorithm. Two B&B algorithms are developed: one for the follower's lower-level problem and one for the leader's upper-level problem. The main idea in this algorithm is recursive partitioning of the original problem into smaller disjoint subproblems until the solution is found. It is guaranteed that a global optimum is found by successively generating new lower and upper bounds that ultimately lie within a given interval.

Tóth and Kovács [48] examine also a similar CFL problem with two differences. First, the problem is formulated on a network where customers represent nodes and facilities can be located along the edges of the network in contrast to the two papers cited above that allow facilities to be opened at candidate sites. Second, operational costs are taken into account for the single facility opened by the leader and the follower. Due to this cost component, the zero-sum property in the objective function, which is used in the B&B algorithm in [45], is violated and hence a more difficult problem is obtained. The authors devise a B&B method to solve the leader's problem that includes another B&B embedded within the former. The calculation of lower and upper bounds involve interval arithmetic and DC (Difference of Convex functions) decomposition.

4   Leader's Decision: Location and Design, Allocation: Deterministic

The only study that belongs to this category is due to Fischer [24]. This study focuses on a discrete Stackelberg-type CFL problem with two competitors. The customers are aggregated at discrete points which make up the markets and each competitor supplies the same product to these markets. Both the leader and the follower want to decide on the locations of a fixed number of new facilities and the price of the product at each market. Since the price at a market is defined by the distance between the facility and the market, the price of the product can differ from market to market (i.e., discriminatory pricing). It is assumed that customers buy the product from the competitor giving the lowest price. Two bilevel models are formulated: a mixed-integer nonlinear bilevel model in which both players fix their locations and prices in the end, and a linear bilevel model with binary variables where price adjustment is possible. A heuristic solution procedure is developed to solve the linear bilevel model, but no computational result is given.

5   Leader's Decision: Location and Design, Allocation: Probabilistic

Hendrix [26] considers a problem where both the leader firm and the follower firm want to open a single facility in the continuous plane. In addition to the location of their facilities, the players also aim to determine the facilities' quality level. The objective functions are represented by the respective profits of the firms, where opening a facility with quality x incurs a linear cost cx. The choice of each customer is modeled using Huff's gravity-based rule based on the ratio of the quality to the Euclidean distance between the location of the customer and facility site. One of the important results is that co-location of the leader's and follower's facilities is not possible. The authors also give conditions under which one of the firms can force the other not to enter the market due to the negative profit.

In Küçükaydın et al. [32], the authors deal with a variant of the CFL problem where multiple facilities are opened by the leader firm among a predetermined set of candidate sites to maximize its profit. The quality levels of these facilities are also decision variables. The reaction of the competitor firm, which is the follower, consists of adjusting the quality of its existing facilities. Employing Huff's gravity-based rule, the authors develop a bilevel mixed-integer nonlinear programming (MINLP) model and solve it using a global optimization method called GMIN-αBB after converting the bilevel model into an equivalent one-level MINLP model. This conversion is possible thanks to the concavity of the objective function of the follower's lower level problem in terms of the quality variables when the leader's decision variables are fixed.

Küçükaydın et al. [33] extend the work in [32] by adding the capability of opening new facilities and/or closing existing ones to the competitor's reaction set. Clearly, the new problem is a more challenging one. First, an exact solution procedure is developed that integrates complete enumeration in terms of competitor's location variables for opening new facilities and keeping/closing existing facilities with GMIN-αBB. Since the exact method can only provide solutions for small problem instances in a reasonable CPU time, three heuristics based on tabu search are proposed. These heuristics perform a search over the location variables of the leader in the upper level problem. Upon fixing these variables, the follower's lower level problem is solved to optimality using a branch-and-bound algorithm with NLP relaxation because it is shown to be a concave optimization problem.

In the CFL problem examined in Redondo et al. [42] the leader firm wants to open a single facility in the continuous plane in addition to its existing facilities in the market. The quality of the new facility is also a decision variable to be set by the firm. The reaction of the follower firm consists of exactly the same decisions as the leader in the sense that the location of a new facility and its quality level has to be determined. This problem is also referred to in the literature as the $(1|1)$-centroid problem. Four heuristics are developed for the solution of the problem. These are a grid search procedure, an alternating procedure and two evolutionary algorithms. The only difference between the problem investigated in Redondo et al. [43] from the one in [42] is that the demand of customer j is not fixed, but varies according to the utility the facility provides to the customer, which is computed according to Huff's gravity-based model. The variable demand also referred to as elastic demand of a customer is assumed to vary between the minimum possible demand and maximum possible demand. Note that most of the literature on CFL uses the assumption of inelastic or fixed demand. In order to solve the medianoid problem, the authors apply an exact interval branch-and-bound method and an evolutionary algorithm called UEGO that uses a Weiszfeld-like procedure and can find the global optimum with a certain reliability. Then, they develop three heuristic methods to solve the centroid problem, i.e. leader's problem: a grid search procedure, a multi-start algorithm, and a subpopulation-based evolutionary algorithm, called TLUEGO. The same problem is then considered by Arrondo et al. [2] to improve the computational performance of TLUEGO. To this end, they suggest three parallelizations of the algorithm, namely a distributed memory programming algorithm, a shared memory algorithm, and a hybrid of these two algorithms.

Drezner et al. [20] introduce a BP model for a discrete CFL problem taking into account the concept of cover. Each existing and new facility possesses a *radius of influence* which defines a threshold distance to the customers, i.e. a customer visits a facility if and only if his/her distance to that facility is less than or equal to the radius of facility. Hence, a facility can capture only the customers who are within its *sphere of influence*. Thus, the radius of each facility determines its attractiveness level. The demand of a customer who is not covered by the sphere of influence of any facility is considered lost. The leader and follower aim to maximize the market share under a certain budget by expanding their own chains conceiving three strategies. The first strategy takes into account only enlarging the sphere of influence of the existing facilities, whereas the second one solely opens new facilities whose location and radii need to be determined. The final strategy combines the first two strategies. The authors define the radius of the facilities as a continuous decision variable. However, they employ discrete design (radius) scenarios in the solution of the model and one of a finite number of available radii is determined for each open facility. First, the follower's problem is solved to optimality using a B&B method. Then a tabu search algorithm is implemented for the solution of leader's problem which uses a greedy-type heuristic to find a starting solution. The computational results indicate that both the leader and follower can extend their market share by acquiring the lost demand.

Kochetov et al. [30] propose a discrete CFL bilevel problem making use of the Huff's gravity-based rule. Both the leader and follower want to maximize their market share by deciding on the location and attractiveness for their new facilities under a budget limitation. As in the study of Drezner et al. [20], a discrete design scenario set is adopted for the attractiveness levels of the facilities. They apply a linear transformation on the follower's problem to turn it into a linear mixed-integer programming problem. Then an alternating matheuristic, derived from Bhadury et al. [9], is employed for the solution of the bilevel model which terminates when either a Nash equilibrium is reached or an already visited solution is obtained.

6 Concluding Remarks

In this chapter, we review and categorize studies on CFL problems formulated as BP models which are published between 2002 and 2016. We first categorize each study according to various features of the CFL problem investigated. After collecting each work in one of the four groups depending on two distinguishing factors of CFL problems, namely the leader's decision(s) and the allocation of customers to facilities, we give detailed information on each work. It must be noticed that the studies in which decisions are made not only about the locations of the facilities but also about their design is outnumbered by the papers containing only location decisions. Thus, future studies can give more emphasis to this issue. Another observation is related to the fact that relocation, redesign, and/or closing of existing facilities concurrently with opening new facilities can be studied more thoroughly with the inclusion of additional features incorporated into the CFL problem. Another fruitful research direction can be the extension of bilevel models in a setting where multiple firms acting simultaneously among themselves react to the leader. Such a setting introduces the requirement of considering Nash games in the lower-level problem of a bilevel programming model.

References

1. Alekseeva, E., Kochetov, Y., Plyasunov, A.: An exact method for the discrete $(r|p)$-centroid problem. J. Glob. Optim. **63**(3), 445–460 (2015)
2. Arrondo, A.G., Redondo, J.L., Fernández, J., Ortigosa, P.M.: Solving a leader-follower facility problem via parallel evolutionary approaches. J. Supercomput. **70**(2), 600–611 (2014)
3. Ashtiani, M.G., Makui, A., Ramezanian, R.: A robust model for a leader-follower competitive facility location problem in discrete space. Appl. Math. Model. **37**, 62–71 (2013)
4. Ben-Akiva, M., Bierlaire, M.: Discrete choice models with applications to departure time and route choice. In: Hall, R (ed.) Handbook of Transportation Science. International Series in Operations Research and Management Science, pp. 7–37. Kluwer Academic Publishers, Dordrecht (1999)
5. Beresnev, V.L.: Upper bounds for objective functions of discrete competitive facility location problems. J. Appl. Ind. Math. **3**(4), 419–432 (2009)
6. Beresnev, V.L.: Local search algorithms for the problem of competitive location of enterprises. Autom. Remote Control **73**(3), 425–439 (2012)
7. Beresnev, V.L.: Branch-and-bound algorithm for a competitive facility location problem. Comput. Oper. Res. **40**(8), 2062–2070 (2013)
8. Beresnev, V.L., Mel'nikov, A.A.: Approximate algorithms for the competitive facility location problem. J. Appl. Ind. Math. **5**(2), 180–190 (2011)
9. Bhadury, J., Eiselt, H.A., Jaramillo, J.H.: An alternating heuristic for medianoid and centroid problems in the plane. Comput. Oper. Res. **30**(4), 553–565 (2003)
10. Biesinger, B., Hu, B., Raidl, G.: An evolutionary algorithm for the leader-follower facility location problem with proportional customer behavior. In: Pardalos, P.M., Resende, M.G., Vogiatzis, C., Walteros, J.L. (eds.) Learning and Intelligent Optimization. Lecture Notes in Computer Science, pp. 203–217. Springer, Heidelberg (2014)
11. Biesinger, B., Hu, B., Raidl, G.: A hybrid genetic algorithm with solution archive for the discrete $(r|p)$-centroid problem. J. Heuristics. **21**(3), 391–431 (2015)
12. Biesinger, B., Hu, B., Raidl, G.: Models and algorithms for competitive facility location problems with different customer behavior. Ann. Math. Artif. Intell. **76**(1), 93–119 (2016)
13. Church, R.L., ReVelle, C.: The maximal covering location problem. Pap. Reg. Sci. **32**(1), 101–118 (1974)
14. Daskin, M.S.: Network and Discrete Location Models, Algorithms, and Applications. Wiley, New York (1995)
15. Davydov, I., Kochetov, Y., Carrizosa, E.: A local search heuristic for the $(r|p)$-centroid problem in the plane. Comput. Oper. Res. **52**, 334–340 (2014)
16. Davydov, I.A., Kochetov, Y.A., Mladenovic, N., Urosevic, D.: Fast metaheuristics for the discrete $(r|p)$-centroid problem. Autom. Remote Control **75**(4), 677–687 (2014)
17. Drezner, T.: Competitive facility location in the plane. In: Drezner, Z. (ed.) Facility Location: A Survey of Applications and Methods, pp. 285–300. Springer, New York (1995)
18. Drezner, T.: A review of competitive facility location in the plane. Logist. Res. **7**, 1–12 (2014)
19. Drezner, T., Drezner, Z.: Retail facility location under changing market conditions. IMA. J. Manag. Math. **13**(4), 283–302 (2002)
20. Drezner, T., Drezner, Z., Kalczynski, P.: A leader-follower model for discrete competitive facility location. Comput. Oper. Res. **64**, 51–59 (2015)
21. Eiselt, H.A., Laporte, G.: Sequential location problems. Eur. J. Oper. Res. **96**(2), 217–231 (1996)
22. Eiselt, H.A., Laporte, G., Thisse, J.F.: Competitive location models: a framework and bibliography. Transport. Sci. **27**(1), 44–54 (1993)
23. Eiselt, H.A., Marianov, V., Drezner, T.: Competitive location models. In: Laporte, G., Nickel, S., Saldanha da Gama, F. (eds.) Location Science, pp. 365–398. Springer International Publishing, Switzerland (2015)

24. Fischer, K.: Sequential discrete *p*-facility models for competitive location planning. Ann. Oper. Res. **111**(1), 253–270 (2002)
25. Hakimi, S.L.: Locations with spatial interactions: competitive locations and games. In: Mirchandani, P.M., Francis R.L. (eds.) Discrete Location Theory, pp. 439–478. Wiley, New York (1990)
26. Hendrix, E.M.T.: On competition in a Stackelberg location-design model with deterministic supplier choice. Ann. Oper. Res. (2015). doi:10.1007/s10479-015-1793-9
27. Hotelling, H.: Stability in competition. Econ. J. **39**, 41–57 (1929)
28. Huff, D.L.: Defining and estimating a trading area. J. Mark. **28**, 34–38 (1964)
29. Huff, D.L.: A programmed solution for approximating an optimum retail location. Land. Econ. **42**, 293–303 (1966)
30. Kochetov, Y., Kochetova, N., Plyasunov, A.: A matheuristic for the leader-follower facility location and design problem. In: Lau, H., Van Hentenryck, P., Raidl, G. (eds.) Proceedings of the 10th Metaheuristics International Conference (MIC 2013), Singapore, pp. 31/1–32/3 (2013)
31. Kononov, A.V., Kochetov, Y.A., Plyasunov, A.V.: Competitive facility location models. Comput. Math. Math. Phys. **49**(6), 994–1009 (2009)
32. Küçükaydın, H., Aras, N., Altınel, İ.K.: Competitive facility location problem with attractiveness adjustment of the follower: a bilevel programming model and its solution. Eur. J. Oper. Res. **208**(3), 206–220 (2011)
33. Küçükaydın, H., Aras, N., Altınel, İ.K.: A leader-follower game in competitive facility location. Comput. Oper. Res. **39**(2), 437–448 (2012)
34. Mel'nikov, A.A.: Randomized local search for the discrete competitive facility location problem. Autom. Remote Control **75**(4), 700–714 (2014)
35. MirHassani, S.A., Raeisi, S., Rahmani, A.: Quantum binary particle swarm optimization-based algorithm for solving a class of bi-level competitive facility location problems. Optim. Method Softw. **30**(4), 756–768 (2015)
36. Moore, J.T., Bard, J.F.: The mixed-integer linear bilevel programming problem. Oper. Res. **38**(5), 911–921 (1990)
37. Panin, A.A., Pashchenko, M.G., Plyasunov, A.V.: Bilevel competitive facility location and pricing problems. Autom. Remote Control **75**(4), 715–727 (2014)
38. Plastria, F.: Static competitive facility location: an overview of optimisation approaches. Eur. J. Oper. Res. **129**(3), 461–470 (2001)
39. Plastria, F., Vanhaverbeke, L.: Discrete models for competitive location with foresight. Comput. Oper. Res. **35**(3), 683–700 (2008)
40. Rahmani, A., Yousefikhoshbakht, M.: Using a mathematical multi-facility location model for the market competition. Int. Res. J. Appl. Basic Sci. **3**(12), 2442–2449 (2012)
41. Ramezanian, R., Ashtiani, M.G.: Sequential competitive facility location problem in a discrete planar space. Int. J. Appl. Oper. Res. **1**(2), 15–20 (2011)
42. Redondo, J.L., Fernández, J., García, I., Ortigosa, P.M.: Heuristics for the facility location and design (1|1)-centroid problem on the plane. Comput. Optim. Appl. **45**(1), 111–141 (2010)
43. Redondo, J.L., Arrondo, A.G., Fernández, J., García, I., Ortigosa, P.M.: A two-level evolutionary algorithm for solving the facility location and design (1|1)-centroid problem on the plane with variable demand. J. Glob. Optim. **56**, 983–1005 (2013)
44. Reilly, W.J.: The Law of Retail Gravitation. Knickerbocker Press, New York (1931)
45. Sáiz, M.E., Hendrix, E.M., Fernández, J., Pelegrín, B.: On a branch-and-bound approach for a Huff-like Stackelberg location problem. OR Spectrum **31**(3), 679–705 (2009)
46. Shiode, S., Yeh, K.Y., Hsia, H.C.: Competitive facility location problem with demands depending on the facilities. Asia Pac. Manage. Rev. **14**(1), 15–25 (2009)
47. Shiode, S., Yeh, K.Y., Hsia, H.C.: Optimal location policy for three competitive facilities. Comput. Ind. Eng. **62**(3), 703–707 (2012)
48. Tóth, B.G., Kovács, K.: Solving a Huff-like Stackelberg location problem on networks. J. Glob. Optim. **64**, 233–247 (2016)
49. von Stackelberg, H.: Marktform und Gleichgewicht. Springer, Vienna (1934)

Partial Cooperation in Location Choice: Salop's Model with Three Firms

Subhadip Chakrabarti and Robert P. Gilles

1 Introduction

The theory of economic competition has been at the centre of economic reasoning since the seminal contributions of Cournot [5], Walras [26][1] and Edgeworth [11]. The notion of imperfect competition was furthered by contributions of Edgeworth [12] on monopoly and Bertrand [1], who founded his approach to price competition as a critique of the work of Cournot [5] and Walras [27]. Later contributions by Chamberlin [4] and Robinson [22] have had less impact through the later rise of game theoretic approaches founded on the ideas put forward in the seminal work by Cournot [5].

Hotelling [13] is considered the main seminal contribution to a game theoretic approach to imperfect competition based on a spatial approach to product differentiation. Hotelling introduced a fictional space in which competing firms choose a *location*. Thus, assuming that the goods sold by these firms are actually homogeneous, the differentiation is represented purely by the firm's relative location. Consumers in this location model now select one firm from which to obtain one unit of the homogeneous good. A cost related to the distance between the location

---

[1] We refer to Walras [28] for a translation of this seminal work.

S. Chakrabarti (✉) • R.P. Gilles
Economics Department, Queen's Management School, Queen's University of Belfast, Riddel Hall, 185 Stranmillis Road, BT9 5EE Belfast, UK
e-mail: s.chakrabarti@qub.ac.uk; r.gilles@qub.ac.uk

© Springer International Publishing AG 2017
L. Mallozzi et al. (eds.), *Spatial Interaction Models*, Springer Optimization and Its Applications 118, DOI 10.1007/978-3-319-52654-6_2

of the consumer and the selected firm in the space now represents the disutility that this consumer incurs from purchasing a product that imperfectly fits with her own location.[2]

Subsequent game theoretic contributions focussed on the exact formulation of the transportation cost function in the Hotelling framework. Hotelling [13] used linear costs and argued that minimal product differentiation would result in such circumstances. d'Aspremont et al. [6] demonstrated, however, that the equilibrium in Hotelling's linear cost model does not in fact exist due to discontinuous reaction functions and advocated the use of quadratic cost functions instead. They showed that such quadratic costs result in maximal product differentiation. Economides [8] subsequently analysed a wider class of transportation cost functions and showed that one can have both maximal and non-maximal product differentiation but never minimal product differentiation.

Salop [24] solved the existence problem by introducing a different topology of the fictional space, realising that the existence problems are related to the existence of a boundary in the product space. Thus, instead of using a bounded space, Salop employed a circular space without a boundary. Firms now select a location on the circle and the disutilities to consumers are a function of the shortest distance between the consumer's location and the firm's location.

Salop [24] did not introduce this spatial device to analyse location choice but rather to analyse *equidistant entry* of firms into the market. Location choice in the Salop city has been analysed by Economides [10] for quadratic costs and by Kats [14] for linear costs. Economides finds that in equilibrium firms locate equidistantly from each other and charge uniform prices, while Kats identifies a continuum of equilibria.

1.1 Partial Cooperation in Competitive Situations

In a recent study, Mallozzi and Tijs [16] introduced the concept of *partial cooperation* in a non-cooperative, normal form game. It is assumed that there is a single coalition of "cooperators" that collectively determines the strategies of its members. This coalition of cooperators acts as a single, large player in the game and its decisions are guided by the aggregated payoffs of the coalition as a whole.

There emerge two equilibrium concepts that properly capture the effects of partial cooperation on the outcome in the game. The first notion is that of *partial cooperative equilibrium* in which the coalition of cooperators acts as a single player in the game and seeks to play a collective best response to the non-cooperators in the

---

[2]Hotelling [13] himself used the metaphor of consumers located along a "high street", who incur costs to travel to the location of the store of each producer. These travel or transportation costs now represent the disutility from consuming a product that is imperfectly fitting with the exact location of the consumer on this high street.

game. As such, the partial cooperative equilibrium is a Nash equilibrium in which the coalition of cooperators acts as a single player.

The second equilibrium concept is that of *leadership equilibrium* in which the coalition of cooperators is assumed to attain a Stackelberg-like leadership position [25] and selects its strategy tuple prior to the non-cooperators in the game. This imposes a two-stage structure on the normal form game in which the coalition cooperators selects its strategy prior to the group of non-cooperators.

Mallozzi and Tijs [16] introduced the leadership equilibrium concept on the restricted class of symmetric potential games [19]. They show that, under well-chosen restrictions on the strategy set and the payoff functions, the non-cooperators select equilibrium strategies which are both symmetric and unique for every strategy tuple that is selected by the coalition of cooperators. This implies that the coalition of cooperators can perfectly anticipate the resulting equilibrium for any of their available strategic choices. This allows the coalition of cooperators to assume a Stackelberg leadership position and to maximise its aggregated payoff function based on both their strategy and the unique Nash equilibrium strategies of the non-cooperators. A maximiser of this function—provided it exists—now determines the corresponding leadership equilibrium strategy tuple of the coalition of cooperators. Mallozzi and Tijs [16] provide conditions for which such an equilibrium exists.

Following this initial contribution, subsequent work in Mallozzi and Tijs [17, 18] extended the partial cooperative framework to games in which the non-cooperating players select from multiple best responses. Mallozzi and Tijs [18] consider symmetric aggregative games [7] and assume that the non-cooperative players coordinate on the symmetric Nash equilibrium that yields the highest payoff to them, and, thus, do not consider any non-symmetric Nash equilibrium that, possibly, might result in higher payoff for all. Mallozzi and Tijs [17], on the other hand, assume that the non-cooperating players coordinate on the Nash equilibrium with the numerically greatest or lowest strategy vector, irrespective of the payoff attained by the non-cooperators in this equilibrium.

The assumptions of Mallozzi and Tijs are rather restrictive and seem inapplicable to most strategic situations. However, at the same time they are hard to do away with. Subsequent work by Chakrabarti et al.[2] suggests a solution by simply letting the coalition of cooperators choose a strategy that maximises its joint, utilitarian payoffs. Chakrabarti et al. [2] assume that the coalition of cooperators is risk-averse and chooses a maximin strategy. Hence, if there are multiple best responses given a strategic agreement of the coalition of cooperators, then the coalition of cooperators takes into account only the worst possible outcome.

Chakrabarti et al. [3] generalise this approach further to situations in which certain strategic elements are subject to partial cooperation, while other strategic elements are considered to be selected in a non-cooperative, competitive fashion. In this general structure, Chakrabarti et al. [3] extend both the partial cooperative as well as the leadership equilibrium concepts to their most general implementations. We utilise these general formulations in our analysis of Salop's location model.

1.2 Partial Cooperation in a Two-Stage Salop Model of Product Differentiation

Chakrabarti et al. [2] were the first to apply the partial cooperative framework to analyse competition between firms in a Salop model of product differentiation using a circular location approach. Their model only addressed the pricing of products in the case that firms are dispersed at equal distances on the Salop circle and that some of these firms cooperate in their pricing strategies. They arrived at two main insights. First, in the partial cooperative equilibrium there is a clear advantage for non-cooperators, as free rider benefits exceed the benefits from cooperation. This corresponds to the so-called "merger paradox", which states that there are no benefits to cartel formation in an imperfectly competitive market [23].

Second, Chakrabarti et al. [2] showed that in the leadership equilibrium the merger paradox vanishes and there are significant benefits from cooperation to the members of the coalition of cooperators. Chakrabarti et al. [3] further corroborate that the benefits from cooperation exceed the free-rider effect for multi-market oligopolies and related competitive environments.

In the present contribution we consider partial cooperation in the first stage of the two-stage Salop model, which concerns the selection of locations by the firms, rather than the pricing strategies. We limit our analysis to the case of three identical firms in which partial cooperation is simplified to the cooperation of two of these three firms. We use the linear cost model as seminally introduced in Hotelling [13] and restrict the choice of locations with the introduction of a minimum distance between the firms.[3]

We analyse three different equilibrium concepts in this context. First, we look at the standard Nash equilibrium concept and show that there emerges a continuum of equilibria, extending the insight of Kats [14] from two firms to three firms. Second, we introduce partial cooperation between two of the three firms in their location choices discussed above. We investigate the partial cooperative equilibrium concept as well as the leadership equilibrium concept.

We show that there do not exist any partial cooperative equilibria in the location stage of the two-stage Salop model with linear costs. This contravenes the existence of standard Nash equilibria for two firms established by Kats [14]. The main difference is that one decision maker selects two locations from which to supply products rather than a single location as is the case in the Kats model. This shows the fragility of the model with regard to the exact assumptions made.

On the other hand, the introduction of leadership under partial cooperation shrinks the set of equilibrium outcomes considerably. We show that if the coalition of cooperators assumes a Stackelberg-like leadership position, there only exist two different equilibrium configurations of locations. Again, this shows that partial

---

[3] The introduction of this minimal distance guarantees the existence of an equilibrium in this model.

cooperation and leadership resolve competitively complex situations into simple, predictable configurations.

Furthermore, we investigate two extensions of the standard model. First, we look at the introduction of an investment opportunity to lower the marginal production costs of the firm. This extends the model to three stages: After the selection of a location, firms invest in their production technology and only then set the price of the good. We show that in this extended three-stage Salop model with linear costs there are no significant changes to any of the established results for the standard two-stage model. Hence, there are no partial cooperative equilibria, but there emerge two leadership equilibrium configurations.

Finally, we consider the two-stage Salop model with quadratic costs. Here, consumers face quadratic transportation costs, which modifies the size of the market captured by the three firms for different price-location configurations. For quadratic costs, we confirm numerically the existence of partial cooperative equilibrium configurations in the location stage of the model. This is rather different from the linear cost case in which such equilibria do not exist.

1.3 Relationship to Some Other Literature

We briefly discuss some other related papers. We point out that if prices are given, the principle of minimal product differentiation holds. However, this is no longer the case with three firms [15]. One situation of interest is what happens in equilibrium when firms locate sequentially rather than simultaneously. Prescott and Visscher [21] discuss sequential location choice where prices are given. Neven [20] extends this to situations where prices are chosen subsequent to entry. Others have discussed multiple dimensions of strategic choice. Economides [9] for instance introduces quality choice in the model of Hotelling [13] and shows that there exists an equilibrium with minimal differentiation with regard to price and quality and maximum differentiation with regard to location.

2 The Two-Stage Salop Model with Linear Costs

We consider a spatial model of imperfect competition with differentiated goods in the sense of Salop [24]. As in Salop's original setup, we consider *three* firms—denoted by 1, 2 and 3, respectively—that compete in a two-stage process. In the first stage, the three firms select a location on a perfect circle of unit length. In the second stage, the three firms simultaneously select a price of their good. Throughout we assume that the three firms compete perfectly in the price setting stage 2 of the game, using best response rationality, resulting in a Nash equilibrium in that stage of the game.

The modelling environment of the unit circle allows the use of distances between the three firms to represent the location choice in the first stage of the game. Indeed, let $0 \leq x_{ij} \leq 1$ denote the distance between the locations of firms i and j such that the third firm h is not located between them. Hence, $x_{12} + x_{13} + x_{23} = 1$. In particular, $x_{ij} = \frac{1}{2}$ if the firms i and j are located on opposite locations on the unit circle. Note that the distance between firms is maximal at $x_{ij} = 1 - \delta$ if firms i and j are located at a distance of $\delta > 0$ with the third firm h located between them.

For reasons of tractability, we assume that two firms cannot locate at the same point but must maintain a distance of at least $\varepsilon > 0$ from each other. This assumption can be justified by minimum space required to set up the physical location. This makes the model in question an abstract economy rather than a game.

Assumption 1 *There exists a parameter $0 < \varepsilon \leq \frac{1}{3}$ such that $x_{ij} \geq \varepsilon$ for any $i, j \in \{1, 2, 3\}$ with $i \neq j$.*

We show below that for ε as introduced in the assumption, there exists an equilibrium in the two-stage model.

For notational completeness, we use $p_i > 0$ as the price set by firm $i \in \{1, 2, 3\}$ in the second stage of the two-stage model.

2.1   Consumers

At every location on the unit circle we assume that there resides a consumer. Each consumer purchases one unit of the good produced by one of the three firms on the circle. If consumer $x \in [0, 1)$ purchases one unit from firm $i \in \{1, 2, 3\}$, then the consumer attains a gross utility level $V_x > 0$ from consuming the good and has disutility from paying the price p_i and from the transportation cost to bridge the distance to collect the good from firm i. The transportation cost is assumed to be *linear* with a cost $t > 0$ per unit of distance. Hence, the utility of consumer x can be determined as

$$U_x(i) = V_x - p_i - t \cdot d(i, x) \tag{1}$$

where $d(i, x) \geq 0$ is the distance from location x to the location of firm i on the unit circle. This linear cost model is the same as seminally studied by Hotelling [13].

Throughout we apply the standard assumption that all consumers $x \in [0, 1)$ optimise the net utility formulated above. Hence, the consumer x's problem is to solve $\max_{i \in \{1, 2, 3\}} U_x(i)$.

2.2 Formalisation of the Two-Stage Salop Model

Using the descriptors of the decision variables of the three firms and the defined behaviour of the consumers, we can now state the model formally. The **two-stage Salop model with linear costs** for three firms consists of two decision stages:

Stage 1: The three firms select locations on the unit circle represented by the distances x_{12}, x_{13} and x_{23} satisfying $x_{ij} > \varepsilon$ for all $i,j \in \{1,2,3\}$ with $i \neq j$.

Stage 2: Given the locations selected in Stage 1, all three firms simultaneously select prices $p_i \geq 0$, $i \in \{1,2,3\}$, for their respective products and sell their products to the consumers on the unit circle according to the linear cost model formulated in (1).

The objective of the firms in this model is to maximise their net profits, defined as their revenues minus their total production costs. The profit function will be determined by the assumptions made about the cost structure in this model.

The two-stage Salop model is now solved through the application of *backward induction* in which we first determine an equilibrium in the second stage of the model for every set of firm locations and, subsequently, solve the location problem in the first stage of the model, given the solution of the second stage.

2.3 The Existence of a Solution in the Linear Cost Model

Consider a consumer located between firms i and j on the unit circle. Assume further that each firm $k \in \{1,2,3\}$ sets a price $p_k \geq 0$ and that this consumer is located at a distance $d \in [0, x_{ij}]$ from firm i and a distance $x_{ij} - d$ from firm j. Then this consumer is indifferent between purchasing the product from firm i and firm j, if

$$p_i + td = p_j + t(x_{ij} - d).$$

This is equivalent to

$$2td = (p_j - p_i) + tx_{ij} \quad \text{implying} \quad d = \left(\frac{p_j - p_i}{2t}\right) + \frac{x_{ij}}{2}.$$

Thus, the demand of firm i can be determined as

$$D_i = \left(\frac{p_j - p_i}{2t}\right) + \left(\frac{p_h - p_i}{2t}\right) + \frac{x_{ij}}{2} + \frac{x_{ih}}{2}. \tag{2}$$

If we assume an uniform marginal production cost of $c \geq 0$ across all firms, the profit of firm i is given by

$$\pi_i = (p_i - c)D_i$$

$$= (p_i - c)\left[\left(\frac{p_j - p_i}{2t}\right) + \left(\frac{p_h - p_i}{2t}\right) + \frac{x_{ij}}{2} + \frac{x_{ih}}{2}\right]. \tag{3}$$

Using this formulation we can now show the following solution to the second stage problem and determine the equilibrium prices for all firms, given their locations.

Proposition 1 *Let the three firms be located on the unit circle represented by the distances x_{12}, x_{13} and x_{23} satisfying $x_{ij} > \varepsilon$ for all $i,j \in \{1,2,3\}$ with $i \neq j$. Then the Nash equilibrium in the second stage pricing game of the two-stage Salop model with linear costs is given by*

$$p_1 = c + \frac{t}{5}(2 - x_{23}); \tag{4}$$

$$p_2 = c + \frac{t}{5}(2 - x_{13}); \tag{5}$$

$$p_3 = c + \frac{t}{5}(2 - x_{12}). \tag{6}$$

resulting in equilibrium profit levels given by

$$\pi_1 = \frac{t}{25}(2 - x_{23})^2; \tag{7}$$

$$\pi_2 = \frac{t}{25}(2 - x_{13})^2; \tag{8}$$

$$\pi_3 = \frac{t}{25}(2 - x_{12})^2. \tag{9}$$

Proof Let $i \in \{1,2,3\}$. Differentiating the computed profit function π_i with respect to p_i, the first order conditions are given by

$$\frac{\partial \pi_i}{\partial p_i} = \left[\left(\frac{p_j - p_i}{2t}\right) + \left(\frac{p_h - p_i}{2t}\right) + \frac{x_{ij}}{2} + \frac{x_{ih}}{2}\right] - \left(\frac{p_i - c}{t}\right) = 0.$$

Hence, we get a linear simultaneous equation system of three equations in three unknowns given by

$$\left[\left(\frac{p_2 - p_1}{2t}\right) + \left(\frac{p_3 - p_1}{2t}\right) + \frac{x_{12}}{2} + \frac{x_{13}}{2}\right] - \left(\frac{p_1 - c}{t}\right) = 0; \tag{10}$$

$$\left[\left(\frac{p_1 - p_2}{2t}\right) + \left(\frac{p_3 - p_2}{2t}\right) + \frac{x_{12}}{2} + \frac{1 - x_{12} - x_{13}}{2}\right] - \left(\frac{p_2 - c}{t}\right) = 0; \tag{11}$$

$$\left[\left(\frac{p_1 - p_3}{2t}\right) + \left(\frac{p_2 - p_3}{2t}\right) + \frac{x_{13}}{2} + \frac{1 - x_{12} - x_{13}}{2}\right] - \left(\frac{p_3 - c}{t}\right) = 0. \tag{12}$$

Solving the system of equations (10)–(12), we get the expressions asserted in the proposition. This completes the proof of Proposition 1. □

Note that from Proposition 1, if all firms are equidistantly located with $x_{ij} = \frac{1}{3}$ for all $i \neq j$, we arrive at $p_i = c + \frac{t}{3}$, which is the result obtained by Salop [24].

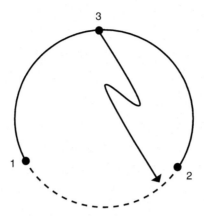

Fig. 1 The competitive relocation among three firms

Given the solution of the pricing stage of the two-stage Salop model with linear costs, we now turn to the solution of the location problem that forms the first stage of this model.

Proposition 2 *Given the equilibrium prices stated in Proposition 1, every location configuration of the three firms on the unit circle represented by the distances x_{12}, x_{13} and x_{23} satisfying $\varepsilon \leqslant x_{ij} \leqslant \frac{1}{2}$ for all $i, j \in \{1, 2, 3\}$ with $i \neq j$ and $x_{12} + x_{13} + x_{23} = 1$ is an equilibrium in the first stage of the two-stage Salop model with linear costs.*

Proof The assertion follows from the fact that, if the pricing rules stated in Proposition 1, the firm's profit is independent of its location from (7)–(9). Furthermore, the condition that $x_{ij} \leqslant \frac{1}{2}$ implies that firm h will not switch its location as illustrated in Fig. 1. On the other hand, such a switch is beneficial if $x_{ij} > \frac{1}{2}$. □

From Proposition 2 it is clear that the two-stage Salop model with linear costs is largely undetermined due to the large class of equilibria in the first stage. This is reduced significantly reduced by the introduction of different forms of competitive behaviour. See also Kats [14].

2.4 Introducing Partial Cooperation in the Linear Cost Model

In the two-stage Salop model with linear costs we now introduce the idea of cooperation by a subgroup of firms in their location decisions. Here we consider that firms 1 and 2 cooperate in the first stage of location selection, while firm 3 acts purely competitively. In the second stage all three firms are assumed to be fully competitive.

Using the concepts of partial cooperative equilibrium and leadership equilibrium developed in Chakrabarti et al. [2], we assume that the coalition $C = \{1, 2\}$

seeks to maximise their total profit by selecting their locations collectively. Hence, the objective function of the coalition C in the location stage of the model is formulated as

$$\Pi_C = \pi_1 + \pi_2 = \frac{t}{25}(1 + x_{12} + x_{13})^2 + \frac{t}{25}(2 - x_{13})^2. \tag{13}$$

Therefore, under the partial cooperative equilibrium concept in the location selection stage, firms 1 and 2 collectively select locations to maximise Π_C, while firm 3 selects a location to maximise its profit function π_3. Thus, the coalition $C = \{1, 2\}$ acts as a single decision maker with objective function Π_C in a two-player normal form game with firm 3.

Proposition 3 *Given the second-stage equilibrium prices stated in Proposition 1, there does not exist a partial cooperative equilibrium for the location stage of the two-stage Salop model with linear costs.*

Proof Consider the collective payoff function Π_C in (13). Then one can easily check that for all location configurations

$$\frac{\partial \Pi_C}{\partial x_{12}} = \frac{2t}{25}(1 + x_{12} + x_{13}) > 0. \tag{14}$$

This implies that the coalition of cooperators $C = \{1, 2\}$ aims to maximise their internal distance x_{12}. Therefore, $x_{12} = 1 - 2\varepsilon$. Therefore,

$$\frac{\partial \Pi_C}{\partial x_{13}} = \frac{2t}{25}(x_{12} + 2x_{13} - 1) = \frac{4t}{25}(x_{13} - 2\varepsilon), \tag{15}$$

implying that selecting a minimal value for x_{13} indeed maximises Π_C. Hence, $x_{12} = 1 - 2\varepsilon$ and $x_{13} = \varepsilon$ represent a best response to the location of firm 3.

Therefore, the cooperators try to locate on either side of firm 3, to which firm 3 responds by switching location to the opposite side of the unit circle—as illustrated in Fig. 1. This implies there emerges a process of perpetual relocation. This excludes the existence of a stable tuple of locations in this game, thereby, establishing the assertion that there is no partial cooperative equilibrium in this situation.   □

On the other hand, if the coalition of cooperators $C = \{1, 2\}$ is given a Stackelberg-like leadership position, our analysis leads to the conclusion that there exist two leadership equilibrium configurations.

Proposition 4 *Given the second-stage equilibrium prices stated in Proposition 1, there exist two leadership equilibrium configurations in the location stage of the two-stage Salop model given by*

$$x_{12}^a = \tfrac{1}{2} \ x_{13}^a = \varepsilon \qquad x_{23}^a = \tfrac{1}{2} - \varepsilon$$
$$x_{12}^b = \tfrac{1}{2} \ x_{13}^b = \tfrac{1}{2} - \varepsilon \ x_{23}^b = \varepsilon$$

Proof The way we have set up the model, it seems that initially player 3's profit depends entirely on the distance between 1 and 2 which 3 does not control under the leadership assumption. So, 3 is a passive best responder to the location decision of the coalition of cooperators $C = \{1, 2\}$. But this is slightly misleading, since 3 can switch location as shown in Fig. 1.

So, as illustrated in Fig. 1, given the locations of 1 and 2, 3 selects either to locate in the opposite side of the circle or to locate in the dotted zone. Given that π_3 is strictly decreasing in x_{12}, firm 3 will not switch locations if and only if $x_{12} \leqslant 1 - x_{12}$. Hence, $x_{12} \leqslant \frac{1}{2}$.

Hence, the above and (14) imply that x_{12} is selected to be maximal given the best response of firm 3, i.e., $x_{12} = \frac{1}{2}$. Furthermore from (15), we can compute for $x_{12} = \frac{1}{2}$ that

$$\frac{\partial \Pi_C}{\partial x_{13}} = \frac{t}{25} (4x_{13} - 1)$$

and, therefore, Π_C is minimised at $x_{13} = \frac{1}{4}$ and maximised at the minimal value of x_{13}, i.e., $x_{13} = \varepsilon$ or at its maximal value, i.e., $x_{13} = \frac{1}{2} - \varepsilon$. Finally, we note that firm 3 is indifferent between locating anywhere between the dispersed locations of firms 1 and 2. Hence, we conclude that there are two leadership equilibrium configurations as asserted in Proposition 4.   □

It is clear that if the identity of the cooperating firms in C is removed, there emerges only a single configuration, where both cooperators locate on opposite sides of the unit circle and the competitive firm locates between them.

In fact, independently of where firm 3 locates, its profit is computed as $\pi_3 = 0.09t$ at an equilibrium price $p_3 = c + 0.3t$. Therefore, firm 3 is completely indifferent where to locate between the cooperating firms 1 and 2, although only the location at minimal distance from either of them is an equilibrium.

3 Extensions of the Two-Stage Salop Model

We consider some extensions of the two-stage Salop model with linear costs that has been discussed in the previous section. We look at two modifications, namely (1) the introduction of the endogenous, strategic determination of the marginal production costs through costly investment in production technology and (2) the case of a quadratic cost structure as considered in the related literature such as Economides [10]. In both extensions we limit our investigations to the consideration of partial cooperation in location choice in a three firm environment.

3.1   Introducing Endogenous Marginal Production Costs

Thus far, we have assumed that there are uniform marginal costs fixed at some $c \geq 0$ across the three firms. However, the marginal costs may be made a function of a costly, strategically selected reduction $Y_i \geq 0$ for each firm i. We introduce a marginal cost function for every firm $i \in \{1, 2, 3\}$ with

$$c_i(Y_i) = \gamma - Y_i \qquad (16)$$

where $\gamma > 0$ is a given upper bound on the firm's marginal cost.

We assume that marginal cost reduction can be accomplished through additional investment in the firm's production technology. In particular, we consider the case that a marginal cost reduction of $Y_i \geq 0$ is given by an investment of the size Y_i^2 in the firm's production technology.

We let these marginal cost reductions be chosen competitively by the firms after the location has been selected and before prices are determined. Hence, we restructure the two-stage Salop model as a three stage model, where first the firms choose location, followed by investment and then prices.

This extends the model as follows. The **three-stage Salop model with linear costs** for three firms consists of three decision stages given by

Stage 1:   The three firms select locations on the unit circle represented by the distances x_{12}, x_{13} and x_{23} satisfying $x_{ij} \geq \varepsilon$ for all $i, j \in \{1, 2, 3\}$ with $i \neq j$. We assume here that there is partial cooperation with $C = \{1, 2\}$ the coalition of cooperators.

Stage 2:   The three firms $i = 1, 2, 3$ simultaneously determine the investment in production technology to reduce its marginal costs with $Y_i \geq 0$.

Stage 3:   Given the locations selected in Stage 1 and the marginal cost reductions established in Stage 2, all three firms simultaneously select prices $p_i \geq 0$, $i \in \{1, 2, 3\}$, for their respective products.

We emphasise that we assume that firms are fully competitive in stages 2 and 3 of this process, while partial cooperation only occurs in the selection of locations of firms 1 and 2 in the first stage.

With reference to the original profit function in (3), we determine that in this modified three-stage model the objective of firm $i \in \{1, 2, 3\}$ is to maximise their net profits now determined as

$$\pi_i = (p_i - \gamma + Y_i) \left[\left(\frac{p_j - p_i}{2t} \right) + \left(\frac{p_h - p_i}{2t} \right) + \frac{x_{ij}}{2} + \frac{x_{ih}}{2} \right] - Y_i^2. \qquad (17)$$

Using the revised formulation of the net profit function (17) we can now state a modification of Proposition 1 for this three-stage Salop model with endogenous marginal cost. A proof is directly derived from the first order conditions from (17) and, therefore, omitted.

Proposition 5 *In the third price-setting stage of the three-stage Salop model for given locations of the firms and given marginal cost reductions Y_1, Y_2 and Y_3, the firms set prices given by*

$$p_1^* = \gamma + \frac{1}{5}\left[t(2 - x_{23}) - (3Y_1 + Y_2 + Y_3)\right]; \tag{18}$$

$$p_2^* = \gamma + \frac{1}{5}\left[t(2 - x_{13}) - (Y_1 + 3Y_2 + Y_3)\right]; \tag{19}$$

$$p_3^* = \gamma + \frac{1}{5}\left[t(2 - x_{12}) - (Y_1 + Y_2 + 3Y_3)\right]. \tag{20}$$

Now, using the results in Proposition 5, we can solve the second stage of the game. We substitute (18)–(20) in (17) and differentiate it with respect to Y_i, $i = 1, 2, 3$ to arrive at the first order conditions for the second stage of the three-stage Salop model with endogenous marginal costs. Solving the resulting system of equations, we get the following:

Proposition 6 *Given the solution of the third stage stated in Proposition 5, the solution of the second investment stage of the three-stage Salop model with endogenous marginal costs is given by*

$$Y_1^* = \frac{2[5t(2 - x_{23}) - 2]}{5(25t - 6)}; \tag{21}$$

$$Y_2^* = \frac{2\left[5t(2 - x_{13}) - 2\right]}{5(25t - 6)}; \tag{22}$$

$$Y_3^* = \frac{2\left[5t(2 - x_{12}) - 2\right]}{5(25t - 6)}. \tag{23}$$

Using the values of investments in (21) to (23) of Proposition 6 into (17) to get the following reduced form profit functions:

$$\pi_1 = \frac{(25t - 4)(5t(2 - x_{23}) - 2)^2}{25(25t - 6)^2};$$

$$\pi_2 = \frac{(25t - 4)(5t(2 - x_{13}) - 2)^2}{25(25t - 6)^2};$$

$$\pi_3 = \frac{(25t - 4)(5t(2 - x_{13}) - 2)^2}{25(25t - 6)^2}.$$

Using these reduced form net profit functions, we can finally determine the first location stage of the three-stage Salop model with endogenous marginal costs under partial cooperation. Indeed,

$$\frac{\partial \Pi_C}{\partial x_{12}} = \frac{2t(25t - 4)(5t(1 + x_{12} + x_{13}) - 2)}{5(25t - 6)^2} > 0 \tag{24}$$

for a sufficiently large t. Hence, in equilibrium x_{12} is again chosen to be maximal, i.e., $x_{12} = \frac{1}{2}$. Therefore,

$$\frac{\partial \Pi_C}{\partial x_{13}} = \frac{2t^2(25t - 4)(2x_{13} + x_{12} - 1)}{5(25t - 6)^2} = \frac{t^2(25t - 4)(4x_{13} - 1)}{5(25t - 6)^2}.$$

As in the proof of Proposition 4 we conclude that $x_{13} = \varepsilon$ or $x_{13} = \frac{1}{2} - \varepsilon$. Hence, the set of leadership equilibria do not change from the established insights for the two-stage Salop model with linear costs.

Proposition 7 *Given the third-stage equilibrium prices and second-stage equilibrium investments stated in Propositions 5 and 6, respectively, there exist two leadership equilibrium configurations in the location stage of the three-stage Salop model with endogenous marginal costs given by*

$$x_{12}^a = \frac{1}{2} \; x_{13}^a = \varepsilon \qquad x_{23}^a = \frac{1}{2} - \varepsilon$$
$$x_{12}^b = \frac{1}{2} \; x_{13}^b = \frac{1}{2} - \varepsilon \qquad x_{23}^b = \varepsilon$$

3.2 Quadratic Cost Structure

Finally we turn to the study of a quadratic cost structure that has been employed in the literature on Hotelling's model to guarantee the existence of a solution. We now impose a cost structure where for every consumer $x \in [0, 1)$ the utility of consuming one unit of the good from firm $i \in \{1, 2, 3\}$ is given by

$$U_x(i) = V_x - p_i - t \cdot d(i, x)^2 \tag{25}$$

where p_i is the price charged by firm i and $t > 0$ is the given transportation cost parameter. This situation has both an advantage and a disadvantage when compared to linear costs. On one hand, it is computationally complex, but on the other hand it guarantees an interior solution to the utility maximisation problem. This implies that we can dispense with the use of the assumption that firms locate at least a distance $\varepsilon > 0$ from each other. This changes the set of solutions to the two-stage Salop model fundamentally.

We denote the two-stage model that emerges under this cost structure as the *two-stage Salop model with quadratic costs* to distinguish from the case of linear costs discussed in the previous section.

As before, we determine the consumer who is indifferent between adjacent firms i and j with $i \neq j$. If a consumer is located at a distance d from firm i, then this consumer is indifferent between purchasing from either i or j if

$$p_i + td^2 = p_j + t(x_{ij} - d)^2. \tag{26}$$

This is equivalent to

$$p_i - p_j = tx_{ij} \left(x_{ij} - 2d \right).$$

Hence, we derive that for the indifferent consumer it holds that

$$d = \frac{p_j - p_i}{2tx_{ij}} + \frac{x_{ij}}{2}. \tag{27}$$

Thus, the demand of firm i is given by

$$D_i = \left(\frac{p_j - p_i}{2tx_{ij}} \right) + \left(\frac{p_h - p_i}{2tx_{ih}} \right) + \frac{x_{ij}}{2} + \frac{x_{ih}}{2}. \tag{28}$$

Therefore, the resulting profit of firm i is derived as

$$\pi_i = (p_i - c)D_i$$
$$= (p_i - c) \left[\left(\frac{p_j - p_i}{2tx_{ij}} \right) + \left(\frac{p_h - p_i}{2tx_{ih}} \right) + \frac{x_{ij}}{2} + \frac{x_{ih}}{2} \right]. \tag{29}$$

Using these findings, we can now formulate the solution for the special case of equidistant location of the three firms on the unit circle that emerges in the second, price setting stage of the two-stage Salop model with quadratic costs:

Proposition 8 *Let the three firms be located equidistantly on the unit circle represented by distances* $x_{12} = x_{13} = x_{23} = \frac{1}{3}$. *Then the Nash equilibrium in the second price setting stage of the two-stage Salop model with quadratic costs is given by*

$$p_1 = p_2 = p_3 = c + \frac{t}{9}. \tag{30}$$

Proof Differentiating profits with respect to p_i, the first order conditions are given by

$$\frac{\partial \pi_i}{\partial p_i} = \left[\frac{p_j - p_i}{2tx_{ij}} + \frac{p_h - p_i}{2tx_{ih}} + \frac{x_{ij}}{2} + \frac{x_{ih}}{2} \right] - \left(\frac{p_i - c}{t} \right) \left(\frac{1}{2x_{ij}} + \frac{1}{2x_{ih}} \right) = 0.$$

Hence, we have a system of three simultaneous equations in three unknowns:

$$\left[\frac{p_2 - p_1}{2tx_{12}} + \frac{p_3 - p_1}{2tx_{13}} + \frac{x_{12}}{2} + \frac{x_{13}}{2} \right] - \left(\frac{p_1 - c}{t} \right) \left(\frac{1}{2x_{12}} + \frac{1}{2x_{13}} \right) = 0; \tag{31}$$

$$\left[\frac{p_1 - p_2}{2tx_{12}} + \frac{p_3 - p_2}{2tx_{23}} + \frac{x_{12}}{2} + \frac{x_{23}}{2} \right] - \left(\frac{p_2 - c}{t} \right) \left(\frac{1}{2x_{12}} + \frac{1}{2x_{23}} \right) = 0; \tag{32}$$

$$\left[\frac{p_1 - p_3}{2tx_{13}} + \frac{p_2 - p_3}{2tx_{23}} + \frac{x_{13}}{2} + \frac{x_{23}}{2} \right] - \left(\frac{p_3 - c}{t} \right) \left(\frac{1}{2x_{13}} + \frac{1}{2x_{23}} \right) = 0. \quad (33)$$

For the case of equidistantly located firms Eqs. (31)–(33) solve to the price levels as asserted in the proposition. □

3.2.1 Nash Equilibrium

As a benchmark in the analysis of partial cooperation in location choice, we replicate the result of Economides [10] that firms in equilibrium are equidistant from each other. If $p_i = c + \frac{t}{9}$ for all i, then

$$\pi_1 = \frac{t}{9} \left(\frac{x_{12}}{2} + \frac{x_{13}}{2} \right) = \frac{t}{18} (1 - x_{23}) \,;$$

$$\pi_2 = \frac{t}{9} \left(\frac{x_{12}}{2} + \frac{x_{23}}{2} \right) = \frac{t}{18} (1 - x_{13}) \,;$$

$$\pi_3 = \frac{t}{9} \left(\frac{x_{13}}{2} + \frac{x_{23}}{2} \right) = \frac{t}{18} (1 - x_{12}) \,.$$

Clearly, starting from a situation of equidistant locations, no movement can make the players better off as $\frac{\partial \pi_1}{\partial x_{12}} = \frac{\partial \pi_2}{\partial x_{23}} = \frac{\partial \pi_3}{\partial x_{13}} = 0$. Hence, we conclude that this indeed forms a Nash equilibrium.

Property 1 Equidistant locations determined by $x_{ij}^* = \frac{1}{3}$ for all $i, j \in \{1, 2, 3\}$ with $i \neq j$ form a Nash equilibrium in the first, location stage of the two-stage Salop model with quadratic costs.

3.2.2 Partial Cooperative Equilibrium[4]

The equation system that describes decisions under the hypothesis of partial cooperation between firms 1 and 2, while firm 3 remains competitive, in the first location stage of the two-stage Salop model with quadratic costs is so complex that an exact solution could not be obtained straightforwardly using calculus. However, we can solve numerically for an approximate solution.

First, solving (31)–(33) to get equilibrium prices, which are subsequently substituted in (29), we can now numerically solve the system of two simultaneous equations given by

$$\frac{\partial \Pi_C}{\partial x_{12}} = 0 \quad \text{and} \quad \frac{\partial \pi_3}{\partial x_{23}} = 0 \quad (34)$$

---

[4]Calculations regarding the reported outcomes here are available upon request.

to arrive at an approximation of a partial cooperative equilibrium in this two-stage Salop model with quadratic costs.

Numerically approximating a solution of the two resulting equations for x_{12} and x_{23}, we arrive at $x_{12} = 0.375$ and $x_{23} = 0.322$. This shows that there is a non-trivial solution and, therefore, we can conclude the following:

Property 2 The location stage of the two-stage Salop model with quadratic costs admits at least one partial cooperative equilibrium.

Acknowledgements We thank an anonymous referee for valuable comments. We are grateful for the support of Emiliya Lazarova and Lina Mallozzi in the development of this paper.

References

1. Bertrand, J.: Book review of 'recherches sur les principes mathématiques de la théorie des richesses'. J. Savants **67**, 499–508 (1883)
2. Chakrabarti, S., Gilles, R.P., Lazarova, E.: Strategic behaviour under partial cooperation. Theor. Decis. **71**(2), 175–193 (2011)
3. Chakrabarti, S., Gilles, R.P., Lazarova, E.: Partial cooperation in strategic multi-sided decision situations. Working Paper, Queen's Management School, Belfast (2016)
4. Chamberlin, E.H.: The Theory of Monopolistic Competition: A Re-orientation of the Theory of Value. Harvard University Press, Cambridge, MA (1933)
5. Cournot, A.: Recherches sur les Principes Mathématiques de la Théorie des Richesses. Hachette, Paris (1838)
6. d'Aspremont, C., Gabszewicz, J.J., Thisse, J.F.: On hotelling's 'stability in competition'. Econometrica **47**, 1145–1150 (1979)
7. Dubey, P., Mas-Colell, A., Shubik, M.: Efficiency properties of strategic market games: an axiomatic approach. J. Econ. Theory **22**, 339–362 (1980)
8. Economides, N.: Maximal and minimal product differentiation in Hotelling's duopoly. Econ. Lett. **21**, 67–71 (1986)
9. Economides, N.: Quality variations and maximal variety differentiation. Reg. Sci. Urban Econ. **19**, 21–29 (1989)
10. Economides, N.: Symmetric equilibrium existence and optimality in differentiated product markets. J. Econ. Theory **47**, 178–194 (1989)
11. Edgeworth, F.Y.: Mathematical Psychics: An Essay on the Application of Mathematics to the Moral Sciences. C. Kegan Paul & Co., London (1881)
12. Edgeworth, F.Y.: The pure theory of monopoly. In: Collected Papers relating to Political Economy, 1925 edn. Macmillan Press/Royal Economic Society, London (1889)
13. Hotelling, H.: Stability in competition. Econ. J. **39**, 41–57 (1929)
14. Kats, A.: More on Hotelling's stability in competition. Int. J. Ind. Organ. **13**, 89–93 (1995)
15. Lerner, A., Singer, H.: Some notes on duopoly and spatial competition. J. Polit. Econ. **45**, 145–186 (1937)
16. Mallozzi, L., Tijs, S.: Conflict and cooperation in symmetric potential games. Int. Game Theory Rev. **10**, 245–256 (2008)
17. Mallozzi, L., Tijs, S.: Partial cooperation and multiple non-signatories decision. Czech Econ. Rev. **2**, 23–30 (2008)
18. Mallozzi, L., Tijs, S.: Coordinating choice in partial cooperative equilibrium. Econ. Bull. **29**, 1–6 (2009)
19. Monderer, D., Shapley, L.S.: Potential Games. Games Econ. Behav. **14**, 124–143 (1996)

20. Neven, D.J.: Endogenous sequential entry in a spatial model. Int. J. Ind. Organ. **5**, 419–434 (1987)
21. Prescott, E.C., Visscher, M.: Sequential location among firms with foresight. Bell J. Econ. **8**, 378–393 (1987)
22. Robinson, J.: The Economics of Imperfect Competition. St. Martin's Press, London (1933)
23. Salant, S., Switzer, S., Reynolds, R.: Losses from horizontal merger: the effects of an exogenous change in industry structure on Cournot-Nash equilibrium. Q. J. Econ. **98**, 185–199 (1983). http://www.jstor.org/stable/1885620
24. Salop, S.: Monopolistic competition with outside goods. Bell J. Econ. **10**(1), 141–156 (1979)
25. von Stackelberg, H.: Marktform und Gleichgewicht. University of Vienna, Habilitation (1934)
26. Walras, L.: Éléments d'économie politique pure; ou, Théorie de la richesse sociale. F. Rouge, Paris (1874)
27. Walras, L.: Théorie mathématique de la richesse sociale. Imprint Corbaz (1883)
28. Walras, L.: Elements of Pure Economics, or the Theory of Social Wealth. Richard D. Irwin Inc., Homewood, IL (1954). Translated by William Jaffe

A Class of Location Games with Type Dependent Facilities

Imma Curiel

1 Introduction

In a location problem one has to find an optimal way of locating a given number of facilities in a given space. Optimality refers here to some preference relation that is attached to the set of feasible ways to locate the facilities. Usually this preference relation is given as a profit or cost function that depends on the distance between the locations of the facilities and the locations of the points of some given, finite set. This is a very general description and any particular location problem studied in the literature has specifications attached to it to define it more precisely. The location problem can be continuous or discrete. In the continuous case the set of feasible locations for the facilities is a metric space (usually R^n or a subset of R^n with $n = 2$ or 3). In the discrete case the set of feasible locations for the facilities is a finite set. In the discrete case the points can be considered to belong to a metric space and the distance between two points is then given by the metric; or they can be the nodes of a graph and the distance between two points is measured using the length of the edges of the graph. Overview of location problems, solving methodologies and applications can be found in [4, 7].

Cooperative game theory is concerned with the analysis of situations in which a group of actors (most commonly called players) can combine forces to achieve a goal. Usually this goal will be to maximize profit or to minimize costs. The idea here is that "the whole is more than the sum of the parts" implying that the profit generated by everybody involved working together will be more than the sum of the profits if everybody works apart or, when the goal is to minimize costs, the total costs if everybody works together is less than the sum of the costs if everybody

I. Curiel (✉)
Anton de Kom University of Suriname, Leysweg 86, Paramaribo, Suriname
e-mail: immacuriel@gmail.com

© Springer International Publishing AG 2017
L. Mallozzi et al. (eds.), *Spatial Interaction Models*, Springer Optimization
and Its Applications 118, DOI 10.1007/978-3-319-52654-6_3

works apart. The question that arises is how the profit or costs generated should be distributed among the players. What are the properties that we want a profit or cost allocation to have? In [9, 10] one can find overviews of cooperative games with their properties and allocation rules with their properties.

The analysis of the properties of cooperative games and their solution concepts in general without referring to the situations that give rise to a particular game constituted the main focus of the research in cooperative game theory after the initial introduction of cooperative games in [12]. However, the last 40 years there has been increasing attention for analysis that takes into account the underlying situation that gives rise to a particular type of cooperative game. Linear production games introduced in [8] and combinatorial optimization games as discussed in [2, 3] are but two examples of such classes of games. Cooperative location games in which a location problem determines the game form another class. Since location problems can vary greatly depending on the specifics of the problem cooperative location games also can vary greatly. An overview of location games can be found in [5].

Non-cooperative game theory studies situations in which the players cannot make binding agreements. Therefore, they cannot coordinate their actions. The players have strategies and the outcome of the game depends on which strategy each player chooses. In a Nash equilibrium of a non-cooperative game each player chooses a strategy and no player has an incentive to unilaterally deviate from his strategy because this will not lead to an improvement of the outcome of the game for him. In [6] a detailed overview of non-cooperative games and Nash equilibria and their refinements is given.

In this paper we consider a new class of location games. The games we introduce arise from location problems on a graph in which more than one facility needs to be located. Differently from the p-facilities games discussed in [1, 2] the facilities in the games we consider here differ with respect to the type of service that they provide. The communities that need the services provided by the facilities can decide to cooperate. We will discuss a cooperative game model and a non-cooperative game model. The remainder of the paper is organized as follows. In Sect. 2 the location problem is introduced and its characteristics are discussed. It is shown that, in general, a cooperative game arising from this location problem can have an empty core. In Sect. 3 properties that are sufficient to guarantee the non-emptiness of the core of these cooperative games are introduced and the proof of the non-emptiness of the core for games that satisfy these properties is given. In Sect. 4 the non-cooperative model is discussed and we consider the nature of the Nash equilibria of the game. We conclude with some ideas for further research in Sect. 5.

2   The Model

We consider a set of communities which are geographically separated and need services that can be provided by facilities that have to be built. The communities belong to a greater administrative entity (county, state, province) which is responsible for building the facility and provides the funding for doing this. This entity has the power to set restrictions with respect to the number of facilities that may be built and the locations where they may be built. It wants to find a balance between providing as much services as possible to the citizens in the communities and preventing waste of resources by having facilities being built that are not fully utilized. Another example of such a governmental structure is given by a community of more or less autonomous islands for which the former colonizing state (=the larger administrative entity) still provides aid in establishing certain service facilities. An example of this last situation is given by the former islands of the Netherlands Antilles and The Netherlands.

Let $N = \{1, 2 \ldots, n\}$ be the set of communities in need of services that can be provided by facilities that are members of the set of facilities $\mathscr{F} = \{F_1, F_2, F_3, \ldots, F_p\}$. Each facility $F_r \in \mathscr{F}$ provides a different type of service and each community is in need of each type of service. The communities are connected by roads (or ferries or air traffic). The set of communities with their connections can be modeled as an undirected graph $G = <N, E>$ where N is the set of nodes of the graph G and equals the set of communities and E is the set of edges of G. Since G is undirected each edge $e \in E$ is a two element subset of N. Each edge $\{i, j\} = e \in E$ has a positive number $l_e = l_{ij}$ associated with it. This number can be viewed as the unit cost of providing the service of a facility located in one of the nodes to a person located in the other node. We assume that G is connected. In graph theoretic terms we consider l_e to be the length of edge $e \in E$. The distance $d(i, j) = d(j, i)$ between two nodes $i, j \in N$ is defined to be the length of a shortest path from i to j. The length of a path is the sum of the lengths of the edges belonging to the path. The administrative entity stipulates that the facilities can be located in the communities (nodes of the graph) with at most one facility in each community (node). The costs for community i related with the service provided by facility F_r is proportional to the distance between i and the node where F_r is located. The proportionality constant is denoted by $w_i(F_r) \geq 0$. The proportionality constant $w_i(F_r)$ can be seen as a measure of how many citizens of i will need the service provided by F_r. With $a(F_r) \in N$ denoting the location of F_r, the costs of i related with the service provided by F_r equals

$$w_i(F_r)d(i, a(F_r)). \tag{1}$$

Each community i has costs associated with not having access at all to the type of service provided by a facility. The costs of community i if it does not have access to the service provided by facility F_r is denoted by $L_i(F_r) \geq 0$. The amount $L_i(F_r)$ can be seen as the loss incurred by community i due to the lack of access to the service provided by facility F_r. A community i is not allowed to decline the services of a facility F_r if F_r is available even if $w_i d(i, a(F_r) > L_i(F_r)$. Each community is allowed to build only one facility and controls access to the facility built in the location of the community. At most one of each type of facility may be built. The communities can and will have to cooperate if they want access to more than one type of service. The problem that arises is how to allocate the total costs among the cooperating communities. Just letting each community be responsible for its own costs may not provide enough incentive for the communities to cooperate. By modeling this situation as a cooperative game we can consider core elements of the game as appropriate cost allocations. Recall that the core $C(c)$ of a cost game $< N, c >$ is defined as

$$C(c) = \{x \in R^n | \sum_{i \in N} x_i = c(N), \sum_{i \in S} x_i \leq c(S) \text{ for all } S \in 2^N \setminus \{\emptyset\}\}. \qquad (2)$$

The following example illustrates this.

Example 1 Let N={1,2,3} be the set of communities. Let $\mathscr{F} = \{F_1, F_2, F_3\}$ be the set of facilities. The proportionality constants are given below.

$$w_1(F_1) = 2 \; w_1(F_2) = 1 \; w_1(F_3) = 1$$
$$w_2(F_1) = 3 \; w_2(F_2) = 3 \; w_2(F_3) = 3$$
$$w_3(F_1) = 1 \; w_3(F_2) = 2 \; w_3(F_3) = 3$$

The losses are

$$L_1(F_1) = 8 \; L_1(F_2) = 6 \; L_1(F_3) = 6$$
$$L_2(F_1) = 7 \; L_2(F_2) = 7 \; L_2(F_3) = 7$$
$$L_3(F_1) = 5 \; L_3(F_2) = 6 \; L_3(F_3) = 6.$$

The graph $< N, V >$ is K_3, the complete graph with 3 nodes. The lengths of the edges are $l_{12} = 1$, $l_{13} = 5$, $l_{23} = 3$. Let $< N, c >$ be the cooperative cost game arising from this situation. The players in this game are the communities. Each player when working alone will build a facility that it needs the most in its node. So player i will build facility F_{i_1} in the node corresponding to i where

$$F_{i_1} = \arg \max_{F_r \in \mathscr{F}} L_i(F_r) \qquad (3)$$

Any tie may be broken arbitrarily. The cost of player i will be the sum of the losses that it incurs by not having access to the facilities beside F_{i_1}. This yields $c(\{1\}) = 12$, $c(\{2\}) = 14$, $c(\{3\}) = 11$. A coalition consisting of two players i and j must decide if it will build zero, one, or two facilities, which one(s) it will build, and where it will put which one. The coalition will choose the option that minimizes it costs. For the coalition consisting of players 1 and 2 this means placing F_1 in node 1 and F_2 or F_3 in node 2. This results in

$$c(\{1, 2\}) = L_1(F_3) + L_2(F_3) + w_1(F_2)d(1, 2) + w_2(F_1)d(1, 2) \qquad (4)$$
$$= 6 + 7 + 1 + 3 = 17.$$

Similarly, we find $c(\{1, 3\}) = 20$, $c(\{2, 3\} = 25$, and $c(\{1, 2, 3\}) = 27$. One may be inclined to divide the costs of 27 equally among the three players. However, this would result in total costs of 18 for players 1 and 2 together which is more than the 17 they incur when working without 3. So, they will not accept this . Instead a core element can be used to divide the costs of 27. The core of this game is $co(\{(2, 14, 11), (9, 7, 11), (3, 14, 10), (10, 7, 10)\})$. Here $co(A)$ is the convex hull of the set A.

In the following we will give a formal definition of a *location game with type dependent facilities*. Let N, G, $l : E \rightarrow R_+$, $\mathscr{F}$, $L_i : \mathscr{F} \rightarrow R_+$ for all $i \in N$, $w_i : \mathscr{F} \rightarrow R_+$ for all $i \in N$ be as described above. For each coalition $S \in 2^N \setminus \{\emptyset\}$ we define the collection $\mathscr{A}_S$ of subsets of $\mathscr{F}$ that S is allowed to build by

$$\mathscr{A}_S = \{A \subset \mathscr{F} \,||A| \leq |S|\}. \tag{5}$$

The definition of $\mathscr{A}$ reflects the fact that S may build at most $|S|$ facilities. For each non-empty coalition S and each element A of $\mathscr{A}_S$ we define a permissible assignment of facilities to nodes belonging to S as a one-to one function

$$a_A^S : A \rightarrow S. \tag{6}$$

Let P_A^S denote the set of all permissible assignments for S and $A \in \mathscr{A}_S$. The cost of a non-empty coalition S in a location game with type dependent facilities $< N, c >$ is given by

$$c(S) = \min_{A \in \mathscr{A}_S, a_A^S \in P_A^S} \sum_{i \in S} (\sum_{F \in \mathscr{F} \setminus A} L_i(F) + \sum_{F \in A} w_i(F) d(i, a_A^S(F)). \tag{7}$$

In Example 1 we discussed a location game with type dependent facilities with a non-empty core. As Example 2 shows, the core of a location game with type dependent facilities can be empty.

Example 2 Let N={1,2,3} be the set of communities. Let $\mathscr{F} = \{F_1, F_2, F_3\}$ be the set of facilities. The proportionality constants are given below.

$$w_1(F_1) = 2 \; w_1(F_2) = 1 \; w_1(F_3) = 1$$
$$w_2(F_1) = 1 \; w_2(F_2) = 1 \; w_2(F_3) = 1$$
$$w_3(F_1) = 1 \; w_3(F_2) = 2 \; w_3(F_3) = 3$$

The losses are

$$L_1(F_1) = 1 \; L_1(F_2) = 1 \; L_1(F_3) = 1$$
$$L_2(F_1) = 2 \; L_2(F_2) = 2 \; L_2(F_3) = 2$$
$$L_3(F_1) = 1 \; L_3(F_2) = 2 \; L_3(F_3) = 2.$$

The graph $< N, V >$ is K_3, the complete graph with 3 nodes. The lengths of the edges are $l_{12} = 1$, $l_{13} = 5$, $l_{23} = 3$. The location game that corresponds to this situation has $c(\{1\}) = 2$, $c(\{2\}) = 4$, $c(\{3\}) = 3$, $c(\{1, 2\}) = 5$, $c(\{1, 3,\}) = 8$,

$c(\{2, 3\}) = 10$, and $c(\{1, 2, 3\}) = 14$. Since it is impossible for $x \in R^3$ to satisfy

$$x_1 + x_2 \leq 5$$
$$x_1 + x_3 \leq 8$$
$$x_2 + x_3 \leq 10$$
$$x_1 + x_2 + x_3 = 14$$

simultaneously, it follows that this game has an empty core.

In the next section we will study conditions that will guarantee that a location game with type dependent facilities will have a non- empty core.

3 Location Games with Type Dependent Facilities with a Non-empty Core

In this section we will study conditions that guarantee the non-emptiness of the core of the location games with type dependent facilities that were introduced in Sect. 2. Let the finite set N, the connected graph $G = < N, E >$, the length function $l : E \rightarrow R_+$, the finite set $\mathcal{F}$, the loss functions $L_i : \mathcal{F} \rightarrow R_+$ for all $i \in N$, and the weight functions $w_i : \mathcal{F} \rightarrow R_+$ for all $i \in N$ give rise to the location game with type dependent facilities $< N, c >$. For all $i \in N$ we denote the maximum distance between i and the other nodes in N by D_i. That is,

$$D_i = \max_{j \in N} d(i, j). \tag{8}$$

We will first consider situations with $|N| = |\mathcal{F}|$. It is clear that if the loss incurred by a player for not having access to a facility is small compared to the costs of using a facility that is located in another node, there will be less incentive for the player to cooperate with the others and the core of the game is likely to be empty. In fact, this was the case in Example 2. This inspires the following definition.

Definition 1 A location game with type dependent facilities $< N, c >$ is said to satisfy the *no access big loss*-property, or shorter, the NABL-property if

$$L_i(F_r) \geq w_j(F_s)D_j \text{ for all } i, j \in N, \ F_r, F_s \in \mathcal{F}. \tag{9}$$

Note that in Definition 1 i may equal j and F_r may equal F_s.

The following property has to do with the way the players rank the facilities with respect to their need for each facility.

Definition 2 A location game with type dependent facilities $< N, c >$ is said to satisfy the *equal ranking*, or ER-property if

$$L_i(F_r) \geq L_i(F_s) \Leftrightarrow L_j(F_r) \geq L_j(F_s) \text{ for all } i,j \in N, \ F_r, F_s \in \mathscr{F}. \tag{10}$$

The ER-property implies that we can rank the facilities from most desired, i.e., the one whose omission would cause the biggest loss for every player, to least desired, the one whose omission would cause the smallest loss for every player. Any ties that occur may be resolved arbitrarily. Without loss of generality and for ease of notation we assume that this ranking is as follows: $F_1, F_2, \ldots, F_n$. Given this ranking we define the *equal ranking with respect to weights* property as follows.

Definition 3 A location game with type dependent facilities $< N, c >$ is said to satisfy the *equal ranking with respect to weights*, or ERW-property if

$$w_i(F_1) \geq w_i(F_2) \geq w_i(F_3) \geq \ldots \geq w_i(F_n) \text{ for all } i \in N. \tag{11}$$

The next property we introduce has to do with the magnitude of the difference between the losses caused by the absence of two facilities that are ranked subsequently.

Definition 4 A location game with type dependent facilities is said to satisfy the *large loss difference* or LLD- property if

$$L_i(F_r) - L_i(F_{r+1}) \geq (w_i(F_r) - w_i(F_{r+1}))D_i \text{ for all } i \in N, \ r \in \{1, 2, \ldots, n\}. \tag{12}$$

The next theorem uses three of these properties to show that every coalition with the same size will build the same facilities.

Theorem 1 *Let $< N, c >$ be a location game with type dependent facilities that satisfies the NABL-, ER-, and LLD-properties. Then for every coalition of size $|S|$, its minimum costs $c(S)$ are achieved by building the facilities $F_1, F_2, \ldots, F_{|S|}$.*

Proof Let $S \subset N$. First we will show that the minimum in (7) is achieved for a set A with $|A| = |S|$. Let $B \subset \mathscr{F}$ with $|B| < |S|$. Let a_B^S be an optimal way of assigning the facilities in B to the nodes in S. Let $F_r \in \mathscr{F} \setminus B$. Place F_r in a node of S that did not get any facility assigned to it by a_B^S. Assume that this is node j. Then the change in costs will equal

$$\sum_{i \in S, i \neq j} w_i(F_r)d(i, j) - \sum_{i \in S} L_i(F_r) < 0. \tag{13}$$

Here the inequality follows from the NABL-property. It follows that the minimum is achieved for a set with cardinality equal to $|S|$.

Suppose A is a subset of $\mathscr{F}$ with $|A| = |S|$, and $A \neq \{F_1, F_2, \ldots, F_{|S|}\}$. Let $F_t \in A$ but not in $\{F_1, F_2, \ldots, F_{|S|}\}$ and let $F_r \in \{F_1, F_2, \ldots, F_{|S|}\}$ but not in A. Replace F_t by F_r. Assume that F_t was located in node $j \in S$. Then the change in costs will equal

$$\sum_{i \in S} (w_i(F_r) - w_i(F_t))d(i,j) - \sum_{i \in S} (L_i(F_r) - L_i(F_t)) < 0. \qquad (14)$$

The inequality follows from the LLD-property. So, for any set with cardinality $|S|$ the minimum costs, $c(S)$, are achieved by building the facilities $F_1, F_2, \ldots, F_{|S|}$. $\square$

In the proof of the non-emptiness of the core of a location game that satisfies the four properties mentioned above we will use the fact that a related *permutation game* has a non-empty core. Permutation games were introduced in [11] where it was shown that they are totally balanced. This implies that every permutation game has a non-empty core. To define a permutation game with n players an $n \times n$-matrix K is given. The cost of a coalition S in a permutation game $< N, c >$ is given by

$$c(S) = \min_{\pi \in \Pi_S} \sum_{i \in S} K_{i\pi(i)}. \qquad (15)$$

Here K_{ij} is the entry in row i and column j of the matrix K and Π_S is the set of permutations of S.

Given the sets and functions that define a location game with type dependent facilities $< N, c >$, we define a $n \times n$-matrix K by

$$K_{ij} = \sum_{k \in N} w_k(F_j)d(k,i). \qquad (16)$$

Recall that $w_k(F_j)d(k,i)$ is equal to the costs for player k to access facility F_j if facility F_j is located in node i. Let $< N, c' >$ be the permutation game associated with this matrix K. The following theorem describes the relationship between $< N, c >$ and $< N, c' >$.

Theorem 2 *Let $< N, c >$ be a location game with type dependent facilities that satisfies the NABL-, ER-, ERW-, and LLD-properties. Let $< N, c' >$ be the permutation game associated with the matrix K as defined in (16). Then*

1. $c(N) = c'(N)$ and
2. $c(S) \geq c'(S)$ for all $S \subset N$.

Proof By Theorem 1 the grand coalition N will build all n facilities. The cost for the grand coalition associated with building facility F_j in node i is given by K_{ij} as defined in (16). So, both $c(N)$ and $c'(N)$ are given by

$$\min_{\pi \in \Pi_N} \sum_{i \in N} K_{i\pi(i)}. \qquad (17)$$

It follows that $c(N) = c'(N)$.

Let S be a subset of N with $S \neq \emptyset, N$. We construct two $|S| \times |S|$-submatrices of the matrix K. The first $|S| \times |S|$-matrix K^S we obtain by deleting every column j with $j \notin S$ and every row i with $i \notin S$ from the matrix K. The second $|S| \times |S|$-matrix K'

we obtain by deleting every column j with $j > |S|$ and every row i with $i \notin S$ from K. From the *ERW*-property it follows that

$$K'_{ij} \geq K^S_{ij} \text{ for all } i,j \in S. \tag{18}$$

In the following $\Pi_{|S|}$ denotes the set of permutations of the set $\{1,2,\ldots,|S|\}$. From the way that c' is defined it follows that

$$c'(S) = \min_{\pi \in \Pi_{|S|}} \sum_{i=1}^{i=|S|} K^S_{i\pi(i)}. \tag{19}$$

Similarly, we define a cooperative cost game $< N,d >$ associated with the matrix K' by

$$d(S) = \min_{\pi \in \Pi_{|S|}} \sum_{i=1}^{i=|S|} K'_{i\pi(i)}. \tag{20}$$

Let the minimum in (20) be attained for π_* and let

$$\sigma : \{1,2\ldots,|S|\} \to S \tag{21}$$

be such that $\sigma(i)$ is that element of S that corresponds to the i-th row in K^S and K'. From (18) it follows that $d(S) \geq c'(S)$. Therefore,

$$c(S) = d(S) - \sum_{j=1}^{|S|} \sum_{k \notin S} w_k(F_j)d(\sigma(\pi_*^{-1}(j)),k) + \sum_{j=|S|+1}^{n} \sum_{k \in S} L_k(F_j) \geq d(S) \geq c'(S). \tag{22}$$

Here the equality follows from the definitions of $c(S)$ and $d(S)$ and the first inequality follows from the NABL-property. $\qquad \square$

The non-emptiness of the core of the location game with type dependent facilities is now easily established.

Theorem 3 *Let* $< N,c >$ *be a location game with type dependent facilities that satisfies the NABL-, ER-, ERW-, and LLD-properties. Then the core* $C(c)$ *of* $< N,c >$ *is not empty.*

Proof Because $< N,c' >$ is a permutation game it has a non-empty core. Because $c(N) = c'(N)$ and $c(S) \geq c'(S)$ for all $S \subset N$ it follows that every core-element of $< N,c' >$ is also a core element of $< N,c >$. $\qquad \square$

If the number of facilities is greater than the number of players, i.e. $|\mathscr{F}| > |N|$, and the game satisfies the NABL-, ER-, ERW-, and LLD-properties it is easy to see that the core will be non-empty. A core element can be constructed as follows. Let $< N,c >$ be the location game with all $|\mathscr{F}|$ facilities and let the truncated location game $< N, c^t >$ be the game we obtain when we consider only the $|N|$ top ranked facilities. From Theorem 3 it follows that $C(c^t) \neq \emptyset$. Let $x^t \in C(c^t)$. Then x defined by

$$x_i = x_i^t + \sum_{r=|N|+1}^{|\mathscr{F}|} L_i(F_r) \text{ for all } i \in N \tag{23}$$

is an element of the core $C(c)$ of the game $< N, c >$.

If the number of facilities is less than the number of players a trivial example that shows that the core can be empty is to consider the case $|\mathscr{F}| = 1$. Then $c(\{i\}) = 0$ for all $i \in N$ while $c(S) > 0$ for all S with at least two members. A less trivial example is given below.

Example 3 Let $N = \{1, 2, 3\}$. Let $G =< N, E >$ be the complete graph on three nodes. The lengths of the edges are $l_{12} = 3$, $l_{13} = 10$, $l_{23} = 5$. Let $\mathscr{F} = \{F_1, F_2\}$. The proportionality constants are given below.

$$w_1(F_1) = 3 \ w_1(F_2) = 2$$
$$w_2(F_1) = 5 \ w_2(F_2) = 4$$
$$w_3(F_1) = 4 \ w_3(F_2) = 3$$

The losses are

$$L_1(F_1) = 45 \ L_1(F_2) = 34$$
$$L_2(F_1) = 54 \ L_2(F_2) = 37$$
$$L_3(F_1) = 60 \ L_3(F_2) = 52.$$

Then the NABL-, ER-, ERW-, and LLD-properties are satisfied and $c(\{1\}) = 34$, $c(\{2\}) = 37$, $c(\{3\}) = 52$, $c(\{1, 2\}) = 21$, $c(\{1, 3\}) = 48$, $c(\{2, 3\}) = 40$, and $c(\{1, 2, 3\}) = 65$. The core of this game is empty since it is impossible for an $x \in R^3$ to satisfy

$$x_1 + x_2 \leq 21$$

$$x_1 + x_3 \leq 48$$

$$x_2 + x_3 \leq 40$$

$$x_1 + x_2 + x_3 = 65$$

simultaneously.

Although in the case that $|\mathscr{F}| < |N|$ adding additional players to a coalition S with $|S| = |\mathscr{F}|$ would not lead to more facilities being built the core of the game can still be non-empty as the following example shows.

Example 4 Let $N = \{1, 2, 3\}$. Let $G =< N, E >$ be the complete graph on three nodes. The lengths of the edges are $l_{12} = 1$, $l_{13} = 6$, $l_{23} = 6$. The proportionality constants are

$$w_1(F_1) = 4 \; w_1(F_2) = 3$$
$$w_2(F_1) = 2 \; w_2(F_2) = 2$$
$$w_3(F_1) = 1 \; w_3(F_2) = 1$$

The losses are the same as in Example 3. The NABL-, ER-, ERW-, and LLD-properties are satisfied. The costs of the coalitions are $c(\{1\}) = 34$, $c(\{2\}) = 37$, $c(\{3\}) = 52$, $c(\{1, 2\}) = 5$, $c(\{1, 3\}) = 24$, $c(\{2, 3\}) = 18$, $c(\{1, 2, 3\}) = 17$. The core of this game is $co(\{(-1, -7, 25), (-1, 6, 12), (12, -7, 12)\})$.

4   A Non-cooperative Approach

In the previous sections we introduced an analyzed situations in which a set of communities that belong to a greater administrative entity can cooperate in the building of facilities that provide services that are needed by the communities. In this section we analyze what happens if the communities can not make binding agreements. We cannot use a cooperative game anymore. Instead we will look at a non-cooperative game. We assume that $|\mathscr{F}| = |N|$. The rules of the game are as follows. The greater administrative entity allows every community to indicate which facility it wants to have built in its geographical location. If only one community chooses facility F_r then facility F_r will be built in the location of that community. If s communities choose facility F_r then a lottery will be conducted to decide where facility F_r will be built. A facility that is not chosen by any player will not be built. Each one of the s communities will have probability $\frac{1}{s}$ of having facility F_r built in its location. The game we will consider is a game of complete information. Each community has all the information about the losses, proportionality constants, and distances of all communities. Each community also knows the decision rule that the administrative entity will apply. The communities have to submit their choice simultaneously. Each community wants to minimize its expected costs. The following example uses the data from Example 2 to generate this non-cooperative game.

Example 5 Let N, $G =< N, E >$, l_e for $e \in E$, $\mathscr{F}$, L_i and w_i for $i \in N$ be as in Example 2. In the non-cooperative location game with type dependent facilities arising from this situation each player has three pure strategies. The costs associated with the 3-tuples of pure strategies are given below.

$$\begin{pmatrix} (\; 5\frac{1}{3}, 5\frac{1}{3}\; , 6\frac{1}{3}\;) & (\; 6, \quad 4, \quad 10\;) & (\; 6, \quad 4, \quad 13\;) \\ (\; 6, \quad 4\frac{1}{2}, 11\frac{1}{2}\;) & (\; 9\frac{1}{2}, 5\frac{1}{2}, 9\;) & (\; 9, \quad 4, \quad 17\;) \\ (\; 6, \quad 4\frac{1}{2}, 7\frac{1}{2}\;) & (\; 9, \quad 4, \quad 18\;) & (\; 9\frac{1}{2}, 5\frac{1}{2}, 11\frac{1}{2}\;) \end{pmatrix}$$

$$
\begin{pmatrix}
(6, & 5\frac{1}{2}, 5\frac{1}{2}) & (3\frac{1}{2}, 4\frac{1}{2} & ,9) & (5, & 4, & 13) \\
(5\frac{1}{2}, 4, & 9) & (3\frac{2}{3}, 5\frac{1}{3}, 5\frac{1}{3}) & & (5\frac{1}{2}, 4, & 12) \\
(6, & 4, & 10) & (3\frac{1}{2}, 4\frac{1}{2}, 16) & (5\frac{1}{2}, 5\frac{1}{2}, 12\frac{1}{2})
\end{pmatrix}
$$

$$
\begin{pmatrix}
(6, 5\frac{1}{2}, 5\frac{1}{2}) & (5, & 4, & 10) & (3\frac{1}{2}, 4\frac{1}{2}, 9\frac{1}{2}) \\
(6, 4, & 11) & (5\frac{1}{2}, 5\frac{1}{2}, 8) & (3\frac{1}{2}, 4\frac{1}{2}, 11\frac{1}{2}) \\
(5, 4, & 11) & (4, & 4, & 13) & (3\frac{2}{3}, 5\frac{1}{3}, 10)
\end{pmatrix}
$$

The first, second, and third matrix corresponds to the pure strategies F_1, F_2, F_3, respectively, of player 3. In each matrix the first, second and third row corresponds to the pure strategies F_1, F_2, F_3, respectively, of player 1. For player 2 the first, second, and third column in each matrix corresponds to the pure strategies F_1, F_2, F_3, respectively. By inspection we see that this game has two Nash equilibria in pure strategies. They are (F_2, F_1, F_2) and (F_3, F_2, F_3).

In a non-cooperative location game with type dependent facilities the set of pure strategies of every player is the set $\mathscr{F}$. Let $x = (x_1, x_2, \ldots, x_n) = (F_{i_1}, F_{i_2}, \ldots, F_{i_n})$ be an n-tuple of pure strategies of the n players. For every $F_r \in \mathscr{F}$ let $|x_{F_r}|$ denote the number of times that F_r occurs in x. Let $P_x(F_r)$ denote the set of players that choose F_r in x. The cost function u_i for player $i \in N$ is given by

$$
u_i(x_1, x_2, \ldots, x_n) = \sum_{r=1, |x_{F_r}| \neq 0}^{n} \frac{1}{x_{F_r}} \sum_{j \in P_x(F_r)} w_i(F_r) d(i, j) + \sum_{r=1, |x_{F_r}| = 0}^{n} L_i(F_r). \tag{24}
$$

In Example 5 none of the two Nash equilibria in pure strategies corresponds to choices of the players that would lead to all three facilities being built. This cannot happen if the game satisfies the strict NABL-property, that is, the NABL-property with the inequality replaced by a strict inequality. This is shown in Theorem 4.

Theorem 4 Let $x = (x_1, x_2 \ldots, x_n)$ be a Nash equilibrium in pure strategies in a non-cooperative game with type dependent facilities that satisfies the strict NABL-property. Then $|x_{F_r}| = 1$ for all $F_r \in \mathscr{F}$.

Proof Let x be such that there is a $F_r \in \mathscr{F}$ such that $|x_{F_r}| \neq 1$. Let F_r be such that $|x_{F_r}| > 1$ and let F_t be such that $|x_{F_t}| = 0$. Consider a player i with $i \in P_x(F_r)$. Define y by

$$
y_j = \begin{cases} x_j & \text{if } j \neq i, \\ F_t & \text{otherwise.} \end{cases}
$$

Then

$$
u_i(x) - u_i(y) = L_i(F_t) - \frac{1}{|x_{F_r}||y_{F_r}|} \sum_{j \in P_x(F_r)} w_i(F_r) d(i, j) > 0. \tag{25}
$$

Here the equality follows from the definition of y and the inequality follows from the strict NABL-property. Since player i can do better by unilaterally changing his strategy it follows that x is not a Nash equilibrium. □

From the proof of Theorem 4 it is easy to see that every $x = (x_1, x_2, \ldots, x_n)$ with $|x_{F_r}| = 1$ for all $F_r \in \mathscr{F}$ is a Nash equilibrium. That leads to the following theorem.

Theorem 5 *The set of Nash equilibria of a non-cooperative location game with type dependent facilities that satisfies the strict NABL-property is equal to the set of n-tuples of strategies in which each player chooses a different facility.*

Proof By Theorem 4 and its proof. □

So if the strict NABL-property is satisfied each Nash equilibrium will lead to every facility being built.

5   Concluding Remarks

In the previous sections we analyzed, in a cooperative and a non-cooperative setting, location situations with type dependent facilities. In both settings a crucial aspect of the analysis was the magnitude of the loss incurred by the players if they didn't have access to a certain facility. We gave sufficient conditions for the core of a cooperative game with type dependent facilities to be non-empty. However, these conditions are not necessary as Example 1 shows. If the facilities under consideration are necessary for the saving of lives (like certain types of medical treatments) the NABL-property is a natural property to occur. The ER- and ERW-properties will also occur naturally in such situations if the medical conditions that require certain treatments have approximately the same distributions in all the communities. The LLD-property is less intuitive. It compares the difference between the losses of a community due to the lack of access to two facilities with the difference between the costs when these facilities are as far as possible from the community. Other conditions that guarantee non-emptiness of the core could be found. It would be especially interesting to find conditions that are necessary and sufficient. We did not consider setup costs for the facilities. All costs involved in building the facilities were assumed to be handled by an administrative entity that was not one of the players but controlled the rules of the game being played. Further research could look at an analysis of the situation when the players themselves are responsible for the setup costs of the facilities. Another research direction is the study of how a game with an empty core could be broken in to smaller games with non-empty cores in a way that is as cost efficient as possible.

References

1. Curiel, I.: Location games. Research Report 90-20, Department of Mathematics and Statistics University of Maryland, Baltimore County (1990)
2. Curiel, I.: Cooperative Game Theory and Applications. Kluwer, Dordrecht (1997)
3. Curiel, I.: Cooperative combinatorial games. In: Chinchuluun, A., Pardalos, P., Migdalas, A., Pitsoulis, L. (eds.) Pareto Optimality, Game Theory and Equilibria, pp. 131–157. Springer, New York (2008)
4. Drezner, Z., Hamacher, H.: Facility Location. Springer, Berlin (2002)
5. Fragnelli, V., Gagliardo, S.: Open problems in cooperative location games. Int. Game Theor. Rev. **15** (2013). doi:10.1142/S021919891340015X
6. Maschler, M., Solan, E., Zamir, S.: Game Theory. Cambridge University Press, Cambridge (2013)
7. Nickel, S., Puerto, J.: Location Theory: A Unified Approach. Springer, Berlin (2005)
8. Owen, G.: On the core of linear production games. Math. Prog. **9**, 358–370 (1975)
9. Owen, G.: Game Theory. Academic press, Orlando (1995)
10. Peleg, B., Sudhölter, P.: Introduction to the Theory of Cooperative Games. Kluwer, Boston (2003)
11. Tijs, S., Parthasarathy, T., Potters, J., Rajendra Prasad, V.: Permutation games: another class of totally balanced games. OR Spektrum **6**, 119–123 (1984)
12. von Neumann, J., Morgenstern, O.: Theory of Games and Economic Behavior. Princeton University Press, Princeton (1944)

Location Methods and Nash Equilibria for Experimental Design in Astrophysics and Aerospace Engineering

Elia Daniele, Pierluigi De Paolis, Gian Luca Greco, and Alessandro d'Argenio

Notation

α_1, α_2	Positive real number
β_1, β_2	Positive real number
δ	Positive real number
Γ	Game
Ω	Rectangular region in two-dimensional space
a	Positive constant
b	Positive constant
c	Positive constant
f	Payoff function
n	Number of test points
x	Player's strategy or cartesian coordinate
y	Cartesian coordinate
AoA	Angle of Attack
DoE	Design of Experiment
ED	Experiments Design
E_i	Experiments
FT	Flight Tests
H	Pressure altitude

E. Daniele (✉)
Fraunhofer IWES Institute for Wind Energy and Energy System Technology, Küpkersweg 70, 26129 Oldenburg, Germany
e-mail: elia.daniele@iwes.fraunhofer.de

P. De Paolis • G.L. Greco • A. d'Argenio
Italian Air Force - Flight Test Centre, Via Pratica Di Mare 45, 00040 Pomezia (RM), Italy
e-mail: pierluigi.depaolis@aeronautica.difesa.it; gianluca.greco@aeronautica.difesa.it; alessandro.dargenio@aeronautica.militare.it

© Springer International Publishing AG 2017
L. Mallozzi et al. (eds.), *Spatial Interaction Models*, Springer Optimization and Its Applications 118, DOI 10.1007/978-3-319-52654-6_4

K	Payoff function
L	Payoff function
M	Mach number
N	Finite players' set
NE	Nash Equilibrium
P, Q	Experiment or test points
StE	Structural Engineering
SyE	Systems Engineering
U_E	Equivalent airspeed
X	Players' strategy set

1 Introduction

In this chapter a survey of location methods based on Nash equilibria for the design of experiments is presented. An experimental design, or design of experiment (DoE), is the detailed planning of testing activities. As the time and cost requirements nowadays could experience no difference between a numerical simulation and an experimental campaign, DoE would refer in general to the design of a test, either being numerical or experimental. The design phase is needed due to maximize the information gathered from tests and to solve some related issues, as their non-repeatability. Experimental design is an effective technique made for maximizing the amount of information gained from a test while minimizing the amount of data to be acquired, ensuring the assessment of valid and reliable conclusions. DoE can be used in several engineering problem-solving areas:

- Sensitivity analysis, whose goal is the determination of the dependency of the whole process on a single factor. It is used to discriminate among relevant and non-relevant factors [36].
- Modeling, whose scope is to find the good fitting between a model and an experimental data set to estimate the parameters characterizing the modeled process [20].
- Optimization, i.e. the determination of the optimal setting to achieve the best results [21].

These areas are strictly connected because usually the optimization process involves modeling and sensitivity analysis to increase the knowledge of the system under optimization. The location of receivers is a problem that can be included in DoE, because the scope is to determine the optimal location of a certain number of receivers [14] to maximize data gathering resulting from test execution. Location problems deal with finding the best location for one or more facilities such that some objectives are optimized [18, 19]. Generally one distinguishes three classes of location models depending on the domain of feasible locations: continuous location problem, network location problem and discrete location problem. If more that one facility has to be located one deals with a multi-facility location problem. Facility location models assume that the closer the facility is to demand, the better the objective is met.

The solution of a facility location problem is restricted to the bi-dimensional case and approached by means of a potential formulation and a Nash non-cooperative game. The most important definitions and proofs are reported in Sect. 2.

A game theoretical approach to a multi-facility continuous location problems was presented in [23], where each competitor controls one or more facilities trying to optimize an economic objective. This model is representative of a classical facility location problem common in Operation Research area. The solution of the location problem is a Nash equilibrium solution of the related game. The game has a peculiar structure, namely it is a potential game where the Nash equilibrium solutions are the minimum points of a function, that is called the potential function or potential [24, 29]. The numerical procedure to compute the *maxima* of the potential is based on a genetic algorithm [5, 8–10, 16, 32, 37]. Another case is also considered, where some experiments are already planned and the problem is to determine the optimal value of the design variables of a certain number of new experiments, that still optimize the same criteria of the previous set, but are taking into account the informations derived from the previous design. This situation corresponds to a multi-facility continuous location problem, where some existing facilities are already in the admissible region providing, within the game theoretical model, an additional term in the objective functions [27].

Two main application fields are employed to stress the capability of an ad hoc numerical methodology involved in the solution of the location problem. The first one refers to optimal (constrained) location of sensors collecting cosmic rays for astrophysics experiments [11, 12, 25, 26], see Sect. 3. This is related to the location of a certain number of receivers on ground, under uniform cosmic source distribution, with a bounded settlement area, further constrained to a limited number of receivers due to a budget cap. Assuming that the capture area of each receiver (e.g. a radar) has a circular shape, this application shares many aspects with the classic sphere packing problem [7, 22, 33] that has been applied in several fields (Fig. 1) and solved with algorithmic optimization procedures [4, 6, 17, 23, 28, 31, 34, 35]. In this case the greater difference between the typical location problem on a bounded domain and the considered location problem is the following: in the latter, the nature of the domain's boundaries is such to act as a cut-off line on which the receiver lost its efficacy or any other measure of profit. In other words, as in the sphere packing problem, the spheres are forced to stay within a limited bounded volume avoiding to consider their elasticity in reduce their size, in the examined location problem the receiver or sensor (for an experiment) loses a portion of its efficacy in collecting the signal (pressure, temperature, etc.) by allowing itself to be pushed on the boundary, because the information is limited within the same boundary (see Sect. 2.2). For this reason, a classical facility location problem with an additional requirement is studied: to locate the facilities far from the boundary of the admissible region.

The second application field presented in this manuscript regards the design of a test matrix for an envelope expansion flight test campaign [13, 27], see Sect. 4. In this case, the test points must be located inside the flight envelope of the airplane,

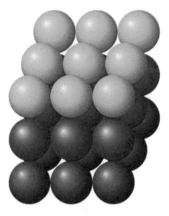

Fig. 1 Sketches for classic sphere packing problem in a cubic domain

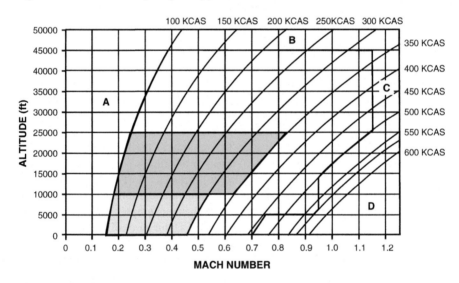

Fig. 2 Fighter type aircraft flight envelope

which is defined as the region where the aircraft is allowed to fly and is limited by the aircraft limits. The aircraft could be conducted outside the envelope flight only under a particular flight clearance and using a controlled test environment. Figure 2 shows a typical flight envelope for a fighter type aircraft with the limiting factors per each area where flying the aircraft is forbidden. Those areas are the following:

- A: Stall limit/High angle of attack (abbreviated as AoA).
- B: Engine performance (related to the service altitude).
- C: Compressibility, or Mach number effects.
- D: Dynamic pressure effects (related to structural limits).

Classical methods used to design test matrices for flight tests are the *Economy Methods* and the *Extensive Methods*. The Economy Methods concentrate on minimizing the number of test points in order to save money. In this case, a subset of flight test conditions (i.e. Mach number and pressure altitude) will be chosen considering a build-up approach in dynamic pressure in order to increase flight test safety. The Extensive Methods attempts to cover the majority of the flight envelope resulting very expensive and time consuming but also giving more data. Generally, due to money constraints, Economy Methods are more used in flight test.

The approach presented in [13, 27] consists of an alternative method to those previously mentioned. The main driving factors in this kind of DoE are the requirements to be demonstrated by two engineering departments: Structural Engineering (StE) and Systems Engineering (SyE). Generally, StE is more interested on true airspeed, load factor and compressibility effects on the aircraft structure, thus, requires optimized test points distribution in Mach number. SyE is instead more focused on environmental effects on the aircraft and its systems, thus, demanding an optimized distribution in pressure altitude.

However, it is fundamental to add an additional requirement that guarantees testing at the conditions corresponding to the maximum equivalent airspeed, which is a function of both Mach number and pressure altitude. This kind of problem can be approached as a particular non-cooperative game, where the two engineering departments represent the two players. The test matrix is designed in order to give the opportunity to both departments to optimize the test points distribution for their objectives considering the problem of a new store certification process.

The structure of this manuscript presents in Sect. 2 a discussion on the facility location problem, while in Sects. 3 and 4 are presented the results for, respectively, sensors location in experiments for astrophysics and test matrix for airplane flight test activities. The conclusions are summarized in Sect. 5.

2   Location Problem in Two Dimensions

In this section the location problem in a bi-dimensional space is discussed. Some preliminaries of Game Theory are presented in Sect. 2.1, where the definition of Nash equilibrium and exact potential game are reported. In Sect. 2.2 the experimental design and the facility location game definitions are recalled. Those informations are later used for the applications described in Sects. 3 and 4.

2.1   Preliminaries of Game Theory

Let consider an n-player normal form game Γ ($n \in \mathcal{N}$, where $\mathcal{N}$ is the set of natural numbers), that consists of a tuple

$$\Gamma =< N; X_1, \ldots, X_n; f_1, \ldots, f_n >$$

where $N = \{1, 2, \ldots, n\}$ is the finite player set, for each $i \in N$ the set of player i's strategies is X_i (i.e. the set of player i's admissible choices) and $f_i : X_1 \times \cdots \times X_n \rightarrow \mathscr{R}$ is player i's payoff function ($\mathscr{R}$ is the set of real numbers). It is supposed that players are cost minimizing, so that player i has a cost $f_i(x_1, x_2, \ldots, x_n)$ when player 1 chooses $x_1 \in X_1$, player 2 chooses $x_2 \in X_2$, and player n chooses $x_n \in X_n$. It is defined $X = X_1 \times \ldots \times X_n$ and for $i \in N$: $X_{-i} = \Pi_{j \in N \setminus \{i\}} X_j$. Let $\mathbf{x} = (x_1, x_2, \ldots, x_n) \in X$ and $i \in N$. In the following it is also used $\mathbf{x} = (x_i, \mathbf{x}_{-i})$, where $\mathbf{x}_{-i} = (x_1, \ldots, x_{i-1}, x_{i+1}, \ldots, x_n)$.

Definition 1 A *Nash equilibrium* [3] for Γ is a strategy profile $\hat{\mathbf{x}} = (\hat{x}_1, \hat{x}_2, \ldots, \hat{x}_n) \in X$ such that for any $i \in N$ and for any $x_i \in X_i$ one has that

$$f_i(\hat{\mathbf{x}}) \le f_i(x_i, \hat{\mathbf{x}}_{-i}).$$

It is denoted by $NE(\Gamma)$ the set of the Nash equilibrium strategy profiles. Any $\hat{\mathbf{x}} = (\hat{x}_1, \ldots, \hat{x}_n) \in NE(\Gamma)$ is a vector such that for any $i \in N$, $\hat{x}_i$ is solution to the optimization problem

$$\min_{x_i \in X_i} f_i(x_i, \hat{\mathbf{x}}_{-i}).$$

A very well known existence result of Nash equilibria is the theorem proved by J. Nash in 1950. Not always a game admits a Nash equilibrium solution. There are special situations in which this is true, for example in potential games. Potential games have been introduced by Monderer and Shapley: the idea is that a game is said potential if the information that is sufficient to determine Nash equilibria can be summarized in a single function on the strategy space, the potential function [24, 29].

Definition 2 A game $\Gamma =< N; X_1, \ldots, X_n; f_1, \ldots, f_n >$ is an *exact potential game* (or simply *potential game*) if there exists a function $V : \Pi_{i \in N} X_i \rightarrow \mathscr{R}$ such that for each player $i \in N$, each strategy profile $x_{-i} \in \Pi_{j \in N \setminus \{i\}} X_j$ of i's opponents, and each pair $x_i, y_i \in X_i$ of strategies of player i:

$$f_i(y_i, \mathbf{x}_{-i}) - f_i(x_i, \mathbf{x}_{-i}) = V(y_i, \mathbf{x}_{-i}) - V(x_i, \mathbf{x}_{-i}).$$

The function V is called an exact potential (or, in short, a *potential*) of the game Γ. If V is a potential function of Γ, the difference induced by a single deviation is equal to that of the deviator's payoff function. Clearly, by definition, the set of all strategy profiles that minimize V (called potential minimizers) is a subset of the Nash equilibrium set of the game Γ:

$$\underset{\mathbf{x} \in X}{argmin}\, V(\mathbf{x}) \subseteq NE(\Gamma).$$

This implies that in a potential game finding a Nash equilibrium means to solve an optimization problem, namely finding the minimum points of the potential function V.

2.2 DoE as a Facility Location Game

Let Ω be a rectangular region of $\mathcal{R}^2$. The model is restricted to the unit square $\Omega = [0, 1]^2$ without leading the generalities (rescaling the variables our results hold). The problem is to decide for two variables x and y the values of n available experiments ($n \in \mathcal{N}$). So one wants to settle n points $P_1, P_2, \ldots, P_n$ in the square in such a way that they are far as possible from the rest of the points and, in some cases (as the optimization of sensor devices location on the layer), from the boundary of the square. This implies to maximize the dispersion of the points. Each point is assigned to a virtual player, whose decision variables are the coordinates and whose pay-off function translates the dispersion in terms of distances.

Problem 1 (Experimental Design (ED)) The problem of deciding the values of two variables for n assigned experiments is to choose $P_1, \ldots, P_n \in \Omega$ maximizing the

$$dispersion(P_1, \ldots, P_n),$$

where the dispersion function is defined in a suitable way [14].

If the dispersion is translated with a suitable objective function, this problem corresponds to a classical facility location problem.

The problem could be stated in different ways, and one of the better known historical formalization are the Weber's problem or *minisum*, that minimizes the sum of weighted distances, and the *minimax* problem (von Neumann) that minimizes the maximal distance between facilities and demand points (Fig. 3). In a n facility location problem, one considers n virtual players each of them optimizing a suitable objective function in the admissible region. As in Problem 1 it is required to maximize the dispersion, imaging that each player wants to stay as far as possible from the opponents (and, in some cases, from the boundary of the region). So, a non-cooperative behavior emerges and a Nash equilibrium solution for this game is considered (Fig. 4).

In the following the focus lies on the situation where there is a competition between the points in the admissible region, because the dispersion depends on the mutual position of all the points and also on the distance with respect to the boundary of the region. It is assumed that there are n (virtual) players, i.e. the n points to determine, competing in order to choose the optimal locations: the location problem can be stated as a Nash equilibrium problem.

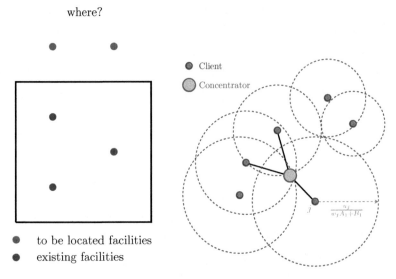

Fig. 3 Location problem statement as a sketch on the *left side*; Weber's problem or *minisum* on the *right side*

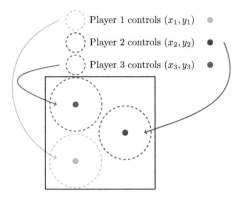

Fig. 4 Location problem as a game

Definition 3 Let $N = \{1, \ldots, n\}$ ($n \in \mathcal{N}$); the *facility location game* is the n-player strategic form game $\Gamma_n^{ED} =< N; \Omega; \{f_i, i \in N\} >$, where Ω and f_i are defined by the following assumptions:

(i) Each player i has to set up a new facility in a point $P^i \in \Omega \subset \mathcal{R}^2$, where Ω is the compact set of the feasible locations for each player, i.e. his strategy set.

(ii) Each point P_i has to be far away as possible from $\partial\Omega$, the boundary of Ω.

(iii) The function $d(P, Q)$ is a measure of the distance between any two points P and Q in $\mathcal{R}^2$.

(iv) For any $i \in N$, $b_i, c_i : [0, +\infty[\rightarrow \mathcal{R}$ are lower semi-continuous and increasing functions.

(v) The new facilities will be locate in $(\hat{P}_1, \ldots, \hat{P}_n) \in \Omega^n$ such that each player i wants to minimize the total cost $f_i : A \to \mathcal{R}$ defined by

$$f_i(P_1, \ldots, P_n) = \sum_{1 \leq j \leq n, j \neq i} b_i \left(\frac{1}{d(P_i, P_j)} \right) + c_i \left(\frac{1}{d(P_i, \partial\Omega)} \right)$$

being $A = \{(P_1, \ldots, P_n) \in \Omega^n : P_i \in (]0, 1[)^2, P_i \neq P_j \, \forall i, j = 1, \ldots, n, \}$ $\{j \neq i\}$ and $d(P, \partial\Omega) = \min_{Q \in \partial\Omega} d(P, Q)$.

The first $n - 1$ terms in the definition of f_i depend on the inverse distance between the point P_i and the rest of the points, where the last term is a decreasing function of the distance of P_i from the boundary of the square.

It may happen that after locating a number k ($k \in \mathcal{N}$) of location points, i. e. facility location positions that correspond to design variable vectors, a budget variation occurs and it is possible to locate n ($n \in \mathcal{N}$) additional facility location points taking into account that in the admissible region there are already k of them (demand points). In this dynamic situation the following location game is defined [27].

Definition 4 Let $k \in \mathcal{N}$ and $\{Q_1, \ldots, Q_k\}$ ($Q_j \in \Omega, j = 1, \ldots, k$); the facility location game $\Gamma_{n,k}^{ED} = <N; \Omega; \{f_i, i \in N\}; k, \{Q_1, \ldots, Q_k\} >$ is defined as follows:

(i) Each player i has to set up a new facility in a point $P^i \in \Omega \subset \mathcal{R}^2$, where Ω is the compact set of the feasible locations for each player, i.e. his strategy set.
(ii) Each point P_i has to be far away as possible from each demand point $Q_j, j = 1, \ldots, k$, and from $\partial\Omega$, the boundary of Ω.
(iii) The function $d(P, Q)$ is a measure of the distance between any two points P and Q in $\mathcal{R}^2$.
(iv) For any $i \in N$, $a_i, b_i, c_i : [0, +\infty[\to \mathcal{R}$ are lower semi-continuous and increasing functions.
(v) The new facilities will be locate in $(\hat{P}_1, \ldots, \hat{P}_n) \in \Omega^n$ such that each player i wants to minimize the total cost $f_i : A \to \mathcal{R}$ defined by

$$f_i(P_1, \ldots, P_n) = \sum_{1 \leq j \leq k} a_i \left(\frac{1}{d(P_i, Q_j)} \right) + \sum_{1 \leq j \leq n, j \neq i} b_i \left(\frac{1}{d(P_i, P_j)} \right) + c_i \left(\frac{1}{d(P_i, \partial\Omega)} \right)$$

being $A = \{(P_1, \ldots, P_n) \in \Omega^n : P_i \in (]0, 1[)^2, P_i \neq P_l \, \forall i \neq l, P_i \neq Q_j\}$ with $i, l = 1, \ldots, n$ and $j = 1, \ldots, k$ and $d(P, \partial\Omega) = \min_{Q \in \partial\Omega} d(P, Q)$.

The game Γ_n^{ED} is a particular case of the game $\Gamma_{n,k}^{ED}$, i.e. without demand points that requires $a_i = 0$ for any $i = 1, \ldots, n$ in Definition 4.

Any Nash equilibrium solution of the game $\Gamma_{n,k}^{ED}$ is an optimal solution of the problem (ED), for which a sketch in Fig. 5 is illustrated.

In the next section are presented some theoretical as well computational results concerning the facility location games Γ_n^{ED} and $\Gamma_{n,k}^{ED}$.

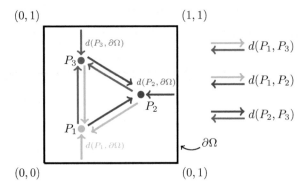

Fig. 5 Sketch for the game of location problem (*ED*)

3 Sensor Location in Experiment for Astrophysics

In the following we denote $P_i = (x_i, y_i)$, $i \in N$. Next definition specifies the solution of the problem (*ED*) in terms of Cartesian coordinates.

Definition 5 Any $(\hat{x}_1, \hat{y}_1, \ldots, \hat{x}_n, \hat{y}_n) \in A$ Nash equilibrium solution of the game Γ_n^{ED} (resp. $\Gamma_{n,k}^{ED}$) is an optimal solution of the problem (*ED*). For any $i \in N$, $(\hat{x}_i, \hat{y}_i)$ is solution to the optimization problem

$$\min_{(x_i, y_i) \in \Omega} f_i(\hat{x}_1, \hat{y}_1, \ldots, \hat{x}_{i-1}, \hat{y}_{i-1}, x_i, y_i, \hat{x}_{i+1}, \hat{y}_{i+1}, \ldots, \hat{x}_n, \hat{y}_n)$$

with $(x_1, y_1, \ldots, x_n, y_n) \in A$, being f_i the cost function specified in Definition 3 (resp. 4).

As proved in [26], Γ_n^{ED} and $\Gamma_{n,k}^{ED}$ are potential games and there exists a solution to the problem (*ED*), namely a Nash equilibrium solution.

3.1 Absence of Demand Points

In this sub-section it is considered the game Γ_n^{ED} as given in Definition 3, with $d(P, Q)$ the Euclidean metric in $\mathscr{R}^2$ and for any $i \in N$

$$b_i(t) = t, \quad c_i(t) = \sqrt{\frac{t}{2}} \quad \forall t \geq 0.$$

In terms of coordinates, if $P_i = (x_i, y_i)$, $i \in N$ the distance of a point $P = (x, y)$ from the set $\partial\Omega$, the boundary of Ω, is

$$d(P, \partial\Omega) = \min_{Q \in \partial\Omega} d(P, Q) = \min\{x, y, 1 - x, 1 - y\}$$

and one has for $(x_1, y_1, \ldots, x_n, y_n) \in A$

$$f_i(x_1, y_1, \ldots, x_n, y_n) = \sum_{1 \leq j \leq n, j \neq i} \frac{1}{\sqrt{(x_i - x_j)^2 + (y_i - y_j)^2}} +$$

$$\frac{1}{\sqrt{2 \min\{x_i, y_i, 1 - x_i, 1 - y_i\}}}. \tag{1}$$

In order to find the Nash equilibrium solutions of the game Γ_n^{ED}, by means of Theorem 2 of [26] it is sufficient to minimize the potential function V, then to solve the following optimization problem:

$$\min_{(x_1, y_1, \ldots, x_n, y_n) \in A} V(x_1, y_1, \ldots, x_n, y_n) =$$

$$\min_{(x_1, y_1, \ldots, x_n, y_n) \in A} \sum_{1 \leq i < j \leq n} \frac{1}{\sqrt{(x_i - x_j)^2 + (y_i - y_j)^2}} + \sum_{1 \leq i \leq n} \frac{1}{\sqrt{2 \min\{x_i, y_i, 1 - x_i, 1 - y_i\}}} \tag{2}$$

being $A = \{(P_1, \ldots, P_n) \in \Omega^n : P_i \in (]0, 1[)^2, P_i \neq P_j \; \forall i, j = 1, \ldots, n, j \neq i\}$. The solution of this problem is achieved through a computational procedure based on a genetic algorithm. The results summarized in this section are purely numerical and they have been computed by a genetic algorithm for several cases, where the number of points to be located is increased.

A genetic algorithm (GA) is an optimization technique based on the Darwin's principles about natural selection. The principal object is a virtual individual (or chromosome) that represents a feasible solution in the search space. In a binary representation it's made by a string of bits, called chromosome, that is the genotype model of problem variables or properties (see, for example, [15]). A population is a finite set of individuals. It is a sampling of the problem domain that evolves generation by generation, exploring zones with an higher probability of minimum cost function. These improvements are achieved by combining the good features of each individual, using crossover and mutation operators. The algorithm consists of several steps:

- Initialization, where in the first step a new population is initialized, generating a set of random solutions in the search space.
- Fitness computation, where fitness means the function that estimates the quality of a chromosome, combining cost function and constraints for each individual.
- Selection, i.e. a probabilistic based selection is performed on sorted population to choose parents for applying genetic operators. The selection does not waste worst chromosomes, useful to move towards unexplored zones of search space.
- Crossover/mutation, where the genotypes of selected parents are mixed to generate new individuals for the following population. To avoid premature stagnation of the algorithm a mutation operator is used, randomly changing a bit of the just created chromosomes.

Table 1 Genetic algorithm characteristics

Parameter	Value or type
Chromosome	Binary string
Crossover	Multi-cat
Mating-pool	50
Mutation probability	0.01%
Population size	100

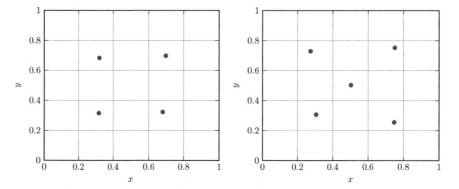

Fig. 6 Cases for $n = 4, 5$

- Population evolution occurs when the whole population built on parents and children is sorted by using a new fitness computation and then it is cut to the initial population size.

The genetic algorithm employed for the solution of the location problem is initialized using parameters that are summarized in Table 1. Each one of the following results is intended to represent only one of the several solutions that differs only for the permutation of design point locations. This reduces the number of evaluation of the location problem solutions proportional to the factorial of the number of points to be located. In Figs. 6 and 7 the results obtained with $n = 4, 5$ and 15, 20 are shown, respectively. The symmetrical nature of the solution for the case with few points is lost when the number of locations increases to 15 and 20 elements.

3.2　Presence of Demand Points

In this sub-section it is considered the game $\Gamma_{n,k}^{ED}$ as given in Definition 4, with $d(P, Q)$ the Euclidean metric in $\mathscr{R}^2$, $Q_1, \ldots, Q_k$ points in Ω and for any $i \in N$

$$a_i(t) = b_i(t) = t, \ c_i(t) = \sqrt{\frac{t}{2}} \ \forall t \geq 0.$$

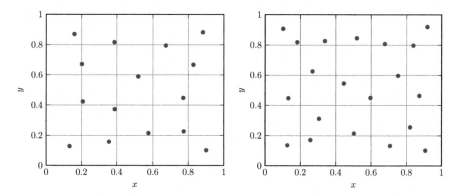

Fig. 7 Cases for $n = 15, 20$

In terms of coordinates one denotes $P_i = (x_i, y_i), i \in N$ and $Q_j = (s_j, t_j), j = 1, \ldots, k$; then

$$f_i(x_1, y_1, \ldots, x_n, y_n) = \sum_{1 \le j \le k} \frac{1}{\sqrt{(x_i - s_j)^2 + (y_i - t_j)^2}} +$$

$$\sum_{1 \le j \le n, j \ne i} \frac{1}{\sqrt{(x_i - x_j)^2 + (y_i - y_j)^2}} + \frac{1}{\sqrt{2} \min\{x_i, y_i, 1 - x_i, 1 - y_i\}}. \qquad (3)$$

As in the previous case, in order to find the Nash equilibrium solutions of the game $\Gamma_{n,k}^{ED}$, it is sufficient to minimize the potential function V, as in the following [26]:

$$\min_{(x_1, y_1, \ldots, x_n, y_n) \in A} V(x_1, y_1, \ldots, x_n, y_n) =$$

$$\min_{(x_1, y_1, \ldots, x_n, y_n) \in A} \sum_{1 \le i \le n} \sum_{1 \le j \le k} \frac{1}{\sqrt{(x_i - s_j)^2 + (y_i - t_j)^2}} +$$

$$\sum_{1 \le i < j \le n} \frac{1}{\sqrt{(x_i - x_j)^2 + (y_i - y_j)^2}} + \sum_{1 \le i \le n} \frac{1}{\sqrt{2} \min\{x_i, y_i, 1 - x_i, 1 - y_i\}}, \qquad (4)$$

being
$A = \{(P_1, \ldots, P_n) \in \Omega^n : P_i \in (]0, 1[)^2, P_i \ne P_l \; \forall i \ne l, P_i \ne Q_j\}$ with $i, l = 1, \ldots, n$ and $j = 1, \ldots, k$. The solution of this problem follows a computational procedure based on a genetic algorithm analogously to the one used in the previous sub-section. In Fig. 8 the results obtained with $(n, k) = (5, 4)$ are shown, compared with the results of the case without demands points: the blue points represent the locations evaluated considering the already present red demand points.

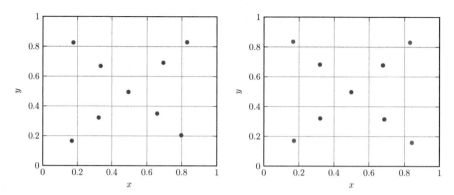

Fig. 8 Solution to Γ_9^{ED}, and to $\Gamma_{5,4}^{ED}$

4 Design of a Test Matrix for an Envelope Expansion Flight Test Activity

The problem of designing a test matrix for an envelope expansion flight test activity is an experimental design problem that we define as a facility location problem.

It is assumed that the requirements of the different engineering departments can be formalized with two different objective functions. The StE, being more interested on combined true airspeed, load factor and compressibility effects on the structures, optimizes distributions in Mach number, while SyE, being more focused on environmental effects on the aircraft and store systems, optimizes distributions in pressure altitude. However, both the engineering departments need to investigate and report, as an additional requirement, the effects of high dynamic pressures on both structure and systems, that means to test at the maximum equivalent airspeed, U_E. The latter is a function of both the Mach number and the pressure altitude (Fig. 9).

A decision should be taken concerning the design variables U_1, U_2, whose values are in suitable real intervals, for n experiments E_1, ..., E_n. The two players StE and SyE decide for each experiment i the values of U_{1i} and U_{2i} in the spatial domain optimizing a payoff function. As already mentioned, the StE optimizes the test points distribution in Mach number, while the SyE optimizes the distributions in pressure altitude, but both of them want to maximize test points density near the maximum equivalent airspeed (U_E) area. Because of this definition, StE and SyE are in a competition. In the next we define the Flight Test Location Game.

Let n be a fixed natural number ($n \geq 1$), that is the number of the prescribed flight test points.

Definition 6 The Flight Test Location Game is the two-player strategic form game $\Gamma^{FT} = \; < 2; X_1^n, X_2^n; f_1, f_2 >$ defined as follows:

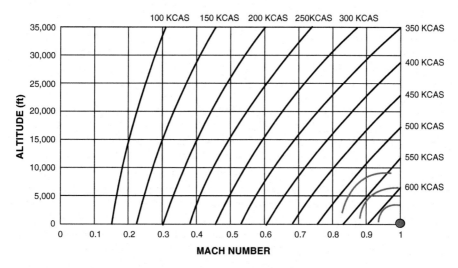

Fig. 9 Flying in the hot spot

(i) Player 1 is the StE team and player 2 is the SyE team.
(ii) The sets X_1, X_2 are real intervals and represent the variable ranges: for each
$i \in [1, n]$ (with $[1, n] = \{1, \ldots, n\}$), player 1 choses the Mach number M_i in the
set $X_1 = [M_L, M_U]$ and player 2 the pressure altitude H_i in $X_2 = [H_L, H_U]$. The
i-th flight test point has coordinates (M_i, H_i) and $(\mathbf{M}, \mathbf{H})$ is the $2n$-dimensional
vector $(M_1, \ldots M_n, H_1, \ldots, H_n)$. Player 1 (resp. player 2) has to choose a n-
dimensional vector $\mathbf{M} \in X_1^n$ (resp. $\mathbf{H} \in X_2^n$).
(iii) The objective functions are real valued functions defined on $X_1^n \times X_2^n$ and
defined by

$$f_1(\mathbf{M}, \mathbf{H}) = \sum_{i=1}^{n} \left[\sqrt{\alpha_1(M_i - M_L)^2 + \alpha_2(M_i - M_U)^2} + \delta(U_U - U_i) - \right.$$

$$\left. \min_{j \in [1,n], j \neq i} \sqrt{\frac{(M_i - M_j)^2}{M_U^2} + \frac{(H_i - H_j)^2}{H_U^2}} \right] \quad (5)$$

and

$$f_2(\mathbf{M}, \mathbf{H}) = \sum_{i=1}^{n} \left[\sqrt{\beta_1(H_i - H_L)^2 + \beta_2(H_i - H_U)^2} + \delta(U_U - U_i) - \right.$$

$$\left. \min_{j \in [1,n], j \neq i} \sqrt{\frac{(M_i - M_j)^2}{M_U^2} + \frac{(H_i - H_j)^2}{H_U^2}} \right] \quad (6)$$

where $\alpha_1, \alpha_2, \beta_1, \beta_2, \delta$ are positive real numbers, $U_U = 400$ KCAS and the equivalent airspeed $U_i = U_E(M_i, H_i)$ is a function of M_i and H_i under the assumption of International Standard Atmosphere [1]. Here the equivalent airspeed is given by $U_E(M_i, H_i) = a\, M_i \sqrt{(1 + b\, H_i)^c}$ with a, b, c positive constants.

The first term of each objective function represents the position of the points with respect to the lower bound and the upper bound of the variable range, the second term is the distance in terms of equivalent airspeed and the last one considers the opposite distance from the closest test point. The objective of each player is to minimize his own objective function in order to obtain an optimal test points distribution. The objective is to distribute the test points maximizing their dispersion inside the flight envelope and at the same time maximizing test points density close to the right lower corner of the flight envelope (maximum equivalent airspeed).

The optimal flight test distribution is obtained using a Nash equilibrium solution of the game Γ^{FT}, i.e. a vector $(\bar{\mathbf{M}}, \bar{\mathbf{H}}) \in X_1^n \times X_2^n$ such that:

$$f_1(\bar{\mathbf{M}}, \bar{\mathbf{H}}) \leq f_1(\mathbf{M}, \bar{\mathbf{H}}), \forall \mathbf{M} \in X_1^n$$

$$f_2(\bar{\mathbf{M}}, \bar{\mathbf{H}}) \leq f_2(\bar{\mathbf{M}}, \mathbf{H}), \forall \mathbf{H} \in X_2^n$$

In terms of facility location problems, the payoff functions of the flight test location game present a *minsum* part as well a *minmax* one [19].

From the analysis of the two objective functions it is possible to observe that our location game is a potential game that reduces its solution to the determination of the minimum of the potential function, which represents a Nash Equilibrium (NE) solution [23, 24, 29]. Here

$$V(\mathbf{M}, \mathbf{H}) = \sum_{i=1}^{n} \left[\sqrt{\alpha_1(M_i - M_L)^2 + \alpha_2(M_i - M_U)^2} + \sqrt{\beta_1(H_i - H_L)^2 + \beta_2(H_i - H_U)^2} + \right.$$

$$\left. \delta(U_U - U_i) - \min_{j \in [1,n], j \neq i} \sqrt{\frac{(M_i - M_j)^2}{M_U^2} + \frac{(H_i - H_j)^2}{H_U^2}} \right].$$

A genetic algorithm has been used in order to find the minimum values of the potential function. Table 2 shows the parameters values used in the genetic algorithm to achieve convergence. Obtained solution of the proposed problem was analyzed in order to evaluate goodness and robustness of the result iterating the process applying step by step minor changes to the setup configuration. Results validation was accomplished comparing the test cases results with the test matrix structure given by other standard empirical testing methods as the Economy method already mentioned in Sect. 1.

Table 2 Genetic algorithm parameters

Parameter	Value or type
Crossover fraction	0.80
Crossover mode	Scattered
Fitness scaling	Rank
Mutation fraction	0.20
Mutation mode	Adaptive feasible
Population size	200
Selection function	Tournament

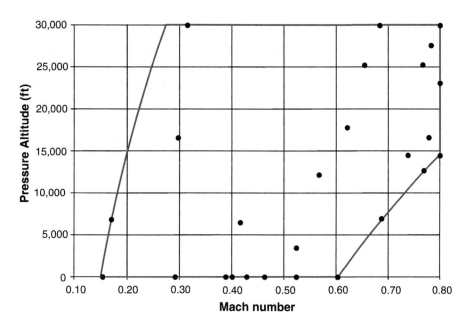

Fig. 10 The optimal distribution for 25 flight test points

A typical store integration test campaign, considering the economical budget usually available and the objectives, requires a test matrix dimension ranging from 10 to 30 test points [2, 30].

Figure 10 shows the test matrix plot in the flight envelope for a 25-test points distribution, where it is possible to verify that the initial requirements in terms of test points distribution have been met by the solution.

5 Conclusions

In this chapter is presented a survey of location methods based on Nash equilibria for the design of experiment. The solution of the location problem is restricted to the bi-dimensional case where it is approached by means of a potential formulation and a Nash game. The most important definitions and proofs are reported in Sect. 2.

Two main application fields are employed to stress the capability of an ad hoc numerical methodology involved in the solution of the location problem. The first one refers to optimal (constrained) location of sensors collecting cosmic rays for astrophysics experiments, see Sect. 3. The DoE problem is solved as a multi-facility continuous location problem and faced by using a game theoretical approach. Each facility is controlled by a competitor that tries to optimize his objective. The solution of the location problem is a Nash equilibrium. A potential approach is presented to reduce computational load due to the numerical evaluation of the Nash equilibria. Furthermore, it is considered the case where some experiments are already planned and the problem is to determine the optimal value of design variables of additional new experiments. Existence results and several test cases are presented in both cases, with and without the presence of demand points, showing the effectiveness of the algorithm. The second application field concerns the design of an experimental test matrix for an envelope expansion flight test activity and a solution is presented. By means of the used genetic algorithm is possible to approach the optimal test points distribution for a test campaign of a new store integration, where optimality is assessed in terms of prescribed objective functions. An alternative approach, based on the concept of potential and repulsive fields has been discussed in [13], where a method to dynamically relocate test points has been presented by considering an initial test points subset, already performed, and the total amount of test points that could be cut (or increased) due to time and cost constraints.

References

1. Anderson, J.D.: Introduction to Flight. McGraw-Hill Series in Aeronautical and Aerospace Engineering, 5th edn. McGraw-Hill, New York (2005)
2. Arnold, R.J., Epstein, C.S.: Store Separation Flight Testing. NATO AGARDograph 300, Flight Test Technique, vol. 5. Neuilly sur Seine, NATO AGARD (1986)
3. Başar, T., Olsder, G.J.: Dynamic Noncooperative Game Theory. Classics in Applied Mathematics, vol. 23. Society for Industrial and Applied Mathematics (SIAM), Philadelphia, PA (1999)
4. Benabbou, A., Borouchaki, H., Laug, P., Lu, J.: Sphere packing and applications to granular structure modeling. In: Garimella, R.V. (eds.) Proceedings of the 17th International Meshing Roundtable, pp. 1–18, Springer, Berlin/Heidelberg (2008). doi:10.1007/978-3-540-87921-3_1
5. Clarich, A., Périaux, J., Poloni, C.: Combining game strategies and evolutionary algorithms for CAD parametrization and multi-point optimization of complex aeronautic systems. In: Proceedings of EUROGEN 2003, Barcelona (2003)
6. Chinchuluun, A., Pardalos, P.M., Huang, H-X.: Multilevel (hierarchical) optimization: complexity issues, optimality conditions, algorithms. In: Gao, D., Sherali, H. (eds.) Advances in Applied Mathematics and Global Optimization, pp. 197–221. Springer, New York (2009). doi:10.1007/978-0-387-75714-8_6
7. Conway, J.H., Sloane, N.J.A.: Sphere Packings, Lattices and Groups. Springer, New York (1998). doi:10.1007/978-1-4757-6568-7
8. D'Amato, E., Daniele, E., Mallozzi, L., Petrone, G.: Equilibrium strategies via GA to Stackelberg games under multiple follower's best reply. Int. J. Intell. Syst. 27(2), 74–85 (2012)

9. D'Amato, E., Daniele, E., Mallozzi, L., Petrone, G., Tancredi, S.: A hierarchical multi-modal hybrid Stackelberg–Nash GA for a leader with multiple followers game. In: Sorokin, A., Murphey, R., Thai, M.T., Pardalos, P.M. (eds.) Dynamics of Information Systems: Mathematical Foundations, pp. 267–280. Springer, Berlin (2012). doi:10.1007/978-1-4614-3906-6_14

10. D'Amato, E., Daniele, E., Mallozzi, L., Petrone, G.: Three level hierarchical decision making model with GA. Engineering Computations: Int. J. Comput.-Aided Eng. Softw. 31(6), 1116–1128 (2014). doi:10.1108/EC-03-2012-0075

11. D'Amato, E., Daniele, E., Mallozzi, L.: Experimental design problems and Nash equilibrium solutions. In: Kalyagin, V., Pardalos, P.M., Rassias, T.M. (eds.) Network Models in Economics and Finance. Springer Optimization and Its Applications, vol. 100, pp. 1–12. Springer, Cham (2014). doi:10.1007/978-3-319-09683-4_1

12. D'Amato, E., Daniele, E., Mallozzi, L.: A genetic algorithm for a sensor device location problem. In: Greiner, D., Galván, B., Périaux, J., Gauger, N., Giannakoglou, K., Winter, G. (eds.) Advances in Evolutionary and Deterministic Methods for Design, Optimization and Control in Engineering and Sciences, pp. 49–57. Springer, Cham (2015). doi:10.1007/978-3-319-11541-2_3

13. D'Argenio, A., de Nicola, C., De Paolis, P., Di Francesco, G., Mallozzi, L.: Design of a flight test matrix and dynamic relocation of test points. J. Algorithms Optim. 2, 52–56 (2014)

14. Dean, A.M., Voss, D.: Design and Analysis of Experiments. Springer, New York (1998). doi:10.1007/b97673

15. Deb, K.: Multi-Objective Optimization Using Evolutionary Algorithms. Wiley, New York (2001). ISBN:978-0-471-87339-6

16. Deb, K., Pratap, A., Agarwal, S., Meyarivan, T.: A fast and elitist multi-objective genetic algorithm: NSGA-II. IEEE Trans. Evolut. Comput. 6(2), 181–197 (2002). doi:10.1109/4235.996017

17. Donev, A., Torquato, S., Stillinger, F.H., Connelly, R.: A linear programming algorithm to test for jamming in hard-sphere packings. J. Comput. Phys. 197(1), 139–166. Academic Press Professional, Inc., San Diego (2004). doi:10.1016/j.jcp.2003.11.022

18. Drezner, Z.: Facility Location – A Survey of Applications and Methods. Springer, New York (1995). ISBN:978-0-387-94545-3

19. Eiselt, H.A., Marianov, V.: Foundations of Location Analysis. International Series in Operations Research and Management Science, vol. 115. Springer, New York (2011). doi:10.1007/978-1-4419-7572-0

20. Gevers, M.: Identification for control – from the early achievements to the revival of experiment design. Eur. J. Control 11(4), 335–352 (2005). doi:10.3166/ejc.11.335-352

21. Giunta, A.A., Balabanov, V., Haim, D., Grossman, B., Mason, W.H., Watson, L.T., Haftka, R.T.: Multidisciplinary optimization of a supersonic transport using design of experiments theory and response surface modeling. Technical Report ncstrl.vatech_cs//TR-97-10, Computer Science, Virginia Polytechnic Institute and State University (1997)

22. Hales, T.C.: The sphere packing problem. J. Comput. Appl. Math. 44(1), 41–76 (1992). doi:10.1016/0377-0427(92)90052-Y

23. Mallozzi, L.: Noncooperative facility location games. Oper. Res. Lett. 35(2), 151–154 (2007). doi:10.1016/j.orl.2006.03.003

24. Mallozzi, L.: An application of optimization theory to the study of equilibria for games: a survey. Cent. Eur. J. Oper. Res. 21(3), 523–539 (2012). doi:10.1007/s10100-012-0245-8

25. Mallozzi, L., D'Amato, E., Daniele, E.: A planar location-allocation problem with waiting time costs. In: Rassias, T.M., Tóth, L. (eds.) Topics in Mathematical and Applications, pp. 541–556. Springer, Cham (2014) doi:10.1007/978-3-319-06554-0_23

26. Mallozzi, L., D'Amato, E., Daniele, E.: Location methods in experimental design. In: Pardalos, P.M, Rassias, T.M. (eds.) Mathematics Without Boundaries: Surveys of Interdisciplinary Research, pp. 429–446. Springer, New York (2014). doi:10.1007/978-1-4939-1124-0_14

27. Mallozzi, L., De Paolis, P., Di Francesco, G., D'Argenio, A.: Computational results for flight test points distribution in the flight envelope. In: Greiner, D., Galván, B., Périaux, J., Gauger, N., Giannakoglou, K., Winter, G. (eds.) Advances in Evolutionary and Deterministic Methods for Design, Optimization and Control in Engineering and Sciences, pp. 401–409. Springer, Cham (2015). doi:10.1007/978-3-319-11541-2_26

28. Migdalas, A., Pardalos, P.M., V'arbrand, P. (eds.): Multilevel Optimization: Algorithms and Applications. Springer, New York (1998). doi:10.1007/978-1-4613-0307-7

29. Monderer, D., Shapley, L.S.: Potential games. Games Econ. Behav. **14**(1), 124–143 (1996). doi:10.1006/game.1996.0044

30. North Atlantic Treaty Organization, Science and Technology Organization: Aircraft/Store Compatibility, Integration and Separation Testing. STO/NATO AGARDograph 300, Flight Test Technique - vol. 29 (2014). ISBN:978-92-837-0214-6

31. Nurmela, K.J.: Stochastic optimization methods in sphere packing and covering problems in discrete geometry and coding theory. Ph.D. thesis, Helsinki University of Technology, printed by Picaset Oy, 1997

32. Periaux, J., Chen, H.Q., Mantel, B., Sefrioui, M., Sui, H.T.: Combining game theory and genetic algorithms with application to DDM-nozzle optimization problems. Finite Elem. Anal. Des. **37**(5), 417–429 (2001). doi:10.1016/S0168-874X(00)00055-X

33. Sloane, N.J.A.: The Sphere Packing Problem. 1998 Shannon Lecture, AT&T Shannon Lab, Florham Park, NJ (1998)

34. Sorokin, A., Pardalos, P.: Dynamics of Information Systems – Algorithmics Approaches. Springer, New York (2013). doi:10.1007/978-1-4614-7582-8

35. Sutou, A., Dai, Y.: Global optimization approach to unequal sphere packing problems in 3D. J. Optim. Theory Appl. **114**(3), 671–694 (2002). doi:10.1023/A:1016083231326

36. Telford, J.K.: A Brief Introduction to Design of Experiments. Johns Hopkins University, Applied Physics Laboratory, Technical Digest **27**(3), 224–232 (2007)

37. Wang, J.F., Périaux, J.: Multi-Point optimization using GAS and Nash/Stackelberg games for high lift multi-airfoil design in aerodynamics. In: Proceedings of the 2001 Congress on Evolutionary Computation CEC2001, pp. 552–559 (2001). doi:10.1109/CEC.2001.934440

Leader-Follower Models in Facility Location

Tammy Drezner and Zvi Drezner

1 Introduction

Facility location models deal, for the most part, with the location of plants, warehouses, distribution centers and other industrial facilities [38, 57, 70, 73, 100, 105]. In this chapter we review the game theoretical concept of the leader-follower in two facilities location models which addresses specific circumstances: the competitive facility location problem and the defensive maximal covering location model on a network. A framework of facility location models that can be investigated in a game theoretical environment is depicted in Fig. 1. The models reviewed in this chapter are marked in boldface italics. Four reliable competitive models (on the plane or network) and one unreliable model on a network with the cover objective are investigated.

There are two well researched two players' games: Nash equilibrium [108] and the leader-follower game also termed the von-Stackelberg equilibrium [136] which in voting theory is known as Simpson's problem [131]. In the Nash equilibrium game no player can improve his objective when the other player does not change his strategy. In many cases no equilibrium exists. In the leader-follower game the leader adopts a strategy and the follower adopts his best strategy knowing the leader's strategy. The follower's goal is to maximize his objective function while the leader's goal is to maximize his objective function *following* the follower's action.

Early contributions to Nash equilibrium location problems include Hotelling [92], Lerner and Singer [103], Eaton and Lipsey [66], Wendell and McKelvey [146]. The leader-follower location problem was introduced by Hakimi [84], and published

T. Drezner (✉) • Z. Drezner
California State University, Fullerton, CA, USA
e-mail: tdrezner@fullerton.edu; zdrezner@fullerton.edu

© Springer International Publishing AG 2017
L. Mallozzi et al. (eds.), *Spatial Interaction Models*, Springer Optimization
and Its Applications 118, DOI 10.1007/978-3-319-52654-6_5

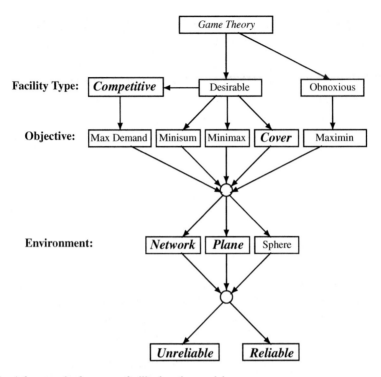

Fig. 1 A framework of common facility location models

in Hakimi [85–87], for location on network nodes using the Hotelling [92] premise that each customer patronizes the closest facility. See also Hansen and Labbè [88].

We first provide an overview the covering and competitive facility location models, and models addressing the location of unreliable facilities. We then review five different leader-follower models:

- Competitive location of two facilities (leader and follower) anywhere on the plane [33]. The proximity rule is assumed.
- Covering a large area by chain facilities so that a future competitor will not be able to attract much demand [59]. The proximity rule is assumed.
- Competitive location of two facilities (leader and follower) applying the gravity (Huff) rule [47].
- Competitive location of multiple facilities (leader's and follower's) using the cover-based rule [65].
- Locating facilities on the nodes of a network to cover as much demand as possible following a disruption of a link by a follower [16].

2 Overview of Covering Models

There are two basic types of covering models: set covering and max-covering. In the set covering model the entire demand is to be covered with the minimum number of facilities [124]. In the max-covering model the objective is to cover as much demand as possible with a given number of facilities [24, 27, 98].

Many extensions to the two basic covering models were proposed. Some papers investigate the hierarchical covering problem where a set of different possible radii is given and the two possible objectives apply [23]. A hierarchy of objectives is considered by Daskin and Stern [31]. The max-covering problem with variable radii is investigated in Berman et al. [18]. The expected maximal covering problem is investigated in Daskin [29], Batta et al. [6], Tavakkoli-Mogahddam et al. [142].

Another stream of research is the gradual covering problem. In one version of the problem the demand covered is a decreasing stepwise function of the radius [12]. In another version there is a minimum and maximum covering distance. The demand is fully covered within the minimum distance and is not covered at all beyond the maximum distance. Between these two distances the coverage is gradually declining either linearly or otherwise [14, 61]. In Drezner et al. [61] the single facility case in the plane using Euclidean distances is optimally solved. The stochastic gradual covering problem is investigated in Drezner et al. [62] and the multiple facilities gradual cover with the maximin objective is investigated in Drezner and Drezner [55].

Carrizosa and Plastria [22] and Plastria and Carrizosa [115] studied covering problems with varying radii. They developed the efficient frontier between the covering radius and the maximal cover with that radius.

Berman et al. [19, 20], Averbakh et al. [5] investigated the cooperative covering location problem. Each facility emits a (possibly non-physical) "signal" which decays over the distance and each demand point receives the aggregate signal emitted by all facilities. A demand point is considered covered if its aggregate signal exceeds a given threshold. Facilities cooperate to provide coverage, compared to the classical coverage location model where coverage is only provided by the closest facility. Examples include: locating warning sirens, heaters or fans in restaurants, light posts in parking lots.

For recent reviews of covering models see Schilling et al. [128], Daskin [30], Current et al. [27], Plastria [114], Snyder [132], García and Marín [76].

3 Overview of Competitive Facility Location Problems

Typical location models do not account for competition or for differences among facilities therefore allocate consumers to facilities by proximity. In reality, retail facilities operate in a competitive environment with an objective of profit and market share maximization. The basic problem is the optimal location of one or more new facilities in a market where competition already exists or will exist in the future. When the budget invested in expanding market share is fixed, profit increases when market share increases, thus maximizing profit is equivalent to maximizing market share. For a discussion of the equivalence of maximization of profit and maximization of market share see Dasci and Laporte [28], Jensen [96], Winerfert [148]. It follows, then, that the location objective is to locate the retail outlet at the location which maximizes its market share. Recent reviews of competitive location models are Berman et al. [17], Drezner [41, 44], Eiselt et al. [71].

Competitive location models are investigated in planar continuous space and in discrete space, in particular in a network environment. Continuous models seek the location of facilities anywhere in the plane thus there is an infinite number of potential locations for the facilities. Discrete models restrict the location of facilities to a pre-specified set of potential locations, typically the nodes of a network.

Most models assume that all the buying power is distributed among the competing facilities. Lost demand is addressed in [2, 11, 52, 54, 63–65, 120].

The underlying theme of competitive models is the existence of an interrelationship among four variables: buying power (demand), distance, facility attractiveness, and market share, with the first three variables being the independent variables and the market share the dependent variable. Many rules were proposed for formulating this relationship:

- Proximity [92].
- Minimum Utility [35, 90].
- Random Utility [45, 102].
- Cover-based [63, 64].
- Gravity (Huff) Model [93, 94].

3.1 The Proximity Rule

Hotelling [92] analyzed location on a line assuming that each consumer patronizes the closest facility. Eiselt [67] and Eiselt and Laporte [69] extended the model to a tree environment. Hotelling [92] showed that when two competitors charge the same price (they do not compete on price) an equilibrium exists. However, when competitors can compete on price, no equilibrium exists. For a discussion of the equilibrium issue the reader is referred to the seminal paper by d'Aspremont et al. [32] and Wong and Yang [150], Yang and Wong [152], Eiselt [68].

3.2 The Minimum Utility Rule

When the facilities are not equally attractive, the proximity premise for allocating consumers to facilities is no longer valid. To account for variations in facility attractiveness, a deterministic utility approach was introduced by Drezner [35]. Hodgson

[90] also suggested to incorporate attractiveness in the competitive location model. A trade-off between distance and attractiveness takes place. It is suggested that a consumer will patronize a better and farther facility as long as the extra distance to it does not exceed its attractiveness advantage [35]. The attractiveness of a facility can be transformed into a distance mark-up. A break-even distance is defined. At the break-even distance the attractiveness of two competing facilities is equal. This break-even distance, therefore, is the maximum distance that a consumer will be willing to travel to a farther facility (new or existing) based on his perception of its attractiveness and advantage relative to other facilities.

3.3   The Random Utility Rule

A random utility model was introduced by Leonardi and Tadei [102] and Drezner and Drezner [45]. The deterministic utility model is extended by assuming that each consumer draws his utility from a random distribution of utility functions. The probability that a consumer will prefer a certain facility over all other facilities is calculated by applying the multivariate normal distribution. Once the probabilities are calculated, the market share captured by a certain facility (new or existing) can be calculated as a weighted sum of the buying power at all demand points. To circumvent the mathematically complicated formulation of the random utility model, Drezner et al. [60] suggested using a simple S-shaped function. The utility declines very slowly for short distances, declines sharply for intermediate distances, and remains around zero for large distances.

3.4   The Cover-Based Rule

Drezner et al. [63, 64] introduced the cover-based approach to estimating market share. Each competing facility has a "sphere of influence" [123] represented by a radius of influence which depends on the facility's attractiveness. A consumer at a distance within the radius of influence is attracted to the facility. Consumers' demand within the sphere of influence of no facility is lost. In Drezner et al. [63] adding additional facilities of a given radius of influence is considered an expansion strategy. In Drezner et al. [64] three market expansion models are analyzed: (1) increasing the radius of influence of existing facilities thereby increasing their attractiveness, (2) adding new facilities (and determining the radius of influence of each), and (3) a combination of both.

3.5   The Gravity (Huff) Rule

The gravity approach is the most commonly used model in recent papers. According to the gravity rule [122] two cities attract retail trade from an intermediate town in direct proportion to the populations of the two cities and in inverse proportion to the square of the distances from them to the intermediate town. Drezner [36, 37] was the first to introduce the gravity model to location analysis. Evaluating market share based on the gravity rule was introduced by Huff [93, 94] and is used by marketers. Huff proposed that the probability a consumer patronizes a retail facility is proportional to its size (floor area) and inversely proportional to a power of the distance to it. At any demand point, the proportion of consumers attracted to each facility is a function of the facility's square footage (attractiveness) and distance. The model finds the market share captured at each potential site, thereby the best location for new facilities whose individual measures of attractiveness are known.

In the original Huff formulation, facility floor area serves as a surrogate for attractiveness. An improvement on Huff's approach was suggested by Nakanishi and Cooper [107], Jain and Mahajan [95] who introduced the multiplicative competitive interaction (MCI) model. The MCI coefficient replaces the floor area with a product of factors, each a component of attractiveness.

In the gravity model a distance decay function $f(d)$ is defined. It represents the decline in facility attractiveness as a function of the distance from the facility and thus the probability that a consumer patronizes that facility. In the original gravity model [122], it is assumed that the distance decay parallels the gravity decay and thus $f(d) = \frac{1}{d^2}$. Huff [93, 94] suggested a decay function of $f(d) = \frac{1}{d^\lambda}$ where the power λ depends on retail category. $\lambda = 3$ was found for grocery stores [94], $\lambda = 3.191$ for clothing stores [93], $\lambda = 2.723$ for furniture stores [93], and $\lambda = 1.27$ for shopping malls [40, 49]. Wilson [147] suggested an exponential decay $e^{-\lambda d}$ which was used in many subsequent papers [1–3, 52, 91]. Drezner [40] compared power and exponential decay on a real data set of shopping malls in Orange County, California and found that exponential decay fits the data better than power decay. The decay function $f(d) = e^{-1.705d^{0.409}}$ was used in Bell et al. [9] who investigated grocery stores. A Logit function $f(d) = \frac{1}{1+e^{\alpha+\beta d+\gamma d^2}}$ was used in Drezner et al. [60]. Goodchild and Noronha [81] applied gravity based models to the location of gas stations, and Drezner [43] applied them to the hotel industry.

There are other location models that apply the gravity rule to various objectives. This is the appropriate approach if demand is not necessarily satisfied by the closest facility because the assignment of demand is not centralized. Examples:

- In the hub location problem [21], a set of hubs needs to be selected from a list of airports so that flyers change planes at a hub on the way to their destination. It is reasonable to assume that customers do not necessarily select the hub that yields the shortest total distance to their destination. The total distance is used in the gravity rule to assess the probability that customers select a particular hub [48].

- In the *p*-median problem [82, 83], *p* facilities are to be located such that each demand point gets its services from the closest facility with the objective of minimizing the total weighted distances for all demand points. When the assumption that each customer gets his service from the closest facility is replaced by the gravity rule we obtain the gravity *p*-median problem [51]. The planar version of this problem is proposed and analyzed in Drezner and Drezner [50].
- The multiple server location problem [10] combines travel time and service time at servers using M/M/k queueing systems. A given number of servers are to be located at nodes of a network. Demand for these servers is generated at each node, and a subset of nodes needs to be selected for locating one or more servers in each. Each customer at a node selects the closest server. The objective is to minimize the sum of travel time and the average time spent at the server, for all customers. The gravity multiple server problem is defined when customers do not necessarily use the server with the shortest travel time plus service time [53].

3.6 Implementation Issues

Once buying power, distance, and attractiveness are known, market share can be calculated by any of the approaches discussed above.

Buying power (demand), sometimes referred to as purchasing power, is available in secondary data sources. Geographic information systems data bases, such as those provided by ESRI,[1] also have data about buying power.

Facility attractiveness is assessed using one of a variety of methods. The attractiveness of a facility is a composite index of a set of attributes. Examples of attractiveness components of shopping malls are: (1) floor area (2) variety of stores, (3) appearance, (4) favorite brand names. Other techniques for inferring or deriving attractiveness levels were proposed in Drezner [40], Drezner and Drezner [49].

The distance between two points can be easily measured. However, since demand points represent areas, the distance correction for an area A and distance d is $\sqrt{d^2 + \alpha A}$ where $\alpha = 0.24$ is recommended by Drezner and Drezner [46].

Plastria and Vanhaverbeke [117], Francis et al. [74] addressed the issue of aggregation and its effect on the optimality of the location solution. Demand points often have to be aggregated due to computational intractability. However, this spatial aggregation typically introduces a bias to the value of the objective function thus the optimality of the solution cannot be guaranteed.

---

[1]Environmental Systems Research Institute, supplier of GIS software such as ArcGIS, ArcView.

3.7   Extensions

Many extensions to the basic model were suggested. The following two extensions are incorporated in this chapter.

Budget Constraints: Combining the location decision with facility design (treating the attractiveness level of the facility as a variable) was recently investigated in [1, 39, 63, 64, 72, 116, 119, 144]. Drezner [39] assumed that facilities attractiveness levels are variables. In that paper it is assumed that a budget is available for locating new facilities and for establishing their attractiveness levels. One needs to determine the facilities attractiveness levels so that the available budget is not exceeded. Plastria and Vanhaverbeke [118] combined the limited budget model with the leader-follower model. Aboolian et al. [1] studied the problem of simultaneously finding the number of facilities, their respective locations and attractiveness (design) levels.

Leader-Follower: The leader-follower model [136] considers a competitor's reaction to the leader's action. The leader decides to expand his chain. The follower is aware of the action taken by the leader and expands his facilities to maximize his own market share. The leader's objective becomes maximizing his market share *following* the follower's reaction [33, 47, 99, 118, 119, 121, 125, 126].

4   Leader-Follower Models in Competitive Models

In this section we review four leader-follower models in a competitive environment.

4.1   The Leader-Follower Model Locating Two Facilities in the Plane

Drezner [33] analyzed two competitive location models in the plane. One is the location of a new facility that will attract the most buying power from an existing facility (the follower's problem). The other is the location of a facility that will secure the most buying power against the best location of a competing facility to be set up in the future (the leader's problem). The proximity rule using Euclidean distances is assumed.

Let n demand points be located on the plane. A weight, or buying power, $b_i > 0$ is associated with demand point i for $i = 1, \ldots, n$. The leader locates his facility at X and the follower locates his facility at Y. Customers will patronize the follower's facility Y if the Euclidean distance between the customer and Y is less than the distance between the customer and X. Two problems are considered:

Problem 1 (the follower's problem): Given the location of an existing facility X serving the demand points, find a location for a new facility Y that will attract the most buying power from demand points.

Problem 2 (the leader's problem): Find a location for X such that it will retain the most buying power against the best possible location for the follower's facility Y.

For given locations X and Y, the distribution of the buying power can be found by constructing the perpendicular bisector to the segment connecting X and Y. This perpendicular bisector divides the plane into two half-planes. All points in the closed half-plane which includes X (including points on the perpendicular bisector itself) will patronize X and all the points in the other open half-plane which includes Y, will patronize Y. This is a generalization of Hotelling's analysis on a line [92].

It is shown in Drezner [33] that one of the optimal locations for Y when X is given is infinitesimally close to X but not on X. It follows that a solution to Problem 1, Y^*, is 'adjacent' (close) to X. The variable yet to be determined is the direction in which Y is 'touching' X. In conclusion, finding an optimal location for Y is equivalent to finding the best line through X such that the open half plane defined by this line contains the most buying power for Y. Finding the best line by simple enumeration is detailed in Drezner [33].

The algorithm that solves Problem 2 is based on the algorithm used for solving Problem 1. It can be found whether attracting a certain market share P_0 or higher by Y is possible by finding whether there is a feasible solution to a linear program. The algorithm is based on a bisection on the value of P_0. Complete details are given in Drezner [33].

1. Calculate all $\frac{1}{2}n(n-1)$ lines through pairs of points and find the market share P_i for each open half-plane defined by the lines.
2. Sort P_i in decreasing order. Set P_{min} and P_{max} to the smallest and largest P_i, respectively.
3. Set P_0 to the median value in the P_i vector for all $P_{min} < P_i < P_{max}$. If the set $P_{min} < P_i < P_{max}$ is empty go to Step 7.
4. Find if there is a feasible point to all half-planes for which $P_i \geq P_0$. This can be done by linear programming.
5. If there is a feasible solution point to the linear program then set P_{max} to P_0 and go to Step 3.
6. Otherwise, set P_{min} to P_0 and go to Step 3.
7. A feasible location for the last P_{max} is an optimal solution. The value of the objective function is P_{min}.

The two problems can be modified by an extra restriction that the follower cannot construct his facility closer than a given distance R from the leader's facility. To solve the modified Problem 1 for a given X it can be shown that the best solution for Y is determined by open half planes defined by tangent lines to the circle centered at X with a radius of $\frac{1}{2}R$ rather than lines through X. The details of the algorithms for solving the modified Problems 1 and 2 are available in Drezner [33].

4.2   A Leader-Follower Model for Covering a Large Area with Numerous Facilities

Drezner and Zemel [59] considered the following problem: a large number of customers are spread uniformly over a given region $A \subseteq \mathbb{R}^2$. What configuration of facilities that cover the area will best protect against a future competing facility? The proximity rule is assumed. Each customer patronizes the closest facility.

Drezner and Zemel [59] found the solution to the problem of covering the whole $\mathbb{R}^2$ plane. Then they analyzed the finite area problem and found bounds on the difference between the configurations as the number of facilities increases.

There are three evenly spread configurations that cover the whole $\mathbb{R}^2$ plane with equilateral polygons depicted in Fig. 2: a triangular grid where facilities are located at the centers of equilateral triangles; a square grid where facilities are located at the centers of squares; and an hexagonal grid (beehive) where facilities are located at the centers of hexagons. No other cover of the plane by identical equilateral polygons exists.

Since customers are attracted to the closest facility, the market share captured by each facility is proportional to the area attracted to the closest facility. This is similar to the Voronoi diagram concept [113, 140, 145]. In the configurations depicted in Fig. 2, the market share attracted by each facility is the area of the polygon. Let A be the area attracted by each facility. It is shown in Drezner and Zemel [59] that:

- For the triangular grid the competitor's facility can attract a maximum of $\frac{2}{3}A = 0.6667A$. The best location for the follower is at a vertex of a triangle (see Fig. 2).
- For a square grid the competitor's facility can attract a maximum of $\frac{9}{16}A = 0.5625A$. The best location for the competitor is at the center of the side of the square.
- For an hexagonal grid the competitor's facility can attract a maximum of $0.5127A$, i.e., 51.27% of the leader facility's market share. All regions inside the equilateral triangles with the leader's facilities at its vertices (see Fig. 2) are equivalent. Also, because of symmetry, the market share at any location, except at the center of the triangle, has two more equivalent locations. The best locations

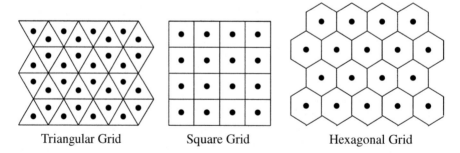

| Triangular Grid | Square Grid | Hexagonal Grid |

Fig. 2   Various configurations

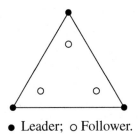

● Leader; ○ Follower.

Fig. 3 Best follower's locations

for the follower inside such triangles are depicted in Fig. 3. The minimum market share of half the area of the hexagon is captured by the follower at the three vertices of the triangle and at its center. The factor 0.512713 has a contrived formula developed in Drezner and Zemel [59]: Define $\theta = \frac{1}{3} \arctan \frac{9\sqrt{1077}}{347}$; $\alpha = \frac{1}{9} \left(16 - \sqrt{94}(\cos\theta + \sqrt{3}\sin\theta) \right)$; then the factor is: $\frac{1}{2} + \frac{\alpha(2-3\alpha)^2}{24(1-\alpha)(2-\alpha)}$. The follower's facility location in Fig. 3 is on the line connecting the vertex of the triangle and its center. Its distance from the leader's facility is $\alpha = 0.2955766$ times the distance between the vertex and the center of the triangle or $\frac{2}{3}\alpha$ of the triangle's height.

The hexagonal pattern provides the best protection from a future competitor. It is interesting that for hexagonal and square grids the competitor captures at least half of A at any point in the plane.

Hexagonal pattern is optimal for many location problems with numerous facilities covering a large area. For example:

- packing the most number of circles in an area [26, 89, 141],
- p-median [112], p-center [138] and p-cover [58],
- p-dispersion [56, 104, 106, 109],
- equalizing the load covered by facilities [139].

It is also the preferred arrangement for a bee-hive in nature which has developed over the years in the evolutionary process.

4.3 The Leader-Follower Problem Using the Gravity (Huff) Rule

It is assumed that a new competing facility will enter the market at some point in the future. The competitor will establish his facility at the location which maximizes his market share given the leader's location. The leader's objective is to find the location that maximizes the market share captured by his facility following the competitor's entry [47]. Ghosh and Craig [77] solved a similar problem by discretizing all

Table 1 Notation for gravity (Huff) based models

n	Number of demand points
b_i	Discretionary buying power at demand point i
k	Number of existing competing facilities
d_{ij}	Euclidean distance between demand point i and existing facility j
X	Location of the leader's new facility
U	Location of the follower's new facility
$d_i(Y)$	Euclidean distance between demand point i and location Y
λ	Power to which the distance is raised
S_j	Measure of attractiveness for existing facility j
S	Measure of attractiveness for the leader's new facility
C	Measure of attractiveness for the follower's facility

variables and assuming a given set of possible locations for both the entrant and the future competitor. They formulated the problem as an integer programming problem which limits the solution procedure to relatively small problems. We briefly summarize the results in Drezner and Drezner [47]. For complete information the reader is referred to that paper.

4.3.1 Calculating the Market Share

Drezner and Drezner [47] applied a power distance decay function as suggested by Huff [93, 94]. As described in Sect. 3, other distance decay functions were also investigated in the literature. The notation is depicted in Table 1.

The market share captured by the leader's new facility as a function of X and U using the gravity model, $M_1(X, U)$, is:

$$
M_1(X, U) = \sum_{i=1}^{n} b_i \frac{\dfrac{S}{d_i^{\lambda}(X)}}{\dfrac{S}{d_i^{\lambda}(X)} + \dfrac{C}{d_i^{\lambda}(U)} + \sum_{j=1}^{k} \dfrac{S_j}{d_{ij}^{\lambda}}} \tag{1}
$$

The market share captured by the follower's facility as a function of X and U, $M_2(X, U)$, is:

$$
M_2(X, U) = \sum_{i=1}^{n} b_i \frac{\dfrac{C}{d_i^{\lambda}(U)}}{\dfrac{S}{d_i^{\lambda}(X)} + \dfrac{C}{d_i^{\lambda}(U)} + \sum_{j=1}^{k} \dfrac{S_j}{d_{ij}^{\lambda}}} \tag{2}
$$

4.3.2   The Optimization

The leader's objective is to maximize his market share $M_1(X, U)$ in the long run by selecting the best location X, taking into consideration that the follower selects its location U so as to maximize his market share $M_2(X, U)$. The best location for the follower, U, is a function of the leader's selected location X. In a mathematical formulation, for a given X, let $U(X)$ be the maximizer of $M_2(X, U)$. The leader's objective is converted to:

$$\max_{X}\{\, M_1(X, U(X)) \,\} \tag{3}$$

subject to: $U(X)$ is the maximizer of $M_2(X, U)$

The leader-follower problem (3) is a complicated problem. The functions $M_1(X, U)$ and $M_2(X, U)$ are not concave and may have many local maxima. The problem may have many local maxima. The constraint of (3) is unusual and cannot be formulated into a mathematical programming formulation. Such a constraint is not in the form of an equality or an inequality.

Drezner and Drezner [47] proposed three heuristic approaches for finding a good solution: brute force, pseudo mathematical programming, and gradient search.

The Brute Force Approach: A grid of locations X that cover the area is generated. For each location X the value of $U(X)$ is found and the value of $M_1(X, U(X))$ is calculated. If the grid is dense enough, the vicinity of the global maximum can be identified. If a more precise location is sought, a finer grid can be evaluated in that vicinity.

The Pseudo Mathematical Programming Approach: If the functions $M_1(X, U)$ and $M_2(X, U)$ were concave, the following mathematical programming formulation (termed the "pseudo" problem) would have solved the problem:

$$\max_{X,U}\{\, M_1(X, U) \,\} \tag{4}$$

subject to:

$$\tfrac{\partial}{\partial U}M_2(X, U) = 0$$

The Gradient Search Approach: A gradient search that directly finds a local maximum for the leader-follower problem (3) is suggested. It guarantees termination at a local maximum of problem (3) once $U(X)$ can be found. It is recommended that this procedure is repeated many times in order to have a reasonable chance of "landing" at the global optimum.

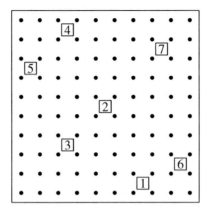

Fig. 4 Problem 1

4.3.3 Computational Experiments

Two sets of problems were tested. The first one (problem 1) consists of 100 demand points evenly distributed in a 10 by 10 miles area, each with buying power of "1" for a total buying power of 100. There are seven existing facilities in the area, each of an attractiveness level of "1" (equally attractive). The data are given in [35–37] and are depicted in Fig. 4. The attractiveness of both one's new facility and the future competing facility are also equal to one.

As explained in Drezner and Drezner [46], the accuracy of the market share estimate is enhanced by replacing the square of the distance d^2 with $d^2 + \alpha A$ where A is the estimated area represented by a demand point. This "distance correction" compensates for the fact that demand points are not mathematical points and if, for example, a facility is located on a demand point, not all customers at that demand point are at distance zero from the facility. It provides for the averaging of the distance between all customers assigned to a demand point and the facility. A distance correction factor of $\alpha A = 0.24$ was used, because the area surrounding a demand point is $A = 1$, as suggested in Drezner and Drezner [46].

The second problem (Problem 2) applies real data. It is based on sixteen communities in Orange County, California, and six existing major shopping malls. The data are given in Table 2. The total buying power for this problem is 1346.5. Both the leader's new facility and the follower's facility have an attractiveness of 5. The distance correction factor [47] is $\alpha A = 1$ because the demand points are farther apart than in Problem 1.

The results reported in Drezner and Drezner [47] indicate that local maxima of problem (3) are encountered more frequently by the gradient search approach than by the pseudo mathematical programming approach. The gradient search approach is recommended as the approach of choice for solving the problem. The pseudo mathematical programming approach is recommended for users who do not wish to code a special program but rather use standard software. The brute force approach is recommended if the value of the market share captured as a function of location is of value to the user.

Table 2 Data for problem 2

Demand points			Shopping malls		
x	y	b_i	x	y	S_j
3	5	163.8	2.7	6.8	7
3	7	28.8	3.9	4.5	3
2	6	39.0	3.6	4.2	7
3	2	77.4	3.2	2.2	10
1	5	42.0	4.0	1.5	7
3	6	107.0	6.1	1.2	3
3	4	64.5			
2	2	250.6			
5	2	101.4			
7	1	57.6			
4	1	132.0			
4	4	77.6			
4	6	29.6			
4	3	67.5			
5	3	50.7			
6	6	57.0			

The sensitivity analysis reported in Drezner and Drezner [47] indicates that the increase in captured market share, compared with the market share at the location that does not anticipate a follower's reaction, increases as competitor's attractiveness increases. In Problem 1 there is a clear shift in location as the attractiveness of the competitor reaches 3.0. The additional market share, using the leader-follower model, reaches an increase of 22.7% in market share. The gain in market share is more modest when solving Problem 2. For an attractiveness of 50, the gain is only 2.4% and the best location does not change significantly. We conclude that in some problems the model significantly improves the location, and therefore the market share captured, when compared with the model that does not consider a follower, and in some problems the improvement is not as significant.

4.4 The Leader-Follower Model Using the Cover-Based Rule

Drezner et al. [65] investigated a leader-follower (Stackelberg equilibrium) competitive location model incorporating facilities' attractiveness (design) subject to limited budgets for both the leader and follower. The competitive model is based on the concept of cover [63, 64]. The leader and the follower, each has a budget to be spent on the expansion of their chains either by improving their existing facilities or constructing new ones. We are interested in the best strategy for the leader assuming that the follower, knowing the action taken by the leader, will react by investing his budget to maximize his market share. The objective of the leader is to maximize his market share following the follower's reaction.

4.4.1 Estimating Market Share

The cover-based competitive location model [63, 64] is applied for estimating market share. Each facility has a sphere of influence (catchment area) for patronage defined by a distance termed "radius of influence". Consumers patronize a facility if they are located within the facility's radius of influence. More attractive facilities have a larger radius of influence, thus they attract consumers from greater distances. Demand at demand points which are not attracted by any facility is lost. When the total captured market share is estimated, it is assumed that if a consumer is attracted to more than one facility, his buying power is equally divided between the attracting facilities.

Models based on this rule are simpler to implement than those based on the gravity rule or the utility-based rule. One only needs to estimate the catchment area of competing facilities which yields their radius of influence. There are established methods for estimating the radius of influence of a facility [8, 143]. For example, license plates of cars in the parking lot are recorded and the addresses of the cars' owners obtained. Drezner [40] conducted interviews with consumers patronizing different shopping malls asking them to provide the zip code of their residence and whether they came from home.

The location of p new facilities with a given radius is sought so as to maximize the market share captured by one's chain. In Drezner et al. [64], three strategies were investigated: In the improvement strategy (IMP) only the improvement of existing chain facilities is considered; in the construction strategy (NEW) only the construction of new facilities is considered; and in the joint strategy (JNT) both improvement of existing chain facilities and construction of new facilities are considered. All three strategies are treated in a unified model by assigning a radius of zero to potential locations of new facilities.

The leader employs one of the three strategies and the follower also implements one of these three strategies. This setting gives rise to nine possible models. Each model is a combination of the strategy employed by the leader and the strategy employed by the follower. For example, the leader employs the JNT model, i.e., considers both improving existing facilities and establishing new ones, while the follower employs the IMP model, i.e., only considers the improvement of his existing facilities. The most logical model is to employ for both the leader and the follower the JNT strategy which yields the highest market share. However, constructing new facilities or improving existing ones may not be a feasible option for the leader or the follower.

The set of potential locations for the facilities is discrete. The notation is given in Table 3.

Note that the radii r_j are continuous variables. However, it is sufficient to consider a finite number of radii in order to find the optimal solution. Consider the sorted vector of distances between facility j and all n demand points. A radius between two consecutive distances covers the same demand points as does the radius equal to the shorter of the two distances yielding the same value of the objective function. Since

Table 3 Notation for cover based models

n	Number of demand points
b_i	Buying power at demand point i, $i = 1, \ldots, n$
L_i	Number of facilities that belong to the leader's chain that attract demand point i
F_i	Number of follower's facilities attracting demand point i
B_L	Budget available to the leader for increasing the attractiveness of existing facilities or constructing new ones
B_F	Budget available to the follower for increasing the attractiveness of existing facilities or constructing new ones
P_L	Set of the existing l of cardinality p_L
P_F	Set of the existing follower's facilities including potential locations for new facilities
d_{ij}	Distance between demand point i and facility j
r_j^o	Present radius of facility j
$f(r)$	Cost of building a facility of radius r (a non decreasing function of r)
r_j	Unknown radius assigned to facility j
R	Set of unknown radii $\{r_j\}$ for $j = 1, \ldots, p$
S_j	Fixed cost if facility j is improved or established, i.e., $r_j > r_j^o$ for existing facilities and $r_j \geq 0$ for establishing new facilities
$C(r_j)$	Cost of improving facility j to a radius r_j. It is zero if $r_j = r_j^o$, and $f(r_j) - f(r_j^o) + S_j$, otherwise

the improvement cost is an increasing function of the radius, an optimal solution exists for radii that are equal to a distance to a demand point.

For demand point i, the numbers L_i and F_i can be calculated by counting the number of leader's facilities that cover demand point i, and the number of follower's facilities that cover it. For a given strategy $R = \{r_j\}$, $j = 1, \ldots, p$ we define these values as $L_i(R)$ and $F_i(R)$. The objective functions for the leader and the follower, before locating new facilities, are:

$$MS_L(R) = \sum_{i=1}^{n} b_i \frac{L_i(R)}{L_i(R) + F_i(R)}, \tag{5}$$

$$MS_F(R) = \sum_{i=1}^{n} b_i \frac{F_i(R)}{L_i(R) + F_i(R)}. \tag{6}$$

Note that if $F_i(R) = L_i(R) = 0$, then in (5) $\frac{L_i(R)}{L_i(R)+F_i(R)} = 0$, in (6) $\frac{F_i(R)}{L_i(R)(F_i(R)} = 0$ and the demand b_i associated with demand point i is lost.

Suppose the leader improves some of his facilities and establishes new ones. Note that $F_i(R)$ does not depend on the actions taken by the leader. The follower's problem is thus well-defined following the leader's action and can be optimally solved by the branch and bound algorithm detailed in Drezner et al. [64].

Once the follower's optimal location is known, the leader's objective function is well defined as his market share is calculated by (5) incorporating changes to the follower's locations and radii.

The leader and follower cannot exceed their respective budgets. For a combined strategy $R = \{r_j\}$ by both competitors the constraints are:

$$\sum_{j \in P_L} C(r_j) \le B_L; \quad \sum_{j \in P_F} C(r_j) \le B_F. \tag{7}$$

Once the leader's strategy is known and thus $L_i = L_i(R)$ are defined, the follower's problem is:

$$\max_{r_j, \, j \in P_F} \left\{ \sum_{i=1}^{n} b_i \frac{F_i(R)}{L_i + F_i(R)} \right\}$$

Subject to:

$$\sum_{j \in P_F} C(r_j) \le B_F. \tag{8}$$

The leader's problem needs to be formulated as a bi-level programming model [75, 137]:

$$\max_{r_j, \, j \in P_L} \left\{ \sum_{i=1}^{n} b_i \frac{L_i(R)}{F_i(R) + L_i(R)} \right\}$$

Subject to: $\qquad\qquad\qquad\qquad\qquad\qquad\qquad\qquad\qquad\qquad\qquad$ (9)

$$\sum_{j \in P_L} C(r_j) \le B_L$$

$$r_j \text{ for } j \in P_F = arg \left[\begin{array}{l} \displaystyle\max_{r_j, \, j \in P_F} \left\{ \sum_{i=1}^{n} b_i \frac{F_i(R)}{F_i(R) + L_i(R)} \right\} \\ \text{subject to: } \sum_{j \in P_F} C(r_j) \le B_F. \end{array} \right]$$

Note that the follower's problem may have several optimal solutions (each resulting in a different leader's objective) and the leader does not know which of these the follower will select. This issue exists in all leader-follower models.

The follower's problems are identical to the three problems analyzed in [64] because market conditions are fully known to the follower. A branch and bound algorithm as well as a tabu search [78–80] were proposed in Drezner et al. [64] for the solution of each of these three strategies. For details the reader is referred to Drezner et al. [64, 65].

4.4.2 Computational Experiments

Drezner et al. [65] experimented with the 40 Beasley [7] problem instances designed for testing p-median algorithms. The problems range between $100 \leq n \leq 900$ nodes. The number of new facilities for these problems was ignored. The leader's facilities are located on the first ten nodes and the follower's facilities are located on the next 10 nodes. The problems are those tested in [64]. The demand at node i is $1/i$ (for testing problems with no reaction by the follower) and the cost function is $f(r) = r^2$. The same radius of influence was used for the existing leader's and follower's facilities. When new facilities can be added (Strategies NEW and JNT), $n - 10$ nodes are candidate locations for the new facilities (nodes that occupy one's facilities are not candidates for new facilities) and are assigned a radius of 0. $r_0^j = 20$, and $S_j = 0$ were applied for expanding existing facilities and the same $S_j > 0$ were used for establishing any new facility.

A branch and bound optimal algorithm was used to solve the follower's location problem. Since it is used numerous times in the solution procedure for the leader's location problem, a budget of 1500 and a set-up cost of 500 were applied to both the leader and the follower. Both the leader and the follower apply the JNT strategy. Tabu search, which does not guarantee optimality, is used to solve the leader's location problem. Therefore, the solution of each problem instance was repeated for at least 20 times to assess the quality of the tabu search solutions. Problems with up to 400 demand points were solved in reasonable run times.

The experiments conducted in Drezner et al. [65] are:

- Solving the leader's location problem when the follower does not react and does not change his facilities. This is performed by the branch and bound rigorous algorithm. When solving the leader-follower problem, this algorithm is performed to find the follower's optimal location solution and consequently the leader's objective function. As expected, one's chain market share increases and the competitor's market share declines. Some of the increase in the leader's market share comes at the expense of the competitor and some comes from capturing demand that is presently lost. It is interesting that the proportion of the additional market share gained from the competitor remains almost constant for all budgets tested. The average for all 40 problems is 44.2% for a budget of 1500, 44.2% for a budget of 2000, 44.9% for a budget of 2500, and 46.7% for a budget of 5000. A larger percentage of market share gained comes from lost demand. These percentages are the complements of the percentages gained form competitors or about 55%.
- The tabu search procedure for finding the leader's best solution after the follower's reaction was coded in C#. Its effectiveness was first tested by optimally solving 160 JNT instances (40 instances for each budget), i.e., assuming no follower's reaction. The tabu search found optimal solutions to 148 out of 160 instances and sub-optimal solutions (avg. error: 0.11%, max error: 0.41%) to the remaining 12 instances. It is also observed that, for the budget of 1500, only one

instance was not optimally solved by tabu search and the error was 0.04%. In the subsequent experiments, a budget of 1500 was used.

- The leader's location solution was extended by adding the branch and bound procedure (the follower's solution) to the tabu search. First, the original Fortran code was recoded statement-for-statement in C# and tested on several JNT instances. It was discovered that it took the C# implementation 10 times longer to find the solution to the follower's problem than the Fortran implementation. This inefficiency was likely caused by pointer arrays used in C#.
- The branch and bound procedure was coded in .NET's C++ language using native C++ arrays. This new implementation was two times slower than its Fortran counterpart. Because the branch and bound procedure is called for each tabu search move, the very efficient Fortran code was compiled to a Digital Link Library (DLL) and was called from the C# code using the .NET inter-operability technology.
- The original parameters used for solving the leader's problem ($b_i = \frac{1}{i}$) did not provide interesting results because the weights declined as the index of the demand point increased and thus both the leader and the follower concentrated their effort on attracting demand from demand points with a low index (high b_i) and "ignored" demand points with higher indices. Therefore, equal weights of "1" were assigned to all demand points. A budget of 1500 and a set-up cost of 500 for both the leader and the follower were used.

Twenty Beasley [7] problem instances with $n \leq 400$ demand points were tested. For 13 instances all results are the same and therefore the minimum and average are 100%. On average, the market share was 99.67% of the best obtained market share and the minimum market share was 98.80% of the best one. The average standard deviation was 0.50%. The numerical simulation experiments were run in parallel as 24 separate threads distributed on 24 virtual CPU cores and four different 64-bit Windows servers. Each server was equipped with 12 to 32 GB RAM and each virtual CPU was approximately equivalent to an Intel Xeon 2 GHz physical processor. These virtual CPUs are on a "cloud" and many applications are executed on them simultaneously by many users. Applications are migrated from one physical CPU to another by the central operating system. Total run time by the "cloud" computers was about 16 months with a maximum of about 2 months for one particular problem.

It was found that the leader has a slight advantage over the follower. The leader's market share increased by an average of 14.80% while the follower's market share increased by 13.95%. However, this difference is not statistically significant. The leader's market share as a percentage of his original market share increased by 84.07% while the follower's increase was only 73.43%. This difference is also not significant.

When the leader takes no action he loses 3.29% of his market share while the follower gains 24.68%. The lost demand (demand not served by either the leader or the follower) also decreases by either action. It drops by 21.4% when the leader takes no action but drops by 28.75% when both the leader and the follower expand. By investing in expanding their chains, the leader attracts 14.80% of the lost demand while the follower attracts 13.95% of the lost demand. Lost demand was reduced on the average by 48.1% of its value prior to the leader's and follower's expansions. The leader and the follower may have attracted demand from one another but these exchanges cancel out. The main source of extra market share for both the leader and the follower was obtained by attracting new consumers that did not patronize any facility prior to the expansion.

Finally, the leader's market share results found by tabu search were compared with a pure random search [153] for the leader's strategy. The budget was allocated to some facilities by the JNT strategy. The follower's problem is then optimally solved by the branch and bound algorithm. The procedure is repeated the same number of times that the follower's problem was solved in the tabu search and the best leader's market share selected. For example, in the tabu search the first problem required about 4490 solutions of the follower's problem in each of the 100 replications. For this problem 449,000 random allocations of the budget by the leader were tested.

The results were quite surprising. For example, for the second problem the best increase in market share found by tabu search is 7 units from 18 to 25. However, the best pure random search increase is only 4 units (from 18 to 22) which is only 57.1% of the potential market share increase. The results indicate that if the leader uses his budget at random, he gets poor results even when such random allocation is repeated hundreds of thousands of times and the best "lucky" result selected! This clearly shows the value of the approach suggested in Drezner et al. [65].

5 Overview of Models Addressing the Location of Unreliable Facilities

The issue of unreliable facilities or inoperable links on the network is discussed in many papers. Wollmer [149] and Wood [151] considered the possibility that some links of the network can be destroyed. Their objective is to maximize the network flow following the loss of some links. It does not involve facility location. Location of unreliable facilities was first introduced by Drezner [34] and extended by Lee [101], Berman et al. [13], Snyder and Daskin [133], Shishebori et al. [130], Shishebori and Babadi [129]. If the closest facility is out of service, then users get their service from the second closest facility. The location of unreliable facilities on a network when the reliability of service declines with the distance from the facility is discussed in Berman et al. [13]. Church and Scaparra [25], Scaparra and Church [127] proposed a model based on the p-median objective. They assume that some of the facilities may be destroyed and a budget is available to protect several facilities. O'Hanley et al. [111] and O'Hanley and Church [110] also suggest models in which one or more facilities are destroyed retaining maximal cover. For a review of unreliable location models the reader is referred to Scaparra and Church [127], Snyder and Daskin [134], Snyder et al. [135], An et al. [4].

6   The Defensive Maximal Covering Location Problem

The defensive maximal covering location model [16] is a leader-follower game theoretic model [87, 136]. The leader locates p facilities on some nodes of a network and at some point in the future a terrorist (the follower) removes one of the links of the network. The follower's objective is to remove the link that causes the most damage. The leader's objective is to cover the most demand *following* a link removal. The model can also be applied to an accident or a natural disaster.

If the disaster is a terrorist attack, it is likely that they will sever the most damaging link. If a link becomes unusable by accident, the leader would like to "protect" himself against the worst possible scenario. This is reminiscent of the minimax regret concept in decision analysis. In both cases, the leader's goal is to maximize coverage *following* the removal of the most damaging link.

The model suggested in Berman et al. [16] is based on demand coverage. For an overview of covering models see Sect. 2. The model is related to location models of unreliable facilities (see an overview in Sect. 5) and can be classified in this category.

We summarize the main results reported in Berman et al. [16]. For complete details the reader is referred to that paper.

The follower's problem is optimally solved while the leader's problem is solved heuristically. Consider a leader's action locating p facilities on a set of p nodes. To solve the follower's (attacker's) problem one needs to find the link whose removal will inflict the maximum decline in coverage. Let F_j be the number of facilities that cover node j when links are intact, and F_j^k be the number of facilities that cover node j following the removal of link k.

By removing link k, the follower may either remove the cover of node j or not. Clearly $F_j^k \leq F_j$. There are four possibilities:

$F_j > 0, F_j^k > 0$ no damage;
$F_j > 0, F_j^k = 0$ lose node j (damage);
$F_j = 0, F_j^k = 0$ no damage;
$F_j = 0, F_j^k > 0$ impossible since $F_j^k \leq F_j$.

Node j loses its cover if and only if $F_j > 0$ and $F_j^k = 0$. F_j is a known number independent of k. Therefore, if $F_j = 0$ no damage can be done to node j because it is not covered before a removal of a link and can be removed from the follower's problem. The follower's problem is solved by efficiently checking the damage for every removed link and selecting the maximum damage rather than solving an integer programming formulation. It is shown in Berman et al. [16] that all links longer than the covering radius can be eliminated from the original problem.

6.1   Properties of the Problem on a Tree

If the network is a tree, some additional properties can be used in the solution approaches. All links longer than the covering radius can be removed from the tree. Each such removal breaks the tree into two sub-trees. It is clear that if K links are removed, there are $K + 1$ disjoint sub-trees. It is shown in Berman et al. [16] that on a tree network, in each direct path the only links that are candidates for removal are links that are adjacent to a facility. In conclusion, for a path network at most $2p$ links are candidates for attack by the follower.

There are other reduction schemes on a path between two facilities. When a direct path connects two facilities, if the distance between each of the end-nodes of a direct path and the node adjacent to the other end-node does not exceed the covering radius, then the whole direct path can be eliminated from consideration by the follower.

6.2   Heuristic Algorithms for the Solution of the Leader's Location Problem

Solving the integer programming formulation for reasonably sized problems requires too much computational effort. Therefore, it is recommended to apply heuristic algorithms in order to solve moderately large problems. Three heuristic algorithms are designed and tested in Berman et al. [16]:

Ascent Algorithm   A set P of p nodes is randomly selected for locating the p facilities. The objective function is the total cover following the follower's move. All possible exchanges between a node in P and a node not in P are evaluated. If an improved exchange is found, the best improved exchange is performed and the next iteration starts. The algorithm terminates when no improving exchange is found. Each iteration requires $p(n - p)$ solutions to follower's problems, each taking $O(n^3 m)$ time for a complexity of $O(n^3 mp(n - p))$.

Simulated Annealing   The simulated annealing [97] simulates the cooling process of hot melted metals. A starting solution is generated, its value of the objective function F is calculated, and the temperature T is set to T_0. The following is repeated for I iterations:

- The solution is randomly perturbed by removing one facility and replacing it with a non-selected facility. The change in the value of the objective function ΔF is calculated.
- If $\Delta F \geq 0$ the search moves to the perturbed solution. If $\Delta F < 0$ the search moves to the perturbed solution with probability $e^{\frac{\Delta F}{T}}$. Otherwise, the search stays at the same solution.
- If the search moves to the perturbed solution, F is changed to the perturbed value of the objective function.
- The time T is changed to αT.

The best encountered solution during the process (usually the last solution) is the result of the algorithm.

Berman et al. [16] followed the approach adopted in Berman et al. [15]. Three parameters are required for the implementation of the simulated annealing algorithm. In Berman et al. [16] the following parameters were used: the starting temperature $T_0 = 10$, the number of iterations $I = 500p(n - p)$, and the factor $\alpha = 1 - \frac{5}{I}$. By this selection of α, the last temperature is $T_0(1 - \frac{5}{I})^I \approx T_0 e^{-5} \approx 0.0067T_0 = 0.067$.

Tabu Search   Tabu search [78–80] proceeds from the ascent algorithm's terminal solution by allowing downward moves, hoping to obtain a better solution in subsequent iterations. A tabu list of forbidden moves is maintained. Tabu moves stay in the tabu list for tabu tenure iterations. To avoid cycling, the forbidden moves are the reverse of recent moves. If a move leads to a better solution than the best solution found so far, this move is executed and the tabu list is emptied. If none of the moves leads to a better solution than the best found solution, the best permissible move (disregarding moves in the tabu list), whether improving or not, is executed.

Each move involves removing a node in P and substituting it with a node not in P. The tabu list consists of nodes recently removed from P so they are not allowed to re-enter P. The maximum possible number of entries in the tabu list is $n - p$. The tabu tenure is randomly generated each iteration between $0.1(n - p)$ and $0.5(n - p)$. Let h be the number of iterations by the ascent algorithm. The tabu algorithm is run for an additional $9h$ iterations.

The easiest way to handle the tabu list is to create a list of all nodes and for each node to store the iteration number at which it entered the tabu list. Emptying the tabu list means assigning a large negative number to every node. A node is in the tabu list if and only if the difference between the current iteration number and the iteration value in the tabu list does not exceed the tabu tenure. This way the length of the tabu list is changing every iteration according to the randomly generated tabu tenure.

6.2.1   Breaking Ties

A useful "trick" for the ascent and tabu algorithms is designed to break ties in the best move. This can be useful in other algorithms for other problems as well. Consider a stream of events (for example a move tying the best move found so far) received over time. It is not known how many events will occur. One event needs to be randomly selected, with equal probability for each event, as a "winner". It is inconvenient to store the list of all events. Thus, the winner cannot be selected at the end of the process.

Drezner [42] suggested the following simple approach: the first event is selected. When the kth event is encountered, it replaces the selected event with a probability of $\frac{1}{k}$. It is shown in Drezner [42] that at the end of the process every event is selected with the same probability. This "trick" simplifies programming algorithms, such as ascent or tabu search, when the events are moves tying for the best one. There is no need to store all tying moves in order to randomly select one.

6.3   Computational Experiments

The three heuristic algorithms were tested on the 40 Beasley [7] problems designed for testing algorithms for the solution of p-median problems. Each problem was solved 100 times by the ascent algorithm and 10 times by simulated annealing and tabu search starting with randomly generated starting solutions leading to similar run times.

We found from the computational results that:

- The ascent algorithm found the best known solution (BK) at least once for 28 of the 40 problems, the tabu search found it in 31 of the 40 problems, while the simulated annealing algorithm found it for all 40 problems. For the 12 problems for which the ascent algorithm did not find the best known solution, the best found solution was, on the average, 1.09% below the best known solution. For the 9 problems for which the tabu search did not find the best known solution, the best found solution was, on the average, 0.50% below the best known solution.
- The ascent algorithm found the BK 33.6% of the 4000 runs, the tabu search found it 56.5% of the 400 runs, while the simulated annealing algorithm found it 83.5% of the 400 runs.
- The average solution of the ascent algorithm was 1.25% below the BK, the tabu search average was 0.34% below the BK, while the average solution of the simulated annealing algorithm was 0.13% below the BK.
- Running 10 replications of the simulated annealing was faster, on the average, than running 100 replications of the ascent algorithm and 10 replications of tabu search.

6.3.1   In Conclusion

- The simulated annealing algorithm performed better than the ascent algorithm and the tabu search.
- The tabu search performed better than the ascent algorithm in a slightly shorter run time.
- The ascent algorithm may be useful for solving large problems for which only one run can be afforded. In such a case, even one run of the simulated annealing or tabu search may take too long.
- The advantage of the simulated annealing over the ascent algorithm and the tabu search is pronounced for larger values of p. For $p = 5$ problems, one hundred runs of the ascent algorithm require about the same time as one run

of the simulated annealing, and the average result is quite comparable (of course, the best result among the 100 runs of the ascent algorithm is better than one run of the simulated annealing). One may consider the ascent approach for small values of p.

7 Summary

The leader-follower model in the context of facility location is an interesting extension to many practical location problems. Four different competitive location models and one defensive maximal covering location model are reviewed.

Many other location problems can be extended to a leader-follower framework and investigating such problems may yield interesting and useful results. For example, the variations of covering objectives in Sect. 2 suggest a plethora of defensive maximal covering location models for future investigation: hierarchical covering [23], variable radii [18], gradual cover [12, 14, 61, 62], cooperative cover [5, 19, 20].

The thirty models (five objectives, each in one of the three environments, and facilities can be reliable or unreliable) depicted in Fig. 1 can also be investigated using a game theoretic framework. For example, unreliable minisum facility location on the sphere (the follower destroys a facility) can be applied in this context and is worth exploring. There are other competitive location models that can be investigated as a leader-follower game. For example, the random utility model [36, 45, 102].

References

1. Aboolian, R., Berman, O., Krass, D.: Competitive facility location and design problem. Eur. J. Oper. Res. **182**, 40–62 (2007)
2. Aboolian, R., Berman, O., Krass, D.: Competitive facility location model with concave demand. Eur. J. Oper. Res. **181**, 598–619 (2007)
3. Aboolian, R., Berman, O., Krass, D.: Efficient solution approaches for discrete multi-facility competitive interaction model. Ann. Oper. Res. **167**, 297–306 (2009)
4. An, Y., Zeng, B., Zhang, Y., Zhao, L.: Reliable p-median facility location problem: two-stage robust models and algorithms. Trans. Res. Part B Methodol. **64**, 54–72 (2014)
5. Averbakh, I., Berman, O., Krass, D., Kalcsics, J., Nickel, S.: Cooperative covering problems on networks. Networks **63**, 334–349 (2014)
6. Batta, R., Dolan, J.M., Krishnamurthy, N.N.: The maximal expected covering location problem: revisited. Trans. Sci. **23**, 277–287 (1989)
7. Beasley, J.E.: OR-library – distributing test problems by electronic mail. J. Oper. Res. Soc. **41**, 1069–1072 (1990). Also available athttp://people.brunel.ac.uk/~mastjjb/jeb/orlib/pmedinfo. html.
8. Beaumont, J.R.: GIS and market analysis. In: Maguire, D.J., Goodchild, M., Rhind, D. (eds.) Geographical Information Systems: Principles and Applications, pp. 139–151. Longman Scientific, Harlow (1991)

9. Bell, D.R., Ho, T.-H., Tang, C.S.: Determining where to shop: fixed and variable costs of shopping. J. Mark. Res. **35**(3), 352–369 (1998)

10. Berman, O., Drezner, Z.: The multiple server location problem. J. Oper. Res. Soc. **58**, 91–99 (2007)

11. Berman, O., Krass, D.: Locating multiple competitive facilities: spatial interaction models with variable expenditures. Ann. Oper. Res. **111**, 197–225 (2002)

12. Berman, O., Krass, D.: The generalized maximal covering location problem. Comput. Oper. Res. **29**, 563–591 (2002)

13. Berman, O., Drezner, Z., Wesolowsky, G.: Locating service facilities whose reliability is distance dependent. Comput. Oper. Res. **30**, 1683–1695 (2003)

14. Berman, O., Krass, D., Drezner, Z.: The gradual covering decay location problem on a network. Eur. J. Oper. Res. **151**, 474–480 (2003)

15. Berman, O., Drezner, Z., Wesolowsky, G.O.: The facility and transfer points location problem. Int. Trans. Oper. Res. **12**, 387–402 (2005)

16. Berman, O., Drezner, T., Drezner, Z., Wesolowsky, G.O.: A defensive maximal covering problem on a network. Int. Trans. Oper. Res. **16**, 69–86 (2009)

17. Berman, O., Drezner, T., Drezner, Z., Krass, D.: Modeling competitive facility location problems: new approaches and results. In: Oskoorouchi, M. (ed.) TutORials in Operations Research, pp. 156–181. INFORMS, San Diego (2009)

18. Berman, O., Drezner, Z., Krass, D., Wesolowsky, G.O.: The variable radius covering problem. Eur. J. Oper. Res. **196**, 516–525 (2009)

19. Berman, O., Drezner, Z., Krass, D.: Cooperative cover location problems: the planar case. IIE Trans. **42**, 232–246 (2010)

20. Berman, O., Drezner, Z., Krass, D.: Continuous covering and cooperative covering problems with a general decay function on networks. J. Oper. Res. Soc. **64**, 1644–1653 (2013)

21. Campbell, J.F.: Integer programming formulations of discrete hub location problems. Eur. J. Oper. Res. **72**, 387–405 (1994)

22. Carrizosa, E., Plastria, F.: Locating an undesirable facility by generalized cutting planes. Math. Oper. Res. **23**, 680–694 (1998)

23. Church, R.L., Eaton, D.J.: Hierarchical location analysis using covering objectives. In: Ghosh, A., Rushton, G. (eds.) Spatial Analysis and Location-Allocation Models, pp. 163–185. Van Nostrand Reinhold Company, New York (1987)

24. Church, R.L., ReVelle, C.S.: The maximal covering location problem. Pap. Reg. Sci. Assoc. **32**, 101–118 (1974)

25. Church, R.L., Scaparra, M.P.: Protecting critical assets: the r-interdiction median problem with fortification. Geogr. Anal. **39**, 129–146 (2007)

26. Coxeter, H.S.M.: Regular Polytopes. Dover Publications, New York (1973)

27. Current, J., Daskin, M., Schilling, D.: Discrete network location models. In: Drezner, Z., Hamacher, H.W. (eds.) Facility Location: Applications and Theory, pp. 81–118. Springer, Berlin (2002)

28. Dasci, A., Laporte, G.: A continuous model for multistore competitive location. Oper. Res. **53**, 263–280 (2005)

29. Daskin, M.S.: A maximum expected covering location model: formulation, properties and heuristic solution. Trans. Sci. **17**, 48–70 (1983)

30. Daskin, M.S.: Network and Discrete Location: Models, Algorithms, and Applications. Wiley, New York (1995)

31. Daskin, M.S., Stern, E.H.: A hierarchical objective set covering model for emergency medical service vehicle deployment. Trans. Sci. **15**, 137–152 (1981)

32. d'Aspremont, C., Gabszewicz, J.J., Thisse, J.-F.: On Hotelling's stability in competition. Econometrica J. Econometric Soc. **47**, 1145–1150 (1979)

33. Drezner, Z.: Competitive location strategies for two facilities. Reg. Sci. Urban Econ. **12**, 485–493 (1982)

34. Drezner, Z.: Heuristic solution methods for two location problems with unreliable facilities. J. Oper. Res. Soc. **38**, 509–514 (1987)

35. Drezner, T.: Locating a single new facility among existing unequally attractive facilities. J. Reg. Sci. **34**, 237–252 (1994)
36. Drezner, T.: Optimal continuous location of a retail facility, facility attractiveness, and market share: An interactive model. J. Retail. **70**, 49–64 (1994)
37. Drezner, T.: Competitive facility location in the plane. In: Drezner, Z. (ed.) Facility Location: A Survey of Applications and Methods, pp. 285–300. Springer, New York (1995)
38. Drezner, Z.: Facility Location: A Survey of Applications and Methods. Springer, New York (1995)
39. Drezner, T.: Location of multiple retail facilities with limited budget constraints – in continuous space. J. Retail. Consum. Serv. **5**, 173–184 (1998)
40. Drezner, T.: Derived attractiveness of shopping malls. IMA J. Manag. Math. **17**, 349–358 (2006)
41. Drezner, T.: Competitive facility location. In: Encyclopedia of Optimization, pp. 396–401, 2nd edn. Springer, New York (2009)
42. Drezner, Z.: Random selection from a stream of events. Commun. ACM **53**, 158–159 (2010)
43. Drezner, T.: Cannibalization in a competitive environment. Int. Reg. Sci. Rev. **34**, 306–322 (2011)
44. Drezner, T.: A review of competitive facility location in the plane. Logist. Res. **7**, 114 (2014). doi:10.1007/s12159-014-0114-z.
45. Drezner, T., Drezner, Z.: Competitive facilities: Market share and location with random utility. J. Reg. Sci. **36**, 1–15 (1996)
46. Drezner, T., Drezner, Z.: Replacing discrete demand with continuous demand in a competitive facility location problem. Nav. Res. Logist. **44**, 81–95 (1997)
47. Drezner, T., Drezner, Z.: Facility location in anticipation of future competition. Locat. Sci. **6**, 155–173 (1998)
48. Drezner, T., Drezner, Z.: A note on applying the gravity rule to the airline hub problem. J. Reg. Sci. **41**, 67–73 (2001)
49. Drezner, T., Drezner, Z.: Validating the gravity-based competitive location model using inferred attractiveness. Ann. Oper. Res. **111**, 227–237 (2002)
50. Drezner, T., Drezner, Z.: Multiple facilities location in the plane using the gravity model. Geogr. Anal. **38**, 391–406 (2006)
51. Drezner, T., Drezner, Z.: The gravity p-median model. Eur. J. Oper. Res. **179**, 1239–1251 (2007)
52. Drezner, T., Drezner, Z.: Lost demand in a competitive environment. J. Oper. Res. Soc. **59**, 362–371 (2008)
53. Drezner, T., Drezner, Z.: The gravity multiple server location problem. Comput. Oper. Res. **38**, 694–701 (2011)
54. Drezner, T., Drezner, Z.: Modelling lost demand in competitive facility location. J. Oper. Res. Soc. **63**, 201–206 (2012)
55. Drezner, T., Drezner, Z.: The maximin gradual cover location problem. OR Spectr. **36**, 903–921 (2014)
56. Drezner, Z., Erkut, E.: Solving the continuous p-dispersion problem using non-linear programming. J. Oper. Res. Soc. **46**, 516–520 (1995)
57. Drezner, Z., Hamacher, H.W.: Facility Location: Applications and Theory. Springer Science and Business Media, New York (2002)
58. Drezner, Z., Suzuki, A.: Covering continuous demand in the plane. J. Oper. Res. Soc. **61**, 878–881 (2010)
59. Drezner, Z., Zemel, E.: Competitive location in the plane. Ann. Oper. Res. **40**, 173–193 (1992)
60. Drezner, Z., Wesolowsky, G.O., Drezner, T.: On the logit approach to competitive facility location. J. Reg. Sci. **38**, 313–327 (1998)
61. Drezner, Z., Wesolowsky, G.O., Drezner, T.: The gradual covering problem. Nav. Res. Logist. **51**, 841–855 (2004)

62. Drezner, T., Drezner, Z., Goldstein, Z.: A stochastic gradual cover location problem. Nav. Res. Logist. **57**, 367–372 (2010)
63. Drezner, T., Drezner, Z., Kalczynski, P.: A cover-based competitive location model. J. Oper. Res. Soc. **62**, 100–113 (2011)
64. Drezner, T., Drezner, Z., Kalczynski, P.: Strategic competitive location: improving existing and establishing new facilities. J. Oper. Res. Soc. **63**, 1720–1730 (2012)
65. Drezner, T., Drezner, Z., Kalczynski, P.: A leader-follower model for discrete competitive facility location. Comput. Oper. Res. **64**, 51–59 (2015)
66. Eaton, B.C., Lipsey, R.G.: The principle of minimum differentiation reconsidered: some new developments in the theory of spatial competition. Rev. Econ. Stud. **42**, 27–49 (1975)
67. Eiselt, H.A.: Hotelling's duopoly on a tree. Ann. Oper. Res. **40**, 195–207 (1992)
68. Eiselt, H.: Equilibria in competitive location models. In: Eiselt, H.A., Marianov, V. (eds.) Foundations of Location Analysis, pp. 139–162. Springer, New York (2011)
69. Eiselt, H.A., Laporte, G.: The existence of equilibria in the 3-facility hotelling model in a tree. Trans. Sci. **27**, 39–43 (1993)
70. Eiselt, H.A., Marianov, V.: Foundations of Location Analysis, vol. 155. Springer, New York (2011)
71. Eiselt, H.A., Marianov, V., Drezner, T.: Competitive location models. In: Laporte, G., Nickel, S., da Gama, F.S. (eds.) Location Science, pp. 365–398. Springer, New York (2015)
72. Fernandez, J., Pelegrin, B., Plastria, F., Toth, B.: Solving a Huff-like competitive location and design model for profit maximization in the plane. Eur. J. Oper. Res. **179**, 1274–1287 (2007)
73. Francis, R.L., McGinnis, L.F. Jr., White, J.A.: Facility Layout and Location: An Analytical Approach, 2nd edn. Prentice Hall, Englewood Cliffs (1992)
74. Francis, R.L., Lowe, T.J., Rayco, M.B., Tamir, A.: Aggregation error for location models: survey and analysis. Ann. Oper. Res. **167**, 171–208 (2009)
75. Gao, Z., Wu, J., Sun, H.: Solution algorithm for the bi-level discrete network design problem. Trans. Res. Part B Methodol. **39**, 479–495 (2005)
76. García, S., Marín, A.: Covering location problems. In: Laporte, G., Nickel, S., da Gama, F.S. (eds.) Location Science, pp. 93–114. Springer, New York (2015)
77. Ghosh, A., Craig, C.S.: Formulating retail location strategy in a changing environment. J. Mark. **47**, 56–68 (1983)
78. Glover, F.: Heuristics for integer programming using surrogate constraints. Decis. Sci. **8**, 156–166 (1977)
79. Glover, F.: Future paths for integer programming and links to artificial intelligence. Comput. Oper. Res. **13**, 533–549 (1986)
80. Glover, F., Laguna, M.: Tabu Search. Kluwer Academic Publishers, Boston (1997)
81. Goodchild, M.F. Noronha, V.T.: Location-allocation and impulsive shopping: the case of gasoline retailing. In: Spatial Analysis and Location-Allocation Models, pp. 121–136. Van Nostrand Reinhold, New York (1987)
82. Hakimi, S.L.: Optimum locations of switching centres and the absolute centres and medians of a graph. Oper. Res. **12**, 450–459 (1964)
83. Hakimi, S.L.: Optimum distribution of switching centers in a communication network and some related graph theoretic problems. Oper. Res. **13**, 462–475 (1965)
84. Hakimi, S.L.: On locating new facilities in a competitive environment. In: Presented at the ISOLDE II Conference, Skodsborg, Denmark (1981)
85. Hakimi, S.L.: On locating new facilities in a competitive environment. Eur. J. Oper. Res. **12**, 29–35 (1983)
86. Hakimi, S.L.: p-Median theorems for competitive location. Ann. Oper. Res. **6**, 77–98 (1986)
87. Hakimi, S.L.: Locations with spatial interactions: Competitive locations and games. In: Mirchandani, P.B., Francis, R.L. (eds.) Discrete Location Theory, pp. 439–478. Wiley-Interscience, New York (1990)
88. Hansen, P., Labbè, M.: Algorithms for voting and competitive location on a network. Trans. Sci. **22**, 278–288 (1988)

89. Hilbert, D., Cohn-Vossen, S.: Geometry and the Imagination. Chelsea Publishing Company, New York (1956). English translation of Anschauliche Geometrie (1932)
90. Hodgson, M.: Toward more realistic allocation in location-allocation models: an interaction approach. Environ. Plan. A **10**, 1273–1285 (1978)
91. Hodgson, M.J.: A location-allocation model maximizing consumers' welfare. Reg. Stud. **15**, 493–506 (1981)
92. Hotelling, H.: Stability in competition. Econ. J. **39**, 41–57 (1929)
93. Huff, D.L.: Defining and estimating a trade area. J. Mark. **28**, 34–38 (1964)
94. Huff, D.L.: A programmed solution for approximating an optimum retail location. Land Econ. **42**, 293–303 (1966)
95. Jain, A.K., Mahajan, V.: Evaluating the competitive environment in retailing using multiplicative competitive interactive models. In: Sheth, J.N. (ed.) Research in Marketing, vol. 2, pp. 217–235. JAI Press, Greenwich (1979)
96. Jensen, M.C.: Value maximization and the corporate objective function. In: Beer, M., Norhia, N. (ed.) Breaking the Code of Change, pp. 37–57. Harvard Business School Press, Boston (2000)
97. Kirkpatrick, S., Gelat, C.D., Vecchi, M.P.: Optimization by simulated annealing. Science **220**, 671–680 (1983)
98. Kolen, A., Tamir, A.: Covering problems. In: Mirchandani, P.B., Francis, R.L. (ed.) Discrete Location Theory, pp. 263–304. Wiley-Interscience, New York (1990)
99. Küçükaydın, H., Aras, N., Altınel, I.: A leader–follower game in competitive facility location. Comput. Oper. Res. **39**, 437–448 (2012)
100. Laporte, G., Nickel, S., da Gama, F.S.: Location Science. Springer, New York (2015)
101. Lee, S.-D.: On solving unreliable planar location problems. Comput. Oper. Res. **28**, 329–344 (2001)
102. Leonardi, G., Tadei, R.: Random utility demand models and service location. Reg. Sci. Urban Econ. **14**, 399–431 (1984)
103. Lerner, A.P., Singer, H.W.: Some notes on duopoly and spatial competition. J. Polit. Econ. **45**, 145–186 (1937)
104. Locatelli, M., Raber, U.: Packing equal circles in a square: a deterministic global optimization approach. Discret. Appl. Math. **122**, 139–166 (2002)
105. Love, R.F., Morris, J.G., Wesolowsky, G.O.: Facilities Location: Models and Methods. North Holland, New York (1988)
106. Maranas, C.D., Floudas, C.A., Pardalos, P.M.: New results in the packing of equal circles in a square. Discret. Math. **142**, 287–293 (1995)
107. Nakanishi, M., Cooper, L.G.: Parameter estimate for multiplicative interactive choice model: least squares approach. J. Mark. Res. **11**, 303–311 (1974)
108. Nash, J.: Non-cooperative games. Ann. Math. **54**, 286–295 (1951)
109. Nurmela, K.J., Oestergard, P.: More optimal packings of equal circles in a square. Discret. Comput. Geom. **22**, 439–457 (1999)
110. O'Hanley, J.R., Church, R.L.: Designing robust coverage networks to hedge against worst-case facility losses. Eur. J. Oper. Res. **209**, 23–36 (2011)
111. O'Hanley, J.R., Church, R.L., Gilless, J.K.: Locating and protecting critical reserve sites to minimize expected and worst-case losses. Biol. Conserv. **134**, 130–141 (2007)
112. Okabe, A., Suzuki, A.: Locational optimization problems solved through Voronoi diagrams. Eur. J. Oper. Res. **98**, 445–456 (1997)
113. Okabe, A., Boots, B., Sugihara, K., Chiu, S.N.: Spatial Tessellations: Concepts and Applications of Voronoi Diagrams. Wiley Series in Probability and Statistics. Wiley, New York (2000)
114. Plastria, F.: Continuous covering location problems. In: Drezner, Z., Hamacher, H.W. (ed.) Facility Location: Applications and Theory, pp. 37–79. Springer, Berlin (2002)
115. Plastria, F., Carrizosa, E.: Undesirable facility location in the Euclidean plane with minimal covering objectives. Eur. J. Oper. Res. **119**, 158–180 (1999)

116. Plastria, F., Carrizosa, E.: Optimal location and design of a competitive facility. Math.
     Program. **100**, 247–265 (2004)
117. Plastria, F., Vanhaverbeke, L.: Aggregation without loss of optimality in competitive location
     models. Netw. Spat. Econ. **7**, 3–18 (2007)
118. Plastria, F., Vanhaverbeke, L.: Discrete models for competitive location with foresight.
     Comput. Oper. Res. **35**, 683–700 (2008)
119. Redondo, J.L., Fernández, J., García, I., Ortigosa, P.M.: Heuristics for the facility location
     and design (1|1)-centroid problem on the plane. Comput. Optim. Appl. **45**, 111–141 (2010)
120. Redondo, J.L., Fernández, J., Arrondo, A., García, I., Ortigosa, P.M.: Fixed or variable
     demand? does it matter when locating a facility? Omega **40**, 9–20 (2012)
121. Redondo, J.L., Arrondo, A., Fernández, J., García, I., Ortigosa, P.M.: A two-level evolutionary
     algorithm for solving the facility location and design (1|1)-centroid problem on the plane with
     variable demand. J. Glob. Optim. **56**, 983–1005 (2013)
122. Reilly, W.J.: The Law of Retail Gravitation. Knickerbocker Press, New York (1931)
123. ReVelle, C.: The maximum capture or sphere of influence problem: Hotelling revisited on a
     network. J. Reg. Sci. **26**, 343–357 (1986)
124. ReVelle, C., Toregas, C., Falkson, L.: Applications of the location set covering problem.
     Geogr. Anal. **8**, 65–76 (1976)
125. Saidani, N., Chu, F., Chen, H.: Competitive facility location and design with reactions of
     competitors already in the market. Eur. J. Oper. Res. **219**, 9–17 (2012)
126. Sáiz, M.E., Hendrix, E.M., Fernández, J., Pelegrín, B.: On a branch-and-bound approach for
     a Huff-like Stackelberg location problem. OR Spectr. **31**, 679–705 (2009)
127. Scaparra, M.P., Church, R.L.: A bilevel mixed-integer program for critical infrastructure
     protection planning. Comput. Oper. Res. **35**, 1905–1923 (2008)
128. Schilling, D.A., Vaidyanathan, J., Barkhi, R.: A review of covering problems in facility
     location. Locat. Sci. **1**, 25–55 (1993)
129. Shishebori, D., Babadi, A.Y.: Robust and reliable medical services network design under
     uncertain environment and system disruptions. Trans. Res. Part E Logist. Trans. Rev. **77**,
     268–288 (2015)
130. Shishebori, D., Snyder, L.V., Jabalameli, M.S.: A reliable budget-constrained FL/ND problem
     with unreliable facilities. Netw. Spat. Econ. **14**, 549–580 (2014)
131. Simpson, P.B.: On defining areas of voter choice: professor tullock on stable voting. Q. J.
     Econ. **83**, 478–490 (1969)
132. Snyder, L.V.: Covering problems. In: Eiselt, H.A., Marianov, V. (eds.) Foundations of
     Location Analysis, pp. 109–135. Springer, New York (2011)
133. Snyder, L.V., Daskin, M.S.: Reliability models for facility location: the expected failure cost
     case. Trans. Sci. **39**, 400–416 (2005)
134. Snyder, L., Daskin, M.: Models for reliable supply chain networks design. In: Murray,
     A., Grubesic, T. (eds.) Reliability and Vulnerability in Critical Infrastructure: A Quantitative
     Geographic Perspective. Springer, Berlin (2006)
135. Snyder, L., Scaparra, M., Daskin, M., Church, R.: Planning for disruptions in supply chain
     networks. Tutor. Oper. Res. 234–257 (2006)
136. Stackelberg, H.V.: Marktform und Gleichgewicht. Julius Springer, Vienne (1934)
137. Sun, H., Gao, Z., Wu, J.: A bi-level programming model and solution algorithm for the
     location of logistics distribution centers. Appl. Math. Model. **32**, 610–616 (2008)
138. Suzuki, A., Drezner, Z.: The *p*-center location problem in an area. Locat. Sci. **4**, 69–82 (1996)
139. Suzuki, A., Drezner, Z.: The minimum equitable radius location problem with continuous
     demand. Eur. J. Oper. Res. **195**, 17–30 (2009)
140. Suzuki, A., Okabe, A.: Using Voronoi diagrams. In: Drezner, Z. (ed.) Facility Location: A
     Survey of Applications and Methods, pp. 103–118. Springer, New York (1995)
141. Szabo, P.G., Markot, M., Csendes, T., Specht, E.: New Approaches to Circle Packing in a
     Square: With Program Codes. Springer, New York (2007)

142. Tavakkoli-Mogahddam, R., Ghezavati, V., Kaboli, A., Rabbani, M.: An efficient hybrid method for an expected maximal covering location problem. In: New Challenges in Applied Intelligence Technologies, pp. 289–298. Springer, New York (2008)

143. Toppen, F., Wapenaar, H.: GIS in business: tools for marketing analysis. Proc. EGIS 1994, EGIS Foundation (1994) www.odyssey.maine.edu/gisweb/spatab/egis/eg94204.html

144. Toth, B., Fernandez, J., Pelegrin, B., Plastria, F.: Sequential versus simultaneous approach in the location and design of two new facilities using planar Huff-like models. Comput. Oper. Res. **36**, 1393–1405 (2009)

145. Voronoï, G.: Nouvelles applications des paramètres continus à la théorie des formes quadratiques. Deuxième mémoire. recherches sur les parallélloèdres primitifs. J. Reine Angew. Math. **134**, 198–287 (1908)

146. Wendell, R., McKelvey, R.: New perspectives in competitive location theory. Eur. J. Oper. Res. **6**, 174–182 (1981)

147. Wilson, A.G.: Retailers' profits and consumers' welfare in a spatial interaction shopping mode. In: Masser, I. (ed.) Theory and Practice in Regional Science, pp. 42–59. Pion, London (1976)

148. Winerfert, B.: The relation between market share and profitability. J. Bus. Strategy **6**, 67–74 (1986)

149. Wollmer, R.: Removing arcs from a network. Oper. Res. **12**, 934–940 (1964)

150. Wong, S.-C., Yang, H.: Determining market areas captured by competitive facilities: a continuous equilibrium modeling approach. J. Reg. Sci. **39**, 51–72 (1999)

151. Wood, R.K.: Deterministic network interdiction. Math. Comput. Model. **17**, 1–18 (1993)

152. Yang, H., Wong, S.: A continuous equilibrium model for estimating market areas of competitive facilities with elastic demand and market externality. Trans. Sci. **34**, 216–227 (2000)

153. Zabinsky, Z.B., Smith, R.L.: Pure adaptive search in global optimization. Math. Prog. **53**, 323–338 (1992)

Asymmetries in Competitive Location Models on the Line

H.A. Eiselt and Vladimir Marianov

1 Introduction

This paper deals with competitive location models in a very simple spatial context. These models are essentially explanatory in nature, i.e., they demonstrate the effects different policies have and how competitors generally behave. Due to the simplicity of the market shape under consideration, it is clearly not designed to provide locating firms with solutions to their specific problems. While problems involving competing firms have long been discussed by economists, Hotelling [41] is generally credited with being the first to consider competition in the spatial context. In a nutshell (and without providing all details at this point), he considered duopolists, who each locate a single branch of their firm in a space that is in the shape of a line segment, which Hotelling referred to as "main street." The demand on the line is assumed to be uniformly distributed. In his original scenario, each of the independently operating firms has two decision variables at its disposal, viz., location and price. Hotelling concluded that an equilibrium is reached when both competitors locate arbitrarily close at the center of the market. He framed this result in the context of product design or brand positioning; more specifically, the two "firms" represent products, while the space symbolizes one quantitatively measurable aspect of the products. In his work, Hotelling used the example of two brands of cider and their respective sweetness. Given his mathematical result, the two products will turn out be become very similar. This result has been dubbed the "principle of minimum differentiation." Similarly, he mapped political candidates

H.A. Eiselt (✉)
Faculty of Business Administration, University of New Brunswick, Fredericton, NB, Canada
e-mail: haeiselt@unb.ca

V. Marianov
Department of Electrical Engineering, Pontificia Universidad Católica de Chile, Santiago, Chile
e-mail: marianov@ing.puc.cl

© Springer International Publishing AG 2017
L. Mallozzi et al. (eds.), *Spatial Interaction Models*, Springer Optimization
and Its Applications 118, DOI 10.1007/978-3-319-52654-6_6

in an "issue space"—in the simplest case, the usual left—right scale—and claimed that his result would prove why that the political platforms tended to be so similar. The analogy of his original model and the political application is not very fitting, though, as this model does not feature prices. This model was later expanded upon to two dimensions by Rusk and Weisberg [65]. It became clear in later years that Hotelling models are notoriously unstable, meaning that even small changes to the model can result in dramatic changes to the solutions (see, e.g., Brown [12], Brenner [11], Yasuda [77], and Marianov and Eiselt [55]). In other words, Hotelling's result with prices may (and does) have very different results from those with fixed prices.

It took 50 years until d'Aspremont et al. [15] proved that Hotelling's original model with linear transportation costs does not have an equilibrium. However, if the transportation costs were assumed to be quadratic, there would be an equilibrium, however one, in which facilities (we will call the entities to be located "facilities," regardless if they represent firms, their branches, brands, politicians, or anything else) locate as far apart as possible from each other. This is yet another indication of the instability of the results.

Lerner and Singer [52] introduced reservation prices into the model, Smithies [70] used elastic demand, and Stevens [71] was the first to formally use game theory in a discretized Hotelling model. Eaton and Lipsey [19] used fixed and equal prices, but generalized Hotelling's model in a different direction by including more than two competitors into the market. They re-established the principle of minimum differentiation for two facilities (even though it can easily be shown that arbitrarily small price changes will destroy the equilibrium, showing again the instability of the model), and demonstrated that the case of $n = 3$ competitors does not have an equilibrium at all. Furthermore, models with more than five firms were shown to have multiple equilibria.

Rothschild [64] appears to have been among the earliest to introduce sequential entry of the firms onto the market into Hotelling's model. This was soon followed by the contribution by Prescott and Visscher [62], who computed locations for two and three firms that sequentially enter the market. Interestingly enough, these authors noticed 2 years before the D'Aspremont et al. [15] paper that there was a problem regarding the nonexistence of price equilibria for firm locations that are close together.

Finally, it was Hakimi [36] and the follow-up contribution [37] that introduced Hotelling models to operations researchers. His arguments place customers and firms on a network, and the main concern are optimal locations (rather than equilibria) and the complexity of the algorithms that compute the optimal locations.

The key results derived from the large variety of Hotelling models in the literature fall into two categories. The first type of result deals with the stability of the model. More specifically, authors use the concept of Nash equilibrium [58] and a later refinement by Selten [67] and his subgame perfect equilibrium. Essentially, Nash essentially states that a situation is in equilibrium, if neither of the decision makers (or players in game-theoretic parlance) can improve his own objective by unilaterally changing his present decisions. There is no concern of how such an equilibrium, assuming it exists, can actually be reached, even though some authors have addressed that issue (see, e.g., Bhadury and Eiselt [7]). Similarly, the existence of an equilibrium is typically dealt with as a binary result: either an equilibrium does exist or it does not. Authors such as Eiselt and Bhadury [24], however, have deviated from this convention and have suggested a continuum to measure stability.

Another solution concept was put forward by the economist von Stackelberg [76]. It subdivides the existing firms or players into two groups, the leaders and the followers. The decision makers of all players will then decide sequentially: first the leaders decide (whatever their decisions may entail, e.g., locations, prices, quality of the products, service level, etc.), knowing fully well that the followers will make their decisions once the leaders have announced their decisions. The followers will wait for the leaders to have made their announcements, and they will then make their decisions. It is important to realize the inherent asymmetry in the leader—follower concept: while the leader must make assumptions concerning the behavior of the follower (e.g., concerning its objectives, perceptions about demand, and other issues), and thus looking over his shoulder, followers do not have any such problems: all they have to do is wait until the leaders make their firm announcements regarding their decisions, take them as given and solve their own conditional problems. This is the concept Rothschild [64], Prescott and Visscher [62] and their many followers have used.

An important concept in sequential facility location is the "first mover advantage." It refers to the (possible) advantage a firm has while being the first to make a location decision. Whereas this concept has been discussed in the marketing literature for a long time, Ghosh and Buchanan [34] introduced it to the discussion of competitive locations. More specifically, the coined the phrase "first entry paradox" for cases, in which the firm that is first to choose its location, does not have an advantage as expected. It turns out (and some of the cases discussed below will witness this) that the "paradox" is more a rule than an exception in competitive location models.

With the multitude of assumptions, rules and concepts pertaining to Hotelling models, it is not surprising that authors have devised taxonomies to classify models, see, e.g., Eiselt et al. [28]. It includes the space, number of players, the pricing policy chosen by the decision makers, the rules of the game (referring to Nash equilibria or von Stackelberg solutions), and the behavior of the customers. However, like most location models, it is assumed that all decision makers perceive reality the same way (e.g., assess demand in the same way; for a model that drops this assumption, see Eiselt [22]), and that all players use the same objective function. While the assumption of symmetry among players may apply in some situations, in others it does not. Firms do not apply the same technologies, so that their cost functions may differ. Upstarts may attempt to establish a foothold in the market (by maximizing sales, while keeping the profit at least at a preset required minimum level), while established firms may attempt to maximize their profit. Similarly, while some firms employ mill pricing as their pricing policy of choice, others, such as internet firms, may use delivered pricing instead. Asymmetries such as these are at the core of this paper.

There is a multitude of introductions, reviews, and surveys to the subject. An early introduction is provided by Gabszewicz and Thisse [33] that focuses on different pricing policies and derives objectives. Plastria's [61] piece concentrates on optimization approaches to competitive location problems. The work by Kress and Pesch [47] is similar, as it also looks at the optimization side of the problem with sequential entry onto the market. Younies and Eiselt [78] rework the fundamental contributions of competitive location models with sequential entry, while Eiselt [23] looks at the early contributions focussing on equilibrium results. In his teaching note, Russell [66] stresses the "limited applicability of the principle under more realistic assumptions." Biscaia and Mota [8] review Hotelling models with their many facets, while the latest review is by Ashtiani [3]. Finally, we do not want to omit mentioning two books devoted to the subject, viz., those by Miller et al. [57] and Karakitsiou [45].

The remainder of this paper is organized as follows. Section 2 introduces the basic model and, in doing so, sets the stage for move involved models. Section 3 considers cases, in which the two firms apply different objectives, while Sect. 4 discusses cases, in which the pricing policies of the firms differ. In the models in Sect. 5 firms have different technological capabilities (reflected in different production and/or transportation costs), and, finally, Sect. 6 summarizes the paper and provides an outlook.

2 The Basic Model

Consider a line segment extending from 0 to ℓ, on which customers are uniformly distributed. There are two competing firms, so that each will locate one facility. The firms (and, if no confusion can arise, their facilities) are referred to as A and B, respectively. Without loss of generality, firm A is located to the left of firm B. Firm A is located a units from the left end (i.e., 0) of the market, while firm B is located b units away from the right side (viz., ℓ) of the market. Given the above assumptions, $a, b \geq 0$ and $a + b \leq \ell$. Given a unified common price p and assuming mill pricing by both firms, customers will purchase the goods they desire from the closer facility, as the good is homogeneous (e.g., a brand name) and transportation costs are the only distinguishing feature. Transportation costs include not only out-of-pocket expenses for the transportation, but also inconvenience, such as time in case where actual shipments occur, or other disutilities such as dislike in case of brand positioning models. Just like Hotelling in his original contribution, we assume in this work that transportation costs are linear in the distance. D'Aspremont et al. [15] introduced quadratic costs, which have very little meaning in case of physical transportation, but are important in case of brand positioning models, in which dislikes of a brand may increase superlinearly in the distance of the brand features and the ideal point that represents a customer's most preferred product design. Following the introduction of linear-quadratic transportation cost functions by Gabszewicz and Thisse [33], Anderson [1] used such functions in his analysis.

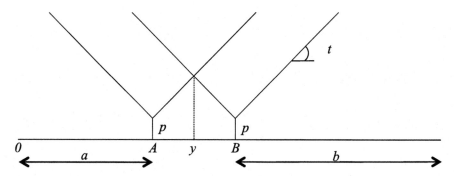

Fig. 1 Facility locations A and B and marginal customer y

He showed that an equilibrium in pure strategies exists only, if the cost function does not include a linear part, however small. Authors such as Hamoudi and Moral [38] expanded the analysis to general convex and concave transportation costs.

For our purposes, we assume that transport costs are linear with unit transportation costs t, so that we can determine the location of the *marginal customer* at point y. The marginal customer is located at a point, at which the costs of purchasing the good from firm A are identical to those of a purchase from firm B. In order to formalize matters, denote by d_{qr} the distance between two suitably defined points q and r. At point y, the cost of purchasing from firm A is $p + td_{Ay}$, while the cost of purchasing from firm B is $p + td_{yB}$. With $d_{yB} = d_{AB} - d_{Ay}$, we obtain $d_{Ay} = \frac{1}{2}d_{AB}$, i.e., the marginal customer is located at the midpoint between firms A and B. This situation is shown in Fig. 1.

We will refer to the area to the left of A as "A's hinterland", the area to the right of firm B is called "B's hinterland", while the area between the two firms is the competitive region. In this situation, firm A will capture all customers between the left end of the market at 0 and the marginal customer at y, while firm B will capture all customers between y and the right end of the market at ℓ. The market areas $M(A)$ and $M(B)$ are then $M(A) = d_{0y} = a + \frac{1}{2}d_{AB} = a + \frac{1}{2}(\ell - a - b)$ and $M(B) = d_{y\ell} = b + \frac{1}{2}d_{AB} = b + \frac{1}{2}(\ell - a - b)$. If either of the two firms were to act unilaterally and relocate, trying to increase its market share, it would move towards its opponent, as for each ϵ units that a firm moves towards its competitor, it gains ϵ units of demand in its hinterland, while losing $\frac{1}{2}\epsilon$ units in the competitive region. Given that moving towards one's competitor is the strategy that maximizes market share, it follows that central agglomeration of the two firms is the only stable solution, i.e., a locational Nash equilibrium. As far as von Stackelberg solutions are concerned, given the leader's location at some point on the line, the follower will always locate right next to the leader on the longer side of the market. The leader anticipates this move, so it will-given that the market size is fixed-minimize the maximum size the follower can get, which is achieved by locating at the center of the market. As demonstrated above, the follower will locate next to the leader, so that again, central agglomeration results. This is Hotelling's result, but it was achieved in a model in which prices are fixed and equal, rather than variable, as in Hotelling's original paper.

Things change dramatically once the prices of the two competitors are still fixed, but no longer equal. Without loss of generality let $p_A < p_B$. This pricing structure allows the lower-priced firm A to follow its competitor B in a predatory fashion by locating arbitrarily close to it, thus cutting him out and capture the entire market. Firm B, on the other hand, will attempt to stay away from A and attempt to capture a small, "out of the way," segment of the market. (This is the fashion of small retailers who, unable to compete in prices with large chain stores, have survived off the main centers.) Clearly, there is no equilibrium. This demonstrates how fickle the equilibrium in the previous case with fixed and equal prices actually is: if one of the two prices changes by an arbitrarily small amount, the equilibrium ceases to exist.

As far as von Stackelberg solutions are concerned, we will consider two cases. In the first case, the higher-priced firm B locates first, followed by the lower-priced firm A. This case is easily dispensed with. Wherever firm B locates, its competitor A can always locate arbitrarily close to B, cut out its competitor, and capture the entire market. The case in which firm A locates first, followed by firm B, is not quite as simple. Given any location of firm A, firm B will locate on the longer side of the market (or on either side, in case A has located at the center of the market), at a distance of $\frac{1}{t}(p_B - p_A) + \epsilon$, which guarantees that firm B is just sufficiently far away from A to avoid being cut out. Its market area is restricted to roughly the area to its right, i.e., $M(B) \approx b$. As firm A anticipates this move, and it obtains the complement of B's market area, i.e., $M(A) = \ell - M(B)$, it will locate at the center of the market with firm B locating $\frac{1}{t}(p_B - p_A) + \epsilon$ away on either side. The two market areas are then $M(A) \approx \frac{1}{2}\ell + \frac{1}{t}(p_B - p_A)$ and $M(B) \approx \frac{1}{2}\ell - \frac{1}{t}(p_B - p_A)$. Notice that (1) the case of equal prices results in central agglomeration as discussed above, and (2) $M(A) > M(B)$ for all cases, in which $p_A < p_B$.

3 Asymmetry of Objectives

This section investigates models, in which the two competitors have different objectives. In particular, we will investigate a model, in which one of the competitors, say firm A, resembles a private firm that maximizes revenue, while its competitor, firm B, is a public firm, whose objective is to maximize access to its facility. Settings such as this are well known in the literature, as they pit a private firm against a public firm. Typical examples are contributions, such as Thisse and Wildasin [74], in which customers decide whether to patronize a centrally located public facility or one of two more peripheral branches of a private firm. Lu [54] has a welfare-maximizing firm compete with a private firm that maximizes its profit, while Herr [39] considers a specific mixed duopoly that examines the potential privatization of a public hospital and whether or not such course of action will increase welfare.

In our analysis, we first consider a model, in which one of the competitors, say firm A, resembles a private firm that maximizes revenue, while its competitor,

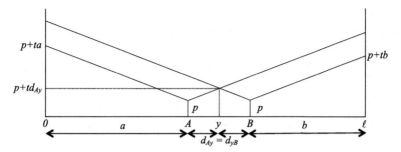

Fig. 2 Delivered prices for two firms with equal prices

firm B, is a public firm, whose objective is to maximize access to its facility. To operationalize, access is expressed as the average distance between a customer and a facility. Again, to simplify matters, assume that the prices at the two firms are equal, so that customers, given that they have no preference for either firm, will patronize the closer facility. Furthermore, let firm A be located to the left of firm B, i.e., $a \leq \ell - b$. The situation is shown in Fig. 2.

At point y, we have $p + td_{Ay} = p + td_{yB}$, resulting in $d_{Ay} = d_{yB} = \frac{1}{2}(\ell - a - b)$. Then firm A's market area (which the firm attempts to maximize), is $M(A) = a + d_{Ay} = a + \frac{1}{2}(\ell - a - b) = \frac{1}{2}(\ell + a - b)$. With the length of the market ℓ and firm B's location being temporarily fixed (as long as firm B does not react to firm A's move), firm A will maximize its sales and revenue, if it chooses the value a as large as possible. Since $a + b \leq \ell$ (firm A locates to the left of its competitor B), i.e., $a = \ell - b$, which is firm A's reaction function. In other words, firm A will locate arbitrarily close to its competitor.

Consider now the public firm B, which attempts to maximize access of its customers to its own facility. The average distance of facility B to its own customers is:

$$D'_B = \frac{1}{M(B)} \left[\int_{x=0}^{\frac{1}{2}(\ell - a - b)} x\,dx + \int_{x=0}^{b} x\,dx \right] = \frac{1}{\frac{1}{2}(\ell - a + b)} \left[\frac{1}{8}(\ell - a - b)^2 + \frac{1}{2}b^2 \right]$$

We already know that firm A will cluster, so that $a = \ell - b$, and $D'_B = \frac{1}{2}b$, which is clearly minimized for $b = 0$. Adding firm A's clustering, we obtain $a = \ell$. This paradoxical result indicates that at equilibrium, the public facility will locate at one end of the market with the private facility right next to it, so that the private firm captures the entire market (certainly its best possible result), while the public facility ends up with no market at all. This is explained by the fact that the public firm considers only its own customers when minimizing average access time, which in this solution equals zero, the lowest possible value. Since neither facility has an incentive to move out of the present situation, this is a Nash equilibrium. Note that the average distance between a customer and its closest facility is $\frac{1}{2}\ell$.

Things change if the public firm considers the average access distance for *all* customers. While firm A's objective and its resulting predatory behavior are still the same, firm B's objective changes. The average customer-facility distance, given that customers patronize the facility closest to them is then

$$D_B = \frac{1}{\ell} \left[\int_{x=0}^{a} x dx + \int_{x=0}^{d_{Ay}=1/2(\ell-a-b)} x dx + \int_{x=0}^{d_{yB}=1/2(\ell-a-b)} x dx + \int_{x=0}^{b} x dx \right]$$

$$= \frac{1}{\ell} \left[\frac{1}{2}a^2 + \frac{1}{4}(\ell - a - b)^2 + \frac{1}{2}b^2 \right]$$

Optimizing this function with respect to b results in firm B's reaction function $b = \frac{1}{3}(\ell - a)$. In other words, the public firm B will disperse and attempt to locate away from its private competitor. Since firm A, on the other hand, needs to agglomerate to achieve its best possible result, we obtain the result:

Theorem 1 *In the competitive model with one firm that maximizes its market capture and another that minimizes the average customer–facility distance, there is no equilibrium.*

Consider now the von Stackelberg solutions in this asymmetric game. There are two versions of it, "first public, then private" as well as "first private, then public." The case of "first public, then private" is easy. The public facility knows that its private competitor will cluster. Thus it locates at the point, at which the average distance is minimized, which is central agglomeration.

Finally in this section, consider the von Stackelberg solution for the "first private, then public" problem. To analyze this case, we first need to derive firm B's reaction function. Evaluating the derivative of $D_B = \frac{1}{\ell} \left[\frac{1}{2}a^2 + \frac{1}{4}(\ell - a - b)^2 + \frac{1}{2}b^2 \right]$ with respect to b, we obtain the reaction function $b = \frac{1}{3}(\ell - a)$. Since the private firm A's objective is to maximize its market, we can rewrite firm A's objective as $M(A) = \frac{1}{2}(\ell + a - b) = \frac{1}{3}(\ell + 2a)$. This is maximized, if a is as large as possible. Note that the maximum value for a is $\frac{1}{2}\ell$. If exceeded, the public facility's best move is to locate on the left side of the private facility (longer side of the market), and by symmetry, firm A would move left to maximize its market. We then obtain $a = \frac{1}{2}\ell$, so that $b = \frac{1}{6}\ell$, firm A's market share is $M(A) = \frac{2}{3}\ell$, and the average distance between a customer and its closest facility is $\frac{1}{6}\ell$. We can summarize our pertinent finding in

Theorem 2 *Consider a competitive Hotelling model on a line segment with a private firm that maximizes its market capture and a public firm that minimizes the average customer–facility distance. The "first public, then private" game results in central agglomeration with $M(A) = \frac{1}{2}\ell$ and an average customer–facility distance of $\frac{1}{4}\ell$, while the "first private, then public" game has a solution at $(a,b) = (\frac{1}{2}\ell, \frac{1}{6}\ell)$ with $M(A) = \frac{2}{3}\ell$ and an average distance of $\frac{1}{6}\ell$.*

It is noteworthy that among the cases discussed above the customer benefits most, if a public facility joins the market after a private firm already exists. This result is somewhat reminiscent of Teitz's [72] "even a small gadfly can keep the big operator 'honest'."

Another "public vs. private" model exists, in which the public facility follows a minimax, rather than a minisum, objective. In other words, while the private facility still maximizes its sales or, proportional to it, its revenue, the public facility attempts to minimize the longest distance any customer has to travel to his closest facility. The private firm A's reaction functions is easily derived: it will again cluster with its public counterpart B by locating on the latter's longer side. Firm B's reaction function can be described as follows. For any $a \leq \frac{1}{4}\ell$, $b = \frac{1}{3}(\ell - a)$, so that the longest distance between any customer and his closest facility is $d_{max} = b = \frac{1}{3}(\ell - a)$ and firm A's market area will be $M(A) = a + b = \frac{1}{3}(\ell + 2a)$. For $a \in \left[\frac{1}{4}\ell, \frac{1}{2}\ell\right]$, firm B will locate at $b \in [\max\{0, \ell - 3a\}, a]$, so that $d_{max} = \max\left\{a, \frac{1}{2}(\ell - a - b), b\right\}$. Given the two bounds for b, we obtain $d_{max} = a$. Furthermore, firm A's market area is $M(A) = a + \frac{1}{2}(\ell - a - b) = \frac{1}{2}(\ell + a - b)$.

In other words, for any set value $a \leq \frac{1}{4}\ell$, firm B locates at a fixed point, while for $a \in \left[\frac{1}{4}\ell, \frac{1}{2}\ell\right]$, firm B has some leeway due to ties. If firm B were to be a "willing cooperator" of B, it would choose among its alternative optima that, which maximizes firm A's objective. This would mean that firm B would always choose $b = \max\{0, \ell - 3a\}$ for $a \in \left[\frac{1}{4}\ell, \frac{1}{2}\ell\right]$. The choice for firm B's secondary objective as a willing cooperator is somewhat arbitrary. It may be justified by considering possible side payments from firm A. Additional possibilities are explored below.

This leads to the following result regarding Nash equilibria. Clearly, for $a \leq \frac{1}{4}\ell$ equilibria cannot exist, as firm A needs to cluster. The only place where clustering is possible is at $a = \frac{1}{2}\ell$, for which $b \in \left[0, \frac{1}{2}\ell\right]$. All of firm B's locations in that interval are optimal, as in all these cases, $d_{max} = a$. The only clustered point in this interval is $b = \frac{1}{2}\ell$, i.e., central clustering. This immediately leads to

Theorem 3 *Consider the problem of one firm that maximizes its market share and a second firm that minimizes the longest distance any customer on the market has to travel to a facility. The unique Nash equilibrium in this model is central agglomeration.*

Consider now von Stackelberg solutions for this model. First we examine the case in which the public firm locates first, followed by the private firm. Given A's reaction function, firm B knows that A will cluster on its longer side. If B locates at the center of the market, so will A, resulting in $d_{max} = \frac{1}{2}\ell$. If B locates anywhere off center at some $b = \frac{1}{2}\ell \pm \epsilon$ with some $\epsilon > 0$, firm A will locate arbitrarily close to B on its longer side, so that $d_{max} = \frac{1}{2}\ell + \epsilon$. Hence, central agglomeration is the result.

Assume now that the private firm locates first, followed by its public counterpart. Here again, we need to specify a tie-breaking rule. (This is the case in many instances when minimax objectives are used. This is due to the fact that the minimax objective focuses exclusively on the longest customer—facility distance, ignoring all other distances). For $a \leq \frac{1}{4}\ell$, firm B's reaction function is as shown above with

$b = \frac{1}{3}(\ell - a)$, resulting in $M(A) = a + b = \frac{1}{3}(\ell + 2a)$. This market area is maximized with the largest possible value of $a = \frac{1}{4}\ell$, resulting in $M(A) = \frac{1}{2}\ell$. For $a \in \left[\frac{1}{4}\ell, \frac{1}{2}\ell\right]$, $d_{max} = a$ and, as a willing cooperator, firm B locates at $b = \max\{0, \ell - 3a\}$, resulting in $M(A) = a + \frac{1}{2}(\ell - a - b) = \frac{1}{2}(\ell + a - b)$. Firm A will maximize this function by choosing a as large as possible and firm B will cooperate by choosing b as small as possible, i.e., $a = \frac{1}{2}\ell$ and $b = 0$. This results in $d_{max} = \frac{1}{2}\ell$ and $M(A) = \frac{3}{4}\ell$, larger than firm A's previous option at $a = \frac{1}{4}\ell$. If firm B had used the minimization of the average customer—facility distance as a secondary objective, it would have used its reaction function $b = \frac{1}{3}(\ell - a)$ derived above and located at $b = \frac{1}{6}\ell$, resulting again in $d_{max} = \frac{1}{2}\ell$ but also in the average distance $\frac{1}{6}\ell$. Here, firm A's market share would have been $\frac{2}{3}\ell$. We can now summarize our pertinent findings in the following

Theorem 4 *Two firms compete on a linear market, a private firm that maximizes its market share and a public firm that minimizes the maximum customer—facility distance. The "first public, then private" game has a unique solution with central agglomeration, while the "first private, then public" game has a solution with the private firm locating at the center, while the public firm's location depends on its tie-breaking rule.*

4   Asymmetry of Pricing Policies

One of the key decisions a firm must make is the choice of a pricing policy. Among the many pricing policies, mill (or fob) pricing, uniform delivered pricing, and spatial price discrimination are the most popular strategies, but certainly not the only ones. Anderson et al. [2] examined the location equilibria for the three aforementioned policies and the social surplus they generate. However, in their study, both firms were using the same strategy. In this paper, we will assume that the two competitors use different exogenously determined pricing policies. Lederer [49] considered the case of brick-and-mortar facilities that compete against an internet e-tailer, which has the choice of pricing policies. Guo and Lai [35] follow a similar route in that they investigate equilibrium conditions for two "regular" retailers and one internet etailer. In contrast, Thisse and Vives [73]—while not considering location choice in their model-let firms first decide which pricing policy they choose, followed by the decision on a price. Our discussion here will follow along the lines of Eiselt [21], who considered the model of one mill-pricing firm competing against one firm that has chosen uniform delivered pricing as their strategy.

Assume that firm A uses mill pricing, while its opponent, firm B, employs uniform delivered pricing. However, this does not mean that firm B's location is irrelevant, as it has to ship the goods to the customers and pay for the shipment. In other words, both firms have to consider transportation costs, albeit in different ways: whereas for firm A, the transportation costs will determine its market area, transportation costs will be paid for by firm B and thus directly affect its profit. The situation is shown in Fig. 3.

The two firms' revenue functions will have a number of breaks. If firm B charges $p_B < p_A$, customers can obtain the good more cheaply from firm B than from firm A anywhere on the market, so that firm B's market area is $M(B) = \ell$. Once firm B's price increases above that of its competitor, its market share starts dropping as firm A gains market symmetrically about its location. After $p_B = p_2$, firm A keeps gaining market shares but only at half the rate as before. Once $p_B = p_3$, firm A captures the entire market, while firm B's market share is reduced to zero. The market shares of firm B can be determined as

$$M(B) = \begin{cases} \ell & p_B < p_1 = p_A \\ \ell - \frac{2}{t}(p_B - p_A) & \text{if} \quad p_B \in [p_1, p_2] \\ \ell - \frac{1}{t}(p_B - p_A) - a & p_B \in [p_2, p_3] \\ 0 & p_B > p_3 \end{cases}$$

The revenue function the above capture function leads to has also four regions: it increases linearly in the first, is concave in the second and third, and equals zero in the fourth region. While the function has nondifferentiabilities at the points p_1, p_2, and p_3, the function does not exhibit jumps.

It is apparent from Fig. 3 that firm A has an incentive to move towards the center of the market, especially when firm B charges a price $p_B \in [p_2, p_3]$. In such a case, firm A can get additional market shares by moving towards the center. More so, firm A can never do better than when located at the center of the market. Hence-even though for a specific combination of prices, there may be alternative optimal solutions for firm A's location-we assume that A will locate, so that $a = \frac{1}{2}\ell$.

In order to determine the market areas of the two facilities, we need to determine the price levels at all points on the market. More specifically, the points at which the prices from the two firms are equal are δ units away from the mill pricing facility A, where $p_A + t\delta = p_B$, i.e., $\delta = \frac{1}{t}(p_B - p_A)$. Then the market areas of the two firms are $M(A) = 2\delta = \frac{2}{t}(p_B - p_A)$ and $M(B) = \ell - M(A) = \ell - \frac{1}{t}p_B$. The profit of firm A is then $P(A) = M(A)p_A = \frac{2}{t}(p_B - p_A)p_A$. Evaluating the derivative of this profit function with respect to p_A and setting it to zero results in $p_A = \frac{1}{2}p_B$.

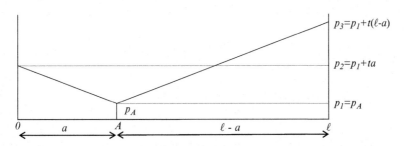

Fig. 3 Determination of breaks in the revenue function

Consider now firm B. First the easy cases: for $p_B > p_A + \frac{1}{2}t\ell$, firm B's market area and profit are zero, while for $p_B < p_A$, firm B's market area is $M(B) = \ell$, its costs, if located at $\frac{1}{2}\ell$, are $C(B) = \frac{1}{4}t\ell^2$, and its profit function is thus $P(B) = \ell p_B - \frac{1}{4}t\ell^2$, so that its optimal price is as large as possible, which is achieved at $p_B = p_A - \epsilon$ for $\epsilon > 0$ but arbitrarily small. Clearly, as firm A's reaction function sets its price at half the level of that of its competitor B, this will not result in a stable situation.

Consider now the case, in which $p_B \in [p_1, p_2] = [p_A, p_A + \frac{1}{2}t\ell]$. With firm A at the center at $a = \frac{1}{2}\ell$, and its price $p_A = \frac{1}{2}p_B$, firm B's market area is symmetric about $\frac{1}{2}\ell$, extending from the two ends of the market to points $d = \frac{1}{2}\ell - \frac{1}{2t}p_B$ from them. It is not difficult to demonstrate (see the Appendix) that as long as firm B locates anywhere in the central interval $[d, \ell - d]$, its costs are $C(B) = (\ell - d)td = \frac{1}{4}t\ell^2 - \frac{1}{t}p_B^2 + \frac{2}{t}p_A p_B - \frac{1}{t}p_A^2$. It can also be shown that firm B's costs are higher for locations closer to the ends of the market.

We can then set up firm B's profit function $P(B) = M(B)p_B - C(B) = \frac{1}{t}p_B^2 + \frac{1}{t}p_A^2 + \ell p_B - \frac{1}{4}t\ell^2$. Differentiation leads to the optimal price for firm B as $p_B = \frac{1}{2}t\ell$, and, subsequently, $p_A = \frac{1}{4}t\ell$. It is noteworthy that with $p_A = \frac{1}{4}t\ell$, the interval $[p_1, p_2] = [\frac{1}{4}t\ell, \frac{3}{4}t\ell]$ does include $p_B = \frac{1}{2}t\ell$, so that the assumptions for this case are indeed satisfied. In other words, firm B's profit is optimized on $[p_1, p_2]$. With equilibrium prices $p_A = \frac{1}{4}t\ell$ and $p_B = \frac{1}{2}t\ell$, the equilibrium profits are $P^*(A) = \frac{1}{8}t\ell^2$ and $P^*(B) = \frac{1}{16}t\ell^2$, giving a strong advantage to the mill pricing facility. However, this is not really surprising as the uniform delivered pricing firm does have to absorb the transportation costs, while firm A lets the customers pay for their own transportation. Note that there are many possible location combinations (a, b) that lead to this solution. We can summarize this result in

Theorem 5 *In a competitive location model with one firm adopting mill pricing while the other uses uniform delivered pricing, there is an equilibrium in which the mill pricing firm charges half the price of its opponent, but obtains twice the profit.*

Consider now the von Stackelberg solutions. Assume that firm A moves first and firm B follows. Optimizing firm B's profit function results in $p_B = \frac{1}{2}t\ell$. Given that firm A locates at the center of the market, $p_A > 0$ implies that firm A will not be a monopolist. Applying firm B's price in its own profit function, we obtain $p_A = \frac{1}{4}t\ell$, leading to $P^A(A) = \frac{1}{8}t\ell^2$ and $P^A(B) = \frac{1}{16}t\ell^2$, where the superscript indicates the leader. These are the same as the equilibrium results obtained earlier.

Now examine the case, in which firm B is the leader, while firm A takes on the role of follower. The leader knows that A locates somewhere near the center of the market and charges a price of $p_A = \frac{1}{2}p_B$. Firm B's profit function can then be written as $P(B) = \ell p_B - \frac{1}{4}t\ell^2 - \frac{3}{4t}p_B^2$. Differentiation results in $p_B = \frac{2}{3}t\ell$, and, given firm A's reaction function, $p_A = \frac{1}{3}t\ell$. Note that both prices are higher than their equilibrium counterparts. The resulting profits are $P^B(A) = \frac{2}{9}t\ell^2$ and $P^B(B) = \frac{1}{12}t\ell^2$, both higher than at equilibrium. Note the very strong advantage for the mill pricing firm. We can summarize these results in

Theorem 6 *In a competitive location model, in which a mill pricing and a uniform delivered pricing facility compete against each other, the solution is a Nash equilibrium if the mill pricing facility moves first. If the uniform delivered pricing facility moves first, its opponent charges half the price and has 8/3 as high a profit as the uniform pricing facility.*

5   Asymmetry of Firm Technology/Ability

This section consists of two parts. The first part assumes that the two firms apply different production technologies, so that their respective costs differ. We will investigate the situation, in which both firms apply spatial price discrimination. The second part has the firms use equal technologies and thus have equal production costs, but their modes of transportation differ, so that their unit transportation costs differ.

5.1   A Duopoly with Different Production Costs

Models in which firms have different production costs have been examined by many authors in the literature. In the context of competitive location models, one of the earlier contributions is that by Hurter and Lederer [44], who located firms in a two-dimensional plane and determine equilibrium conditions. Ziss [80] considers a three-stage game: enter/do not enter the market, choose a location, determine the price. The issue of entry deterrence is an important feature in this mode, which does allow for different production costs of firms. A similar game is examined by Vogel [75], albeit on a circular market. Liang and Mai [53] investigate a product design model (in the usual features space) in order to decide on export taxes or subsidies. Matsumura et al. [56] show how technology transfer can guarantee the existence of an equilibrium in case of different production cost. Ledvina and Sircar [51] research the case of asymmetric costs in the energy industry. Cai and Kobayashi [13] examine two firms with different production costs, one of which already present in the market, while another wants to enter, how lobbying influences entry regulations. Pierce and Sen [59] look at firms with different production costs in the context of outsourcing, and Eleftheriou and Michelacakis [29] prove that firms with different production costs and spatial price discrimination result in locational arrangements that are always socially optimal, given a large variety of different assumptions.

Our discussion stays close to Hotelling's model. We begin by considering the case of firms A and B locating a single facility each with unit production costs c_A and c_B, respectively. Without loss of generality, suppose that $c_A < c_B$. Based on firm A's lower per-unit production costs, we will refer to it as the "efficient firm," while firm B will be called the "inefficient firm." Both firms have their respective locations and prices as variables. As usual, we assume that the costs of delivery of the goods

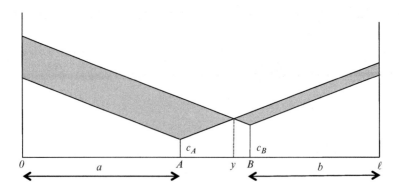

Fig. 4 Delivered cost functions of two firms with different production costs

are linear in the distance and the quantity. Hence, the cost functions of the two firms for delivering the goods are shown in Fig. 4 as the "Y"-shaped functions above the locations of firm A and B, respectively. Note that in contrast to the usual mill pricing Hotelling models, these Y-shaped functions represent costs to the firms, rather than costs to the customers. In other words, at any point on the market, the full cost of delivering one unit of the good from the firm to that point is represented by the curve of that facility. The argument is now that the price that a customer pays is indicated by the upper envelope of the two Y-shaped functions. Note that at any point on the market, the lower-cost firm will capture the demand at that point, but at a price that equals the cost of the higher-priced firm (minus some ϵ).

Furthermore, we define a separation point y as the point on the market, at which the full costs of the two firms are equal. Note the similarity of the marginal customer in the previous sections and the separation point y, one defined on the basis of full prices and the other on the basis of full costs. Formally, the separation point is defined as the point y, at which $c_A + td_{Ay} = c_B + td_{By}$. As a result, firm A's costs are lower than that of its competitor to the left of the separation point y, while to its right, firm B's costs are lower. This means that to the left of y, firm A can set a price slightly lower than the cost of its competitor at $p_x = c_B + td_{xB} - \epsilon$, so that, assuming that firm B does not use a suicidal strategy (i.e., attempt to capture demand regardless of its loss), firm B will not be able to compete and firm A will capture the demand. A similar argument applies all points to the right of the separation point y, where B will be able to undercut its competitor A.

The results of this argument are twofold. First, the prices at some point x on the market are

$$p_x = \begin{cases} c_B + t(\ell - b - d_{0x}) \\ c_A + t(d_{0x} - a) \end{cases} \quad \text{if} \quad \begin{aligned} d_{0x} &\leq d_{0y} \\ d_{0x} &\geq d_{0y} \end{aligned}$$

and secondly the market areas captured by the two facilities are $M(A) = d_{0y}$ and $M(B) = (\ell - d_{0y})$. Note that if $c_A + td_{AB} \leq c_B$, then the efficient firm A will be able to supply the entire market more cheaply than the inefficient firm, so that firm B will be cut out. This will destroy the duopoly and, at least temporarily, create a monopoly. Rewriting the condition that avoids such an occurrence is the regularity condition $c_B - c_A < t(\ell - a - b)$. This condition requires firm B's production costs to be lower than firm A's production cost plus its transportation costs to the location of firm B.

One somewhat compelling reason for the prices to increase as we move away from the separation point y is this. At the separation point, the level of competition is largest. As we move away from that point, the level of competition decreases and with it the price increases. In Fig. 4, the shaded area above point A denotes firm A's profit, while the shaded area above firm B denotes firm B's profit. Also notice that the force that pulls the two facilities together in the original Hotelling model is absent in this model: on the contrary, there is a force that pushes the two facilities apart, as that way, the prices and with it the two firms' profits increase.

Formally, firm A's market area is $M(A) = d_{0y} = a + d_{Ay} = \frac{1}{2t}(c_B - c_A) + \frac{1}{2}(\ell + a - b)$ and its profit is $P(A) = a[c_B - c_A + t(\ell - a - b)] + \frac{1}{4t}[c_B - c_A + t(\ell - a - b)]^2$. Similarly, the expression of firm B's profit is $P(B) = b[c_A - c_B + t(\ell - a - b)] + \frac{1}{4t}[c_A - c_B + t(\ell - a - b)]^2$. The respective derivatives result in the reaction functions of the two firms, which are $a = \frac{1}{3t}(c_B - c_A + t\ell - tb)$ and $b = \frac{1}{3t}(c_A - c_B + t\ell - ta)$. These two equations result in a Nash equilibrium with locations $a^* = \frac{1}{4}\ell + \frac{1}{2t}(c_B - c_A)$ and $b^* = \frac{1}{4}\ell - \frac{1}{2t}(c_B - c_A)$. In other words, with very similar operating costs, the two firms will locate at the first and third quartiles of the market, respectively (this is Hotelling's "social optimum"), and as the differences between the unit operating costs increase, both firms move to the right by equal distances. These equilibrium locations will apply, as long as $a \geq 0$ and $b \geq 0$. This is ensured by the regularity condition. The equilibrium profits of the two firms are $P^*(A) = \frac{3}{4t}(c_B - c_A + t\ell)^2$ and $P^*(B) = \frac{3}{4t}(c_A - c_B + \frac{1}{2}t\ell)^2$, respectively.

While it is possible for firm A to undercut its competitor, one can show that it is not profitable to do so. This leads immediately to

Theorem 7 *As long as* $c_B - c_A \leq \frac{1}{2}t\ell$ *the Hotelling model with delivered prices has a stable equilibrium with locations* $a^* = \frac{1}{4}\ell + \frac{1}{2t}(c_B - c_A)$ *and* $b^* = \frac{1}{4}\ell - \frac{1}{2t}(c_B - c_A)$. *If* $c_B - c_A > \frac{1}{2}t\ell$, *firm A becomes a monopolist.*

Consider now the von Stackelberg solutions. We first investigate the case, in which the efficient firm A locates first, followed by its inefficient counterpart B. Inserting firm B's reaction function into firm A's profit function yields $P(A) = \frac{4}{3}a[c_B - c_A + \frac{1}{2}t\ell - \frac{1}{2}ta] + \frac{4}{9t}[c_B - c_A + \frac{1}{2}t\ell - \frac{1}{2}ta]^2$, which, after taking the derivative with respect to a and setting it equal to zero results in the leader's location at $a' = \frac{4}{5t}(c_B - c_A + \frac{1}{2}t\ell) = \frac{4}{5t}(c_B - c_A) + \frac{2}{5}\ell$. Firm B will then react by choosing its location at $b' = \frac{3}{5t}(c_A - c_B + \frac{1}{3}t\ell) = \frac{3}{5t}(c_A - c_B) + \frac{1}{5}\ell$. Here, $b \geq 0$ requires that $c_B - c_A \leq \frac{1}{3}t\ell$, which is the same as the regulatory condition above. We can also show that for $c_B - c_A > \frac{1}{3}t\ell$, concavity of the profit function $P(B)$ implies that firm B will stay as close to the (now unattainable) optimum, so that it will set $b' = 0$ and thus locate at the right boundary of the market. The profits of the two firms are $P(A) = \frac{4}{5t}(c_B - c_A + \frac{1}{2}t\ell)^2$ and $P(B) = \frac{27}{25t}(c_A - c_B + \frac{1}{3}t\ell)^2$, respectively, so that the leader's profit is $P(A)/P(B) = 66.67\%$ higher than that of the follower. As firm B's cost increase, firm A's profit also increases. This is due to two forces that apply simultaneously: on the one hand, the more efficient firm's relative competitiveness increases (and with it its market share), and secondly, the price of the product in firm A's market area is determined by its less efficient competitor.

Consider now the von Stackelberg game, in which the inefficient firm B locates first, followed by its efficient counterpart A. The same procedure used in the previous analysis results in optimal locations $b = \frac{4}{5t}(c_A - c_B + \frac{1}{2}t\ell)$ for the leader and $a = \frac{2}{3}(c_B - c_A + \frac{1}{3}t\ell)$ for the follower. The profits of the two firms are $P(A) = \frac{27}{25t}(c_B - c_A + \frac{1}{3}t\ell)^2$ for the follower and $P(B) = \frac{20}{25t}(c_A - c_B + \frac{1}{2}t\ell)^2$ for the leader, respectively. In the boundary case of $c_A = c_B$, the strong leader advantage of $5/3$ applies again. We can summarize these results in

Theorem 8 *In a competitive location model with two competitors that have different production costs, consider the von Stackelberg solution. In case the efficient firm locates first, it has a $\frac{2}{3}$ profit advantage, while if the inefficient firm locates first, it may have an advantage if the costs are not too different. In other words, there is always a first mover advantage.*

5.2   A Duopoly with Different Transportation Cost

There are relatively few authors who have looked at models with different transportation costs. Among them are Lederer and Hurter [50], whose firms may have different production costs, face different transportation costs, and use spatial price discrimination in a "first location, then price" game, in which the firms make simultaneous choices in each of the two phases. The authors prove that a location-price equilibrium exists in this case. Konur and Geunes [46] use a model with differential transportation costs and simultaneous location choices. They use variational inequalities to determine optimal locations.

Our model in this section deals with a standard Hotelling model, in which there are different transportation costs associated with each of the two firms. This assumption may be justified in different ways. On the one hand, there is the usual mill-pricing model, in which customers pick up the goods from the stores and pay their own transportation costs. Consider the case, in which one store is located in an area that is harder to get to, e.g., more congested, such as a downtown location. In such a case, a shopping trip to this facility will take more resources, i.e., be more expensive per mile than a trip to an alternative facility, which is, say, located next to the highway, can easily be reached, and has ample and free parking. This argument is somewhat reminiscent of competitive location models with attraction functions, see, e.g., the references in Sect. 6 of this paper. However, this argument does not apply in cases, in which the two firms locate close together. Alternatively, we can think of a model, in which goods are delivered to customers, who are subsequently charged

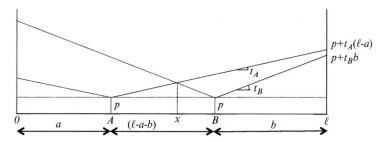

Fig. 5 One marginal customer at x in the case of firms with different transportation costs

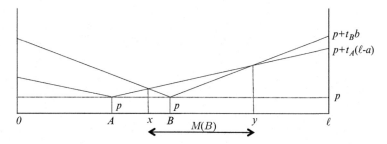

Fig. 6 Two marginal customers at x and y in the case of firms with different transportation costs

for the delivery, based on the distance between them and the facility. Different unit transportation costs can then result from the fact that the two firms employ different modes of transportation with different costs.

Formally, we have again the standard Hotelling model of length ℓ, two firms A and B, which locate at a and b distance units away from the left and right edge of the market (we assume that $a + b \leq \ell$), both firms charge a price p, and their respective unit transportation costs are t_A and t_B, respectively. Without loss of generality, we assume that $t_A \leq t_B$. Again, we define a marginal customer as a customer who faces the same costs from both facilities and thus is indifferent to the choice. Figures 5 and 6 demonstrate that in the case of different unit transportation costs, it is possible that one or two marginal customers exist.

We will discuss these two cases separately. More specifically, Case 1 deals with the situation in Fig. 5, in which there exists a single marginal customer at x, while in Case 2 (Fig. 6), there exist two marginal customers at points x and y.

Case 1: There exists a single marginal customer at x. The condition for this to happen is $p + t_A(\ell - a) \geq p + t_B b$. Given some standard algebraic transformations, we obtain the condition $a \leq \ell - \frac{t_B}{t_A} b$. The distance between firm A and the marginal customer x can then be written as $d_{Ax} = \frac{t_B(\ell - a - b)}{t_A + t_B}$, so that firm A's market share is $M(A) = a + d_{Ax}$, which can be rewritten as $M(A) = \frac{t_A a + t_B(\ell - b)}{t_A + t_B}$. Clearly, this function is linearly increasing in a, so that firm A will choose a location that is as far as possible from the left end of the market. In other words, firm A will move towards firm B. A similar analysis for firm B reveals that $M(B) = \frac{\ell t_A - a t_A + b t_B}{t_A + t_B}$, so that again, this measure will increase with increasing values of b, so that firm B will also gain market shares and profit by moving towards its opponent.

Case 2: The condition for this case is $a \geq \ell - \frac{t_B}{t_A} b$. As shown in Fig. 6, there are now two marginal customers x and y. The distances between firm B and x and y are $d_{Bx} = \frac{t_A}{t_A + t_B}(\ell - a - b)$ and $d_{By} = \frac{t_A}{t_B - t_A}(\ell - a - b)$, respectively. Firm B's market area is then $M(B) = \frac{2 t_A t_B}{t_B^2 - t_A^2}(\ell - a - b)$. The term in brackets expresses the distance between the two firms, whose coefficient is positive, so that, firm B has an incentive to move away from its opponent. On the other hand, firm A's market area is $M(A) = \left[t_B^2 - t_A^2\right]^{-1}\left[\left(t_A^2 - t_B^2 - 2 t_A t_B\right)\ell + (2 t_A t_B)(a + b)\right]$. Here, the coefficient of a, the only variable under firm A's jurisdiction, is positive, so that A has an incentive to move towards its opponent. In the extreme, firm A co-locates with its opponent and it will thus capture the entire market.

The above discussion now allows us to analyze possible solutions. First consider Nash equilibria. Suppose that Case 1 applies. In such a case, both firms have an incentive to move towards each other, so that at some point, the situation becomes Case 2 and there is no equilibrium for Case 1. Assume now that we are in Case 2. Here, firm A is predatory, while its opponent will move away from A, until Case 1 applies. It is apparent that no equilibrium can exist.

Theorem 9 *In a competitive location model, in which firms have different transportation cost functions, there is no equilibrium.*

Consider now von Stackelberg solutions. The case of "first firm B then firm A" is easy. Wherever firm B locates, its opponent will locate arbitrarily close, so that firm B's market share is reduced to virtually zero. A more interesting case is the case, in which firm A locates first, followed by firm B. As firm B will profit on the longer side, i.e., farther away from firm A, firm A will attempt to prevent this by locating at the center of the market, i.e., $a = \frac{1}{2}\ell$. As firm B's reaction function in both cases is $b = \frac{t_A}{t_B}(\ell - a)$, we obtain $b = \frac{1}{2}\frac{t_A}{t_B}\ell$. This means that for any fixed location of firm A, firm B will locate, such that the right end of its market area coincides with the right end of the market. The market areas in the "First A, then B" scenario are then $M(A) = \frac{t_B}{t_A + t_B}\ell$ and $M(B) = \frac{t_A}{t_A + t_B}\ell$, meaning that the ratio of the two market areas is inversely proportional to the ratio of the unit transportation costs, i.e., $\frac{M(A)}{M(B)} = \frac{t_B}{t_A}$. We summarize the result in

Theorem 10 *Consider a competitive location model with firms that are associated with different unit transportation costs. If the higher-cost firm moves first, it will be cut out and be eliminated from competition. If the lower-cost firm moves first, the ratio of its market share is inversely proportional to the ratio of the unit transportation rates.*

6 Conclusions

In this paper we have examined a number of competitive location models, in which asymmetries exist. Special attention has been paid not only to the existence of locational Nash equilibria, but also to von Stackelberg solutions and the possible existence of first (or second) mover advantages.

Many possible extensions exist, some of which have already been suggested in the literature. One issue, suggested by many authors, deals with more than two competitors. Among those who have worked in the field and suggested pertinent models are Shaked [68, 69], DePalma et al. [16] as well as Eiselt and Laporte [27] for three firms, and some more recent contributions are those by Brenner [10], Ben-Porat and Tennenholtz [4], and Fournier and Scarsini [32]. However, all of these contributions have investigated models with identical facilities. There are many asymmetries that could be thought of that exist in real life.

Another type of model that has been widely researched—albeit not with asymmetries—includes attraction functions. Such functions typically associate a weight to each facility, which is a unidimensional measure that expresses the desirability of that facility. As mentioned above, this can and will typically involve the size of the facility, the breadth of goods and services the facility offers, the ease of accessibility, and others. In the context of competitive location, attraction functions were suggested by Eiselt and Laporte [25, 26] and Eiselt [21]. These models have been pioneered by Reilly [63] and Huff [43]. Drezner [17] picked up the issue in her model in the two-dimensional plane. More recent contributions that involve facilities with different attractivenesses include papers by Blanquero et al. [9], Küçükaydin et al. [48], and Fernández and Hendrix [31]. Drezner [18] surveys competitive location models with facilities that have different attractivenesses.

Another large and important field involves uncertainty and imperfection of information. Again, much work has been done in the area-including the specific aspect of competitive locations-but not with asymmetries. Bhadury [6] deals with uncertainty of costs, an issue picked up again by Huang et al. [42] and Pinto and Parreira [60]. Imperfect information was used in a model by Esteves [30]. Eiselt [22] included different perceptions of some features in a competitive location scenario in his model.

Finally, there is a number of issues that have been introduced into some competitive location models, but have not really been examined closely, and most certainly not in the context of asymmetries. Among them are flow capturing models, pioneered by Hodgson [40] and Berman et al. [5], in which customers patronize facilities not by making special trips, but as they drive by them on their regular route, e.g., between home and work. Another interesting issue involves the popular price matching. In his model, Zhang [79] describes a three-stage competitive location model, the second phase of which features the decision of whether or not to adopt a price matching policy. Finally, the issue of taxation was brought up by Casado-Izaga [14] in his competitive location model that does allow for some asymmetries.

Acknowledgements This work was in part supported by the Institute Complex Engineering Systems, through grants ICM-MIDEPLAN P-05-004-F and CONICYT FB0816.

Appendix

Given that firm A locates at the center of the market, i.e., $a = \frac{1}{2}\ell$, firm B's market area is symmetric about $\frac{1}{2}\ell$. Suppose that firm B's market area is d units near both ends of the market. (We deviate from some of the notation in the paper in order to simplify matters.) As usual, B is located b units from the right end of the market. This situation is shown in Fig. 7.

Firm B's transportation costs are then the two trapezoids (D, E) and (F, G). Elementary algebra indicates that the areas of D, E, F, and G are $\frac{1}{2}td^2$, $(\ell - b - d)dt$, $\frac{1}{2}td^2$, and $(b-d)dt$, respectively, so that the total cost (the total area) is $(\ell - d)td$. It is apparent that these costs are independent of b, the location of firm B. The scenario does not change as long as $b \in [d, \ell - d]$.

Consider now the case, in which locates at a point $b \le d$ (or, alternatively, $b \ge \ell - d$). This situation is shown in Fig. 8.

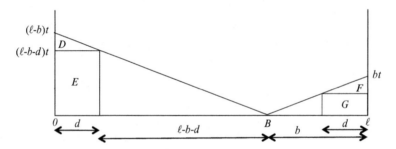

Fig. 7 Transportation cost of a firm that uses delivered prices and whose market share extends d from both ends of the market, the form locates outside its market area

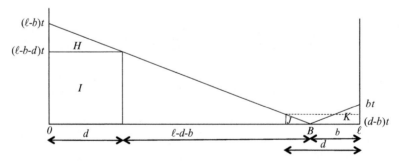

Fig. 8 Transportation cost of a firm that uses delivered prices and whose market share extends d from both ends of the market, the form locates inside its market area

Similar to the above analysis, we have the four areas H, I, J, and K, the sum of whose areas determine the transportation cost incurred by firm B. The areas are $\frac{1}{2}d^2t$, $(\ell - b - d)td$, $\frac{1}{2}(d-b)^2t$, and $\frac{1}{2}tb^2$, respectively, so that the area is $-2btd + \ell td + tb^2$, which is dependent on b. The minimum is found at $b = d$, indicating that firm B best locates somewhere in its opponent's market area.

References

1. Anderson, S.P.: Equilibrium existence in the linear model of spatial competition. Economica **55**(220), 479–491 (1988)
2. Anderson, S.P., De Palma, A., Thisse, J.-F.: Social surplus and profitability under different spatial pricing policies. South. Econ. J. **58**, 934–949 (1992)
3. Ashtiani, M.G.: Competitive location: a state-of-art review. Int. J. Ind. Eng. Comput. **7**, 1–18 (2016)
4. Ben-Porat, O., Tennenholtz, M.: Multi-unit facility location games. http://arxiv.org/pdf/1602.03655.pdf (2016). Accessed 9 Sept 2016
5. Berman, O., Hodgson, M.J., Krass. D.: Flow-interception problems. In: Drezner, Z. (ed.) Facility Location: A Survey of Applications and Methods, pp. 389–426. Springer, New York (1995)
6. Bhadury, J.: Competitive location under uncertainty of costs. J. Reg. Sci. **36**(4), 527–554 (1996)
7. Bhadury, J., Eiselt, H.A.: Reachability of locational Nash equilibria. Oper. Res. Spektrum **20**(2), 101–107 (1998)
8. Biscaia, R., Mota, I.: Models of spatial competition: a critical review. Pap. Reg. Sci. **92**(4), 851–871 (2013)
9. Blanquero, R., Hendrix, E.M.T., Carrizosa, E.: Locating a competitive facility in the plane with a robustness criterion. Eur. J. Oper. Res. **215**(1), 21–24 (2011)
10. Brenner, S.: Hotelling games with three, four, and more players. J. Reg. Sci. **45**(4), 851–864 (2005)
11. Brenner, S.: Location (Hotelling) games and applications. In: Wiley Encyclopedia of Operations Research and Management Science (2011). doi:10.1002/9780470400531.eorms0477
12. Brown, S.: Retail location theory: the legacy of Harold Hotelling. J. Retail. **65**(4), 450–470 (1989)
13. Cai, D., Kobayashi, S.: Lobbying on entry under spatial competition: the case of asymmetric production costs. http://www.webmeets.com/files/papers/earie/2013/257/Cai_2013.pdf (2013). Accessed 9 Sept 2016
14. Casado-Izaga, E.J.: Tax effects in a model of spatial price discrimination: a note. J. Econ. **99**(3), 277–282 (2010)
15. D'Aspremont, C., Gabszewicz, J.J., Thisse, J.-F.: On Hotelling's Stability in competition. Econometrica **47**, 1145–1150 (1979)
16. DePalma, A., Ginsburgh, V., Thisse, J.-F.: On existence of locational equilibria in the 3-firm Hotelling problem. J. Ind. Econ. **36**, 245–252 (1987)
17. Drezner, T.: Locating a single new facility among existing, unequally attractive facilities. J. Reg. Sci. **34**(2), 237–252 (1994)
18. Drezner, T.: A review of competitive facility location in the plane. Logist. Res. 7/1, 1–12 (2014)
19. Eaton, B.C., Lipsey, R.G.: The principle of minimum differentiation reconsidered: some new developments in the theory of spatial competition. Rev. Econ. Stud. **42**(1), 27–49 (1975)
20. Eiselt, H.A.: Different pricing policies in Hotelling's duopoly model. Cah. CERO **33**, 195–205 (1991)
21. Eiselt, H.A.: A Hotelling model with different weights and moving costs. Belg. J. Oper. Res. Stat. Comput. Sci. **30**(2), 3–20 (1990)

22. Eiselt, H.A.: Perception and information in a competitive location model. Eur. J. Oper. Res. **108**(1), 94–105 (1998)
23. Eiselt, H.A.: Equilibria in competitive location models, Chap. 7. In: Eiselt, H.A., Marianov, V. (eds.) Foundations of Location Analysis. Springer, Berlin/Heidelberg (2011)
24. Eiselt, H.A., Bhadury, J.: Stability of Nash equilibria in locational games. Recherche opéra-tionnelle/Oper. Res. **29**(1), 19–33 (1995)
25. Eiselt, H.A., Laporte, G.: Trading areas of facilities with different sizes. Recherche Opera-tionnelle/Oper. Res. **22**(1), 33–44 (1988)
26. Eiselt, H.A., Laporte, G.: Location of a new facility on a linear market in the presence of weights. Asia Pac. J. Oper. Res. **5**(2), 160–165 (1988)
27. Eiselt, H.A., Laporte, G.: The existence of equilibria in the 3-facility Hotelling model on a tree. Transp. Sci. **27**(1), 39–43 (1993)
28. Eiselt, H.A., Laporte, G., Thisse, J.-F.: Competitive location models: a framework and bibliography. Transp. Sci. **27**(1), 44–54 (1993)
29. Eleftheriou, K., Michelacakis, N.: A unified model of spatial price discrimination. https://mpra. ub.uni-muenchen.de/72106/1/MPRA_paper_72106.pdf (2016). Accessed 9 Sept 2016
30. Esteves, R.-B.: Price discrimination with private and imperfect information. Scand. J. Econ. **116**(3), 766–796 (2014)
31. Fernández, J., Hendrix, E.M.T.: Recent insights in Huff-like competitive facility location and design. Eur. J. Oper. Res. **227**, 581–584 (2013)
32. Fournier, G., Scarsini, M.: Hotelling games on networks: existence and efficiency of equilibria. http://arxiv.org/abs/1601.07414 (2016). Accessed 9 Sept 2016
33. Gabszewicz, J.J., Thisse, J.-F.: Spatial competition and the location of firms. Fundam. Pure Appl. Econ. **5**, 1–71 (1986)
34. Ghosh, A., Buchanan, B.: Multiple outlets in a duopoly: a first entry paradox. Geogr. Anal. **20**, 111–121 (1988)
35. Guo, W.-C., Lai, F.-C.: Spatial competition with quadratic transport costs and one online firm. Ann. Reg. Sci. **52**(1), 309–324 (2014)
36. Hakimi, S.L.: On locating new facilities in a competitive environment. Eur. J. Oper. Res. **12**, 29–35 (1983)
37. Hakimi, S.L.: Locations with spatial interactions: competitive locations & games. In: Francis, R.L., Mirchandani, P.B. (eds.) Discrete Location Theory. Wiley, New York (1990)
38. Hamoudi, H., Moral, M.J.: Equilibrium existence in the linear model: concave versus convex transportation costs. Pap. Reg. Sci. **84**(2), 201–219 (2005)
39. Herr, A.: Quality and welfare in a mixed duopoly with regulated prices: the case of a public and a private hospital. Ger. Econ. Rev. **12**(4), 422–437 (2011)
40. Hodgson, M.J.: The location of public facilities intermediate to the journey to work. Eur. J. Oper. Res. **6**, 199–204 (1981)
41. Hotelling, H.: Stability in competition. Econ. J. **39**, 41–57 (1929)
42. Huang, R., Menezes, M,B.C., Kim, S.: The impact of cost uncertainty on the location of a distribution center. Eur. J. Oper. Res. **218**(2), 401–407 (2012)
43. Huff, D.L.: Defining and estimating a trade area. J. Mark. **28**, 34–38 (1964)
44. Hurter, A.P. Jr., Lederer, P.J.: Spatial duopoly with discriminatory pricing. Reg. Sci. Urban Econ. **15**, 541–553 (1985)
45. Karakitsiou, A.: Modeling Discrete Competitive Facility Location. Springer, Cham (2015)
46. Konur, D., Geunes, J.: Competitive multi-facility location games with non-identical firms and convex traffic congestion costs. Transp. Res. E **48**(1), 373–385 (2012)
47. Kress, D., Pesch, E.: Sequential competitive location on networks. Eur. J. Oper. Res. **217**(3), 483–499 (2012)
48. Küçükaydin, H., Aras, N., Altinel, I.K.: Competitive facility location problem with attractive-ness adjustment of the follower: a bilevel programming model and its solution. Eur. J. Oper. Res. **208**(3), 206–220 (2011)
49. Lederer, P.J.: Competitive delivered pricing by mail-order and internet retailers. Netw. Spat. Econ. **11**(2), 315–342 (2011)

50. Lederer, P.J., Hurter, A.P. Jr.: Competition of firms: discriminatory pricing and location. Econometrica **54**(3), 623–640 (1986)
51. Ledvina, A., Sircar, R.: Oligopoly games under asymmetric costs and an application to energy production. Math. Financ. Econ. **6**(4), 261–293 (2012)
52. Lerner, A.P., Singer, H.W.: Some notes on duopoly and spatial competition. J. Polit. Econ. **45**, 145–186 (1937)
53. Liang, W.-J., Mai, C.-C.: Optimal trade policy with horizontal differentiation and asymmetric costs. Rev. Dev. Econ. **14**(2), 302–310 (2010)
54. Lu, Y.: Hotelling's location model in mixed duopoly. Econ. Bull. **8**(1), 1–10 (2006)
55. Marianov, V., Eiselt, H.A.: Agglomeration in competitive location models. Ann. Oper. Res. **246**, 31–55 (2016)
56. Matsumura, T., Matsushima, N., Stamatopoulos, G.: Location equilibrium with asymmetric firms: the role of licensing. J. Econ. **99**(3), 267–276 (2010)
57. Miller, T.C., Friesz, T.L., Tobin, R.L.: Equilibrium Facility Location on Networks. Springer, Heidelberg (2013)
58. Nash, J.F. Jr.: Equilibrium points in n-person games. Proc. Natl. Acad. Sci. U. S. A. **36**(1), 48–59 (1950)
59. Pierce, A., Sen, D.: Outsourcing versus technology transfer: Hotelling meets Stackelberg. J. Econ. **111**(3), 263–287 (2014)
60. Pinto, A.A., Parreira, T.: Maximal differentiation in the Hotelling model with uncertainty. In: Pinto, A.A., Zilberman, D. (eds.) Modeling, Dynamics, Optimization, and Bioeconomics I. Proceedings in Mathematics & Statistics, vol. 73, pp. 585–600. Springer, Cham (2014)
61. Plastria, F.: Static competitive facility location: an overview of optimization approaches. Eur. J. Oper. Res. **129**, 461–470 (2001)
62. Prescott, E.C., Visscher, M.: Sequential location among firms with foresight. Bell J. Econ. **8**(2), 378–393 (1977)
63. Reilly, W.J.: The Law of Retail Gravitation. Knickerbocker, New York (1931)
64. Rothschild, R.: A note on the effect of sequential entry on choice of location. J. Ind. Econ. **24**(4), 313–320 (1976)
65. Rusk, J.G., Weisberg, H.F.: Perceptions of presidential candidates: implications for electoral change, Chap. 21. In: Niemi, R.G., Weisberg, H.F. (eds.) Controversies in American Voting Behavior, pp. 370–388. WH Freeman and Co., San Francisco (1976)
66. Russell, J.E.: Using a retail location game to explore Hotelling's principle of minimum differentiation. Bus. Educ. Innov. J. **5**(2), 48–52 (2013)
67. Selten, R.: Reexamination of the perfectness concept for equilibrium points in extensive games. Int. J. Game Theory **4**, 25–55 (1975)
68. Shaked, A.: Non-existence of equilibrium for the two-dimensional three-firms location problem. Rev. Econ. Stud. **42**(1), 51–56 (1975)
69. Shaked, A.: Existence and computation of mixed strategy Nash equilibrium for 3-firms location problem. J. Ind. Econ. **31**(1–2), 93–96 (1982)
70. Smithies, A.: Optimum location in spatial competition. J. Polit. Econ. **49**(3), 423–439 (1941)
71. Stevens, B.H.: An application of game theory to a problem in location strategy. Pap. Proc. Reg. Sci. Assoc. **7**(1), 143–157 (1961)
72. Teitz, M.B.: Locational strategies for competitive systems. J. Reg. Sci. **8**(2), 135–138 (1968)
73. Thisse, J.-F., Vives, X.: On the strategic choice of spatial price policy. Am. Econ. Rev. **78**(1), 122–137 (1988)
74. Thisse, J.-F., Wildasin, D.E.: Optimal transportation policy with strategic locational choice. Reg. Sci. Urban Econ. **25**, 395–410 (1995)
75. Vogel, J.: Spatial price discrimination with heterogeneous firms. Working Paper 14978. National Bureau of Economic Research, Cambridge, MA. http://www.nber.org/papers/w14978 (2009). Accessed 9 Sept 2016
76. von Stackelberg, H.: The Theory of the Market Economy. Translated from the original German work Grundlagen der theoretischen Volkswirtschaftslehre, 1943, by Peacock AT. William Hodge, London (1952)

77. Yasuda, Y.: Instability in the Hotelling's non-price spatial competition model. Theor. Econ. Lett. **3**, 7–10 (2013)
78. Younies, H., Eiselt, H.A.: Sequential location models, Chap. 8. In: Eiselt, H.A., Marianov, V. (eds.) Foundations of Location Analysis. Springer, Berlin/Heidelberg (2011)
79. Zhang, Z.J.: Price-matching policy and the principle of minimum differentiation. J. Ind. Econ. **43**, 287–299 (1995)
80. Ziss, S.: Entry deterrence, cost advantage and horizontal product differentiation. Reg. Sci. Urban Econ. **23**, 523–543 (1993)

Huff-Like Stackelberg Location Problems on the Plane

José Fernández, Juana L. Redondo, Pilar M. Ortigosa, and Boglárka G.-Tóth

1 Introduction

Locating a new facility usually requires a massive investment. In order to guarantee the survival of the facility, especially in a competitive environment (where other facilities offering the same product or service exist), the locating firm tries to take all the factors which may affect the market share captured by the facility (or its profit) into account. A well-known aphorism states that 'the most important attributes of stores are location, location and location'. The literature about facility location corroborates that point as the number of papers devoted to that topic is huge. Mathematical location models try to combine all the factors of interest for the facility into neat equations which try to faithfully represent (a simplified version of) reality. The location decisions provided by the location models can be of invaluable help to the decision-maker, as the location of a facility cannot be easily altered.

Depending on the location space, competitive facility location models can be subdivided, as any other type of location problems, into three main categories: (1) continuous problems, where the set of feasible locations for the new facility (or

J. Fernández (✉)
Department of Statistics and Operations Research, University of Murcia, Campus de Espinardo,
30100 Espinardo, Murcia, Spain
e-mail: josefdez@um.es

J.L. Redondo • P.M. Ortigosa
Department of Informatics, University of Almería, ceiA3, Ctra. Sacramento s/n,
La Cañada de San Urbano, 04120 Almería, Spain
e-mail: jlredondo@ual.es; ortigosa@ual.es

B.G.-Tóth
Department of Computational Optimization, University of Szeged, H-6720 Szeged,
Arpád tér 2, Hungary
e-mail: boglarka@inf.u-szeged.hu

© Springer International Publishing AG 2017
L. Mallozzi et al. (eds.), *Spatial Interaction Models*, Springer Optimization
and Its Applications 118, DOI 10.1007/978-3-319-52654-6_7

facilities) is (a subset of) the plane; (2) network problems, where any point in a network (on an edge or a vertex) is a possible location, and (3) discrete problems, when the set of potential locations is reduced to a finite set of points. In this chapter we restrict ourselves to continuous models, as this is the main research field of the authors, but the interested reader can find many references on network and discrete competitive location models in literature, see for instance [3, 4, 16, 29, 30, 45] and references therein.

In competitive models there is a demand which has to be, or may be, served by the facilities. This demand is commonly assumed to be concentrated at a finite set of points, called *demand points* (also referred to as *customers*). In most of the research works it is assumed that the demand is *fixed*, regardless the conditions of the market (price, distance to the facilities,...). This implicitly assumes that goods are 'essential' to the customers. It is only recent that the case of 'inessential' goods has been addressed [28, 41]. In those models it is assumed that the demand *varies* depending on the location of the facilities.

The *attraction* of a customer towards a facility depends on both the location and the characteristics of the facility. Usually the characteristics are combined into a single figure which represents the *quality* of the facility. The closer the facility to the customer and the higher its quality, the higher the attraction of the customer towards the facility. Although there are many ways to model the attraction (see [34]), the formula quality divided by a function of the distance (already proposed in [22]) is the most popular in literature, and the one followed in this chapter.

The *patronizing behavior* of customers, which establishes how customers split their demand among the available facilities, is another key factor of the model. Two rules dominate literature. In the *deterministic rule* it is assumed that customers only buy at a single facility, the one to which they are attracted most [8, 33]. However, this hypothesis has not found much empirical support, except in areas where shopping opportunities are limited and transportation is difficult. On the contrary, in the *probabilistic rule* customers patronize all the facilities. However, the demand served at each facility is not the same: it is proportional to the attraction. Hence, more attractive facilities capture more demand than less attractive facilities. The probabilistic rule was already suggested in [22] to estimate the market share captured by competing facilities, and first used in a location model in [9]. In that paper, as in most of the ones using the probabilistic rule, the quality of the facility to be located was fixed, given beforehand. It was in [18] when quality was first considered an additional variable to the problem to be determined. In fact, it was empirically proved that both the location and the quality of the facility to be located have to be found simultaneously, as the location influences the quality, and vice-versa. In general, the probabilistic rule has proved to approximate the market share captured by the facilities more accurately than other alternatives, and it will be the one used in the models in this chapter.

Another point to be taken into account is the possible *reaction* of the competitors. In most competitive location models it is assumed that the competition is *static*. This means that competitors are already present in the market, the locating chain knows their characteristics and no reaction to the location of the new facility (or facilities) is expected from them. However, there are situations where the competitors do react to the location of the new facilities. In those cases, it is very important to *foresee* those reactions, as the market share and profit obtained by the locating chain may vary substantially. Although there are *dynamic* location models, where competitors can change their decisions indefinitely, and then the existence of *equilibrium* situations is of major concern (see for instance [6, 19, 27]), in this chapter the focus is on the so-called 'leader-follower' (or Stackelberg) problems. The scenario considered in that type of problems is that of a duopoly. A chain, the *leader*, makes the first movement, and locates p new facilities in the market, where similar facilities of a competitor (the *follower*), and possibly of its own chain, already exist. Then, the follower, as a reaction, decides to locate r new facilities. Hakimi [20] seems to be the first considering this type of two-level optimization problems. He introduced the term $(r|X_p)$ *medianoid* to refer to the follower's problem of locating r facilities in the presence of the p new leader's facilities located at the set of points X_p. And the term $(r|p)$ *centroid* problem to refer to the leader's problem of locating p new facilities, knowing that the follower, as a reaction, will locate r new facilities by solving the corresponding $(r|X_p)$ medianoid problem. In this chapter only the $(1|1)$ centroid problem will be considered, i.e., it is assumed that the leader will locate only one new facility, and the follower's reaction consists of the location of a new single facility too.

Even in this simple case the leader-follower problem is very hard to solve. In fact, the follower's problem is already a highly nonlinear global optimization problem (see [9, 18]). The literature on leader-follower location problems is scarce (see [15] for a review on the topic until 1996). And this shortage is even more pronounced in the case of continuous problems, largely due to the complexity of this type of bilevel programming problems. Drezner [7] solved the $(1|1)$ centroid problem for the Hotelling model and Euclidean distances exactly, through a geometric-based approach. Bhadury et al. [2] considered the $(r|p)$ centroid problem also for the Hotelling model with Euclidean distances, and gave an alternating heuristic to cope with it. In [11] Drezner and Drezner considered the Huff model, and proposed three heuristic approaches for handling the $(1|1)$ centroid problem (see also [12]).

More recently, the authors of this chapter have worked and extended the Huff-like Stackelberg problems. In [44] an exact branch-and-bound method is proposed for a model closely related to that in [11]. This model was later extended in [39] to consider the quality of the new facilities as additional variables of the model, and also changing the objective from market share maximization to profit maximization; both sequential and parallel heuristics were proposed to cope with it (see [39, 40]). Finally, in [42], the model was extended to take into account the possibility of the variability of the demand (see also [1]); again, sequential and parallel heuristic procedures were proposed. The goal of this chapter is to make a critical review of those papers and to point lines for future research. First, in the next section, the basic notation is introduced, and then, in the three following sections, the three aforementioned models are reviewed. Finally, in the last section we point out an idea which may be used to develop exact methods for the last two models.

Although here we only consider that, as a reaction, the follower will locate an additional facility too, other alternatives have been recently proposed in literature. They all consider that the follower can change the quality of its existing facilities. In particular, in [43], the leader locates one single facility in a region of the plane, and then the follower may *increase* the quality of some of its facilities. The follower does not locate any new facility. In [25] the leader enters the market by locating several facilities at some of the points of a finite set of feasible locations (discrete problem), and then, the reaction of the competitor is to adjust (i.e., increase or decrease) the attractiveness of its existing facilities so as to maximize its own profit. However, it cannot open new facilities and/or close existing ones, either. The model is extended in [26], where the follower can also open new facilities or close some existing ones. The probabilistic rule is used in the three aforementioned papers. A different approach is followed in [14] (see also [13]) where a discrete location model based on the concept of coverage is presented. Each facility attracts consumers within a *sphere of influence* defined by a radius. The leader and the follower, each has a budget to be spent on the expansion of their chains either by *improving* their existing facilities or constructing new ones.

2 Notation

A chain, the *leader*, wants to locate a new single facility in a given area of the plane, where m facilities offering the same goods or product already exist. The first k (≥ 0) of those m facilities belong to the chain, and the other $m - k$ (> 0) to a competitor chain, the *follower*. The leader knows that the follower, as a reaction, will subsequently position a new facility too.

The following notation will be used throughout this chapter:

Indices
i Index of demand points, $i = 1, \ldots, n$.
j Index of existing facilities, $j = 1, \ldots, m$. The first k of those m facilities
       belong to the leader's chain, and the rest to the follower's.
l Index for the new facilities, $l = 1$ for the leader, $l = 2$ for the follower.

Variables
$z_l = (x_l, y_l)$ Location of the new leader's ($l = 1$) or follower's ($l = 2$)
                      facility.
α_l Quality of the new leader's ($l = 1$) or follower's ($l = 2$)
                      facility (in case the quality is to be determined by the model).
$nf_l = (z_l, \alpha_l)$ Variables of the new leader's ($l = 1$) or follower's ($l = 2$)
                      facility.

Input data

p_i Location of the ith demand point.

$\widehat{w}_i$ Fixed demand (or purchasing power) at p_i, $\widehat{w}_i > 0$ (when the demand is assumed to be fixed).

w_i^{min} Minimum possible demand at p_i, $w_i^{min} > 0$ (when the demand is assumed to be variable).

w_i^{max} Maximum possible demand at p_i, $w_i^{max} \geq w_i^{min}$ (when the demand is assumed to be variable).

f_j Location of the jth existing facility.

d_{ij} Distance between p_i and f_j, $d_{ij} > 0$.

β_j Quality of f_j, $\beta_j > 0$.

d_i^{min} Minimum distance from p_i at which the new facilities can be located, $d_i^{min} > 0$.

S_l Location space where the leader ($l = 1$) or the follower ($l = 2$) will locate its new facility.

α_l^{min} Minimum level of quality for the new leader's ($l = 1$) or follower's ($l = 2$) facility, $\alpha_l^{min} > 0$ (when the quality is a variable of the model).

α_l^{max} Maximum level of quality for the new leader's ($l = 1$) or follower's ($l = 2$) facility, $\alpha_l^{max} \geq \alpha_l^{min}$, (when the quality is a variable of the model).

Miscellaneous

$g_i(\cdot)$ A non-negative, non-decreasing function, which modulates the decrease in attractiveness as a function of distance.

$d_i(z_l)$ Distance between p_i and z_l, $l = 1, 2$.

u_{i,nf_l} Attraction that p_i feels for nf_l, $l = 1, 2$,
$$u_{i,nf_l} = \alpha_l / g_i(d_i(z_l))$$

$U_i(nf_1, nf_2)$ Total utility perceived by a customer at p_i provided by all the facilities.

$w_i(U_i(nf_1, nf_2))$ Actual demand at p_i (when the demand is assumed to be variable).

Computed parameters

u_{ij} Attraction that p_i feels for f_j (or utility of f_j perceived by the people at p_i),
$$u_{ij} = \beta_j / g_i(d_{ij}).$$

Market share and profit functions

$M_l(nf_1, nf_2)$ Market share obtained by the leader ($l = 1$) or the follower ($l = 2$) after the location of the new facilities.

$\Pi_l(nf_1, nf_2)$ Profit obtained by the leader ($l = 1$) or the follower ($l = 2$) after the location of the new facilities.

The profit functions Π_1 and Π_2 vary in each of the problems analyzed, and are detailed in the corresponding sections.

In all the models in this chapter it is assumed that the patronizing behavior of customers is probabilistic, that is, demand points split their buying power among *all* the facilities proportionally to the attraction they feel for them. Using these assumptions, the market share attracted by the leader's chain after the location of the leader and the follower's new facilities is

$$M_1(nf_1, nf_2) = \sum_{i=1}^{n} w_i \frac{u_{i,nf_1} + \sum_{j=1}^{k} u_{i,j}}{u_{i,nf_1} + u_{i,nf_2} + \sum_{j=1}^{m} u_{i,j}}, \tag{1}$$

where w_i stands for $\widehat{w}_i$ when the demand is fixed, and for $w_i(U_i(nf_1, nf_2))$ when the demand is variable. Analogously, the market share attracted by the follower's chain is

$$M_2(nf_1, nf_2) = \sum_{i=1}^{n} w_i \frac{u_{i,nf_2} + \sum_{j=k+1}^{n} u_{i,j}}{u_{i,nf_1} + u_{i,nf_2} + \sum_{j=1}^{m} u_{i,j}}. \tag{2}$$

Given nf_1, the problem for the follower is the $(1|nf_1)$ medianoid problem:

$$(FP(nf_1)) \quad \begin{cases} \max \ \Pi_2(nf_1, nf_2) \\ \text{s.t.} \quad z_2 \in S_2 \\ \quad\quad d_i(z_2) \geq d_i^{\min}, i = 1, \dots, n \\ \quad\quad \alpha_2 \in [\alpha_2^{\min}, \alpha_2^{\max}] \end{cases} \tag{3}$$

whose objective is the maximization of the profit obtained by the follower (once the leader has set up its new facility at nf_1). In case the problem $(FP(nf_1))$ has multiple optimal solutions, then it is assumed that the follower selects an optimal solution which provides the worst possible objective function value for the leader (the so-called *pessimistic approach* in bilevel programming [5]).

Let us denote with $nf_2^*(nf_1)$ an optimal solution of $(FP(nf_1))$ for which the objective value of the leader is minimum. The problem for the leader is the $(1|1)$ centroid problem:

$$(LP) \quad \begin{cases} \max \ \Pi_1(nf_1, nf_2^*(nf_1)) \\ \text{s.t.} \quad z_1 \in S_1 \\ \quad\quad d_i(z_1) \geq d_i^{\min}, i = 1, \dots, n \\ \quad\quad \alpha_1 \in [\alpha_1^{\min}, \alpha_1^{\max}] \end{cases} \tag{4}$$

As we can see, the leader problem (LP) is much more difficult to solve than the follower problem $(FP(nf_1))$. Notice, for instance, that to evaluate its objective function Π_1 at a given point nf_1, we have to first solve the corresponding medianoid problem $(FP(nf_1))$ to obtain $nf_2^*(nf_1)$.

3   A Model Without Costs

3.1   The Model

The first model we will describe is that in [44]. Essential goods are considered. Therefore, the demand has to be served by the facilities. The demand quantities are assumed to be known and fixed. Also the quality values of the new facilities to be located, α_1 and α_2, are assumed to be given, i.e., *they are not variables of the model*. As the qualities are fixed, no cost related to the achievement of a given level of quality is considered. No cost related to the setting-up of the facilities at a given location is considered either. Then, taking into account that the profit obtained by a player is an increasing function of the market share it captures, the objective functions considered in [44] were

$$\Pi_l(nf_1, nf_2) = M_l(nf_1, nf_2), \quad l = 1, 2.$$

In addition to this, the location space is the same for the leader and the follower, i.e., $S_1 = S_2$. No other constraints are considered in the model. The corrected Euclidean distance [10] was used as distance function.

Since the demand is fixed and has to be served, then

$$M_1(nf_1, nf_2) + M_2(nf_1, nf_2) = \sum_{i=1}^{n} \widehat{w}_i. \tag{5}$$

In particular, what is a gain for one chain is a loss for the other. This zero-sum concept is the key used in [44] to develop a Branch-and-Bound (B&B) procedure to solve the leader problem rigorously, to have a guarantee on the reached accuracy.

3.2   A B&B Algorithm for the Follower Problem

Branch-and-bound (B&B) algorithms recursively decompose the original problem into smaller disjoint subproblems until the solution is found. The method avoids visiting those subproblems which are known not to contain a solution. The initial set $C_1 = S_1(= S_2)$ is subsequently partitioned in more and more refined subsets (branching). At every iteration, the method has a list Λ of subsets C_k of C_1. The method stops when the list is empty. For every subset C_k in Λ, upper bounds UB^k of the objective function on C_k are determined. Moreover, a global lower bound GLB is updated. If $UB^k < GLB$ for a given subset C_k, it can be removed from the list, since it cannot contain a maximum.

Algorithm 1: B&B algorithm for the *(reverse) follower problem*: function *FunctB&B(M, nf, C, ϵ_f)*

1: $\Lambda := \emptyset$.
2: $C_1 := C$.
3: Determine an upper bound UB^1 on C_1.
4: Compute $nfa^1 :=$ midpoint(C_1), $BestPoint := nfa^1$.
5: Determine lower bound: $LB^1 := M(nfa^1)$, $GLB := LB^1$.
6: Put C_1 on list Λ, $r := 1$.
7: **while** $\Lambda \neq \emptyset$ **do**
8:     Take subset C from list Λ and bisect into C_{r+1} and C_{r+2}.
9:     **for** $t := r + 1$ to $r + 2$ **do**
10:       Determine upper bound UB^t.
11:       **if** $UB^t > GLB + \epsilon_f$ **then**
12:         Compute $nfa^t :=$ midpoint(C_t) and $LB^t := M(nfa^t)$.
13:         **if** $LB^t > GLB$ **then**
14:           $GLB := LB^t$, $BestPoint := nfa^t$ and remove all C_i from Λ with $UB^i < GLB$.
15:           **if** $UB^t > GLB + \epsilon_f$ **then**
16:             save C_t in Λ.
17:     $r := r + 2$.
18: OUTPUT: {$BestPoint, GLB$}.

The steps of the method can be seen in Algorithm 1. In the solution procedure for the leader problem, a similar problem to that of the follower, in which the leader wants to locate a new facility at nf_1, given the location and the quality of all the facilities of the competitor (the follower), has to be solved. In this case, the leader has to solve a medianoid problem in which the roles of leader and follower are interchanged. We will call this problem a *reverse medianoid problem*. To take both the medianoid and the reverse medianoid problems into account, in Algorithm 1 the new facility of the competitor is denoted by nf, the objective function by $M(nfa)$ (where $M(nfa) = M_2(nf, nfa)$ when solving a medianoid problem and $M(nfa) = M_1(nfa, nf)$ when solving a reverse medianoid problem), and the feasible set by C.

The B&B method introduced in [44] uses boxes (2-dimensional intervals) as subsets of the initial region and the subdivision rule bisects a box C over its longest edge. Several selection rules of the next box to be selected (Step 8 of Algorithm 1) were tested in [44], see Sect. 3.4.

Concerning the computation of bounds, the global lower bound is updated by evaluating the objective function at some points (the centers of the boxes). As for the upper bounds, four variants were proposed in [44]. The simplest one (which turned out to be competitive with the other three more elaborated bounds based on D.C. decompositions of the objective function) is based on the underestimation of the distance from demand point p_i to facilities in a box C. Since the new facility is only located at one point within the box, we obtain an overestimation (upper bound) of the market captured by the new facility. The idea developed in [44] is similar to that in [32].

The demand points p_i within box C have a distance $\Delta_i(C) = 0$ from C. For demand points out of box C, $p_i \notin C$, the shortest distance $\Delta_i(C)$ of p_i to the box is calculated, $\Delta_i(C) = \min_{x \in C} d(x, p_i)$. The distance $\Delta_i(C)$ can be determined as follows. Box C is defined by two points: lower-left point $LL = (ll_1, ll_2)$ and upper-right point $UR = (ur_1, ur_2)$. The shortest distance from demand point p_i to the box C can be computed by

$$\Delta_i(C) = \begin{cases} 0 & \text{if } p_i \in C \\ \sqrt{\Delta_{i1}^2 + \Delta_{i2}^2} & \text{if } p_i \notin C \end{cases}$$

where

$$\Delta_{i1} = \max\{ll_1 - p_{i1}, p_{i1} - ur_1, 0\}$$
$$\Delta_{i2} = \max\{ll_2 - p_{i2}, p_{i2} - ur_2, 0\}$$

Notice that this distance calculation can be extended to higher dimensions.

The output of Algorithm 1 is the best point found during the process and its corresponding function value. The best point is guaranteed to differ less than ϵ_f in function value from the optimal solution of the problem.

Another B&B algorithm which can be used to solve the follower problem is described in [18]. It uses interval analysis tools (see [47]) and can also handle the follower problems in the next two sections.

3.3 A B&B Algorithm for the Leader Problem

The corresponding B&B method for the leader problem is given in pseudocode form in Algorithm 2. The branching and selection rules used were the same as in Algorithm 1, as well as the computation of the global lower bound.

The key point in the algorithm is computation of the upper bounds. Let $C \subseteq \mathbb{R}^2$ denote a subset of the search region of the leader problem (LP). An upper bound of the objective function $M_1(nf_1, nf_2^*(nf_1))$ over C can be obtained by having the leader solve the reverse medianoid problem, as the following lemma proves.

Lemma 1 *Let nf_2 be a given solution for the new follower's facility. Then*

$$UB(C, nf_2) = \max_{nf_1 \in C} M_1(nf_1, nf_2)$$

is an upper bound of $M_1(nf_1, nf_2^*(nf_1))$ over C.

Proof According to (5), maximizing the market share captured by the follower given nf_1 is equivalent to finding the facility nf_2 that minimizes the market share captured by the leader. Hence, $M_1(nf_1, nf_2^*(nf_1)) \leq M_1(nf_1, nf_2)$ such that

$$\max_{nf_1 \in C} M_1(nf_1, nf_2^*(nf_1)) \leq \max_{nf_1 \in C} M_1(nf_1, nf_2) = UB(C, nf_2).$$

Algorithm 2: B&B algorithm for the *leader problem*

1: $\Lambda := \emptyset$.
2: $C_1 := S$.
3: Compute $nf_1^1 :=$ midpoint(C_1), $BestPoint := nf_1^1$.
4: Solve the problem for the follower: $\{nf_2^1, lbobj\} := FunctB\&B(M_2, nf_1^1, C_1, \epsilon_f)$.
5: Determine an upper bound UB^1 on C_1 solving a reverse medianoid problem:
    $\{nfa, UB^1\} := FunctB\&B(M_1, nf_2^1, C_1, \epsilon_l)$.
6: Determine lower bound: $LB^1 := M_1(nf_1^1, nf_2^1)$, $GLB := LB^1$.
7: Put C_1 on list Λ, $r := 1$.
8: **while** $\Lambda \neq \emptyset$ **do**
9:      Take subset C from list Λ and bisect into C_{r+1} and C_{r+2}.
10:     **for** $t := r + 1$ to $r + 2$ **do**
11:         Compute $nf_1^t =$ midpoint(C_t).
12:         Solve the problem for the follower: $\{nf_2^t, lbobj\} := FunctB\&B(M_2, nf_1^t, C_1, \epsilon_f)$.
13:         Determine upper bound UB^t solving a reverse medianoid problem:
            $\{nfa, UB^t\} := FunctB\&B(M_1, nf_2^t, C_t, \epsilon_l)$
14:         **if** $UB^t > GLB + \epsilon_l$ **then**
15:             Determine $LB^t := M_1(nf_1^t, nf_2^t)$.
16:             **if** $LB^t > GLB$ **then**
17:                 $GLB := LB^t$, $BestPoint := nf_1^t$, and remove all C_i from Λ with $UB^i < GLB$.
18:             **if** $UB^t > GLB + \epsilon_l$ **then**
19:                 save C_t in Λ.
20:     $r := r + 2$.
21: OUTPUT: $\{BestPoint, GLB\}$.

$\square$

For a given box C_t, the choice of nf_2^t for the upper bound calculation is done as follows. First, the midpoint of C_t is computed, and considering it as the new leader's facility, nf_1^t, the corresponding follower's problem is solved, $(FP(nf_1^t))$, obtaining nf_2^t. Then, the upper bound is obtained by solving the reverse medianoid problem up to an accuracy ϵ_l

$$UB^t = UB(C_t, nf_2^t) = \max_{nf_1 \in C_t} \{M_1(nf_1, nf_2^t)\} = FunctB\&B(M_1, nf_2^t, C_t, \epsilon_l).$$

Again, the output of the B&B method (see Algorithm 2) is the best point found during the process and its corresponding function value, which differs less than ϵ_l from the optimum value of the problem.

3.4   Computational Studies

A random problem with $n = 10$ demand points and $m = 4$ existing facilities was first solved to illustrate the algorithm. The number k of facilities belonging to the leader's chain was varied from $k = 0$ to 4. The other parameters of the problem were chosen from uniform distributions (see [44]). Table 1 shows the resulting optimal

Table 1 Optimal locations and market capture for different number of leader facilities, $k = 0, \ldots, 4$; locations and market captures are rounded to two decimals

		$k = 0$	$k = 1$	$k = 2$	$k = 3$	$k = 4$
Optima location	Leader	$\begin{pmatrix} 2.44 \\ 3.97 \end{pmatrix}$	$\begin{pmatrix} 5.03 \\ 0.69 \end{pmatrix}$	$\begin{pmatrix} 5.33 \\ 4.34 \end{pmatrix}$	$\begin{pmatrix} 5.33 \\ 4.34 \end{pmatrix}$	$\begin{pmatrix} 5.03 \\ 0.69 \end{pmatrix}$
	Follower	$\begin{pmatrix} 2.44 \\ 3.97 \end{pmatrix}$	$\begin{pmatrix} 5.03 \\ 0.69 \end{pmatrix}$	$\begin{pmatrix} 1.41 \\ 4.65 \end{pmatrix}$	$\begin{pmatrix} 1.75 \\ 3.79 \end{pmatrix}$	$\begin{pmatrix} 1.75 \\ 3.79 \end{pmatrix}$
Market Capture	Leader	186.29	367.87	497.70	611.07	773.44
	Follower	813.71	632.13	502.30	388.93	226.56
Gain or loss for the leader		186.29	100.67	14.17	−72.46	−226.56

locations and market capture of both chains. In the last line, the gain or loss for the leader, to be understood as the difference between the market captured by the leader after and before the location of the facilities, is given. The accuracy for Algorithms 1 and 2 were set both to $\epsilon_l = \epsilon_f = 10^{-2}$.

One can observe a characteristic of the problem, where leader and follower tend to *co-locate* when the number of existing facilities of the leader is low. Notice also that when the leader is dominant in the market then the leader suffers a decrease in market share after the location of the two new facilities (see the negative values in the last line of Table 1). This is because in those cases the follower increases its market share more than the leader.

Concerning the efficiency of the selection rule of the next box to be processed, breadth-first and best-bound strategies were researched. The results in [44] concluded that best-bound strategy is the one providing the best results, as in average, the number of iterations employed by Algorithm 1 was reduced significantly. The influence in the number of iterations of Algorithm 2 was not so clear when using the upper bound described in Sect. 3.2, but when additional bounds are employed the best-bound selection rule was also clearly the best for Algorithm 2.

As for the memory requirement, it is known that branch-and-bound algorithms are usually hindered by huge search trees that need to be stored in memory. This complexity usually increases rapidly with dimension and with accuracy. Interestingly, this does not seem to be the case for this problem. There are never more than 30 boxes in the storage tree. And the same remains valid when the accuracy is increased up to 0.0001 for both Algorithms 1 and 2.

The last set of experiments done in [44] studied whether larger problems could be solved in reasonable time. To this aim, random problems were generated varying the number of demand points ($n = 20, 30, \ldots, 110$), number of existing facilities ($m = 5, 10, 15$) and number of those facilities belonging to the leader's chain ($k = [m/2]$). For each (n, m) setting, ten problems were generated by randomly selecting the parameters of the problem from uniform distributions. The results can be seen in Fig. 1. It can be seen that increasing the number of demand points does not make the problem more complex in terms of the memory requirement. The leader problem neither needs more iterations, although the follower problem needs more iterations

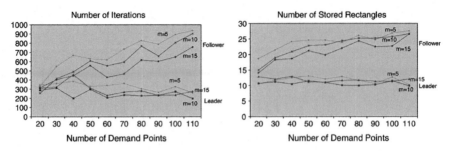

Fig. 1 Average number of iterations and memory requirement (*rectangles*) over ten random cases varying number of demand points $n = 20, \ldots, 110$, existing facilities $m = 5, 10, 15$ and $k = [m/2]$. $\epsilon_l = \epsilon_f = 0.01$

on average. Hence, the results suggest that no exponential effort is required to solve the problems with increasing number of demand points, confirming the viability of the approach.

4 A Model with Costs Assuming Fixed Demand

4.1 The Model

The scenario considered in this section (see [39]) is similar to the one previously described. The demand is again supposed to be fixed and known. But now, both the location and the quality (design) of the new facilities have to be found and several types of costs are considered.

The objective function Π_2 for the follower problem [see Eq. (3)], is now formulated as the difference between the revenues obtained from the captured market share minus the operating costs of the new facility:

$$\Pi_2(nf_1, nf_2) = F_2(M_2(nf_1, nf_2)) - G_2(nf_2). \tag{6}$$

Similarly, the profit obtained by the leader [see Eq. (4)] is given by:

$$\Pi_1(nf_1, nf_2^*(nf_1)) = F_1(M_1(nf_1, nf_2^*(nf_1))) - G_1(nf_1). \tag{7}$$

Functions F_l, $l = 1, 2$, are strictly increasing differentiable functions that transform the market share into expected sales. In the computational studies in [39], they are linear, $F_l(M_l) = c_l \cdot M_l$, where c_l is the income per unit of goods sold.

Functions $G_l, l = 1, 2$, are the operating costs functions. G_l should increase as z_l gets closer to any demand point, since it is rather likely the operating costs of the facility will be higher as the facility approaches the demand points. Furthermore, G_l should be a nondecreasing and convex function in the variable α_l, since the more quality the facility requires, the higher the costs will be, at an increasing rate. In [39] it is assumed that functions G_l consist of the sum of the location costs and the costs needed to achieve a given level of quality, i.e. $G_l(nf_l) = G_l^a(z_l) + G_l^b(\alpha_l)$. In the computational experiments the following choices were made: $G_l^a(z_l) = \sum_{i=1}^n \Phi_l^i(d_i(z_l))$, with $\Phi_l^i(d_i(z_l)) = \widehat{w}_i/((d_i(z_l))^{\phi_l^{i0}} + \phi_l^{i1})$, $\phi_l^{i0}, \phi_l^{i1} > 0$ and $G_l^b(\alpha_l) = \exp(\alpha_l/\xi_l^0 + \xi_l^1) - \exp(\xi_l^1)$, with $\xi_l^0 > 0$ and $\xi_l^1 \in \mathbb{R}$ given values. See [18] for a detailed explanation of these functions, as well as other possible expressions for F_l and $G_l(nf_l)$.

Notice that the key to solving the problem of the previous section with precision was that *what is a gain for one chain is a loss for the other*, see (5). This is no longer true for this model: notice that now $\Pi_1(nf_1, nf_2) + \Pi_2(nf_1, nf_2)$ is not necessarily constant due to the cost functions. This fact impedes using the methodology employed in the previous section to develop a B&B method for the new leader's problem (Lemma 1 does not hold any more). That is why heuristic procedures are proposed in [39] to cope with the new problem. However, other strategies are possible, as described in Sect. 6.

4.2 Solving the Medianoid Problem

The algorithm UEGO is used here to deal with the medianoid problem. UEGO, which stands for Universal Evolutionary Global Optimizer, is a memetic multi-modal global optimization method especially suitable to be parallelized and highly adaptable to different problems [24, 31, 35–38].

The key concept of UEGO is that of species, which is defined by a center and a radius. The center is a solution, and the radius is a positive number that defines an attraction area and hence, multiple solutions. In particular, for the medianoid problem, a species is an array of the form $(nf_2, \Pi_2(nf_1, nf_2), R)$ (we also store information about the objective value at the center of the species). During the optimization procedure, UEGO works with a set of species stored in the *species_list*.

The adaptability of UEGO mainly relies on being defined in two levels, global an local. In the global level, UEGO defines an iterative and progressively cooled management process over a set of available species, and this process is the same for all the problems to which UEGO is applied. In the local one, a particular local optimizer is selected for the studied problem at the context defined by every species. For the current problem, a Weiszfeld-like method (WLM) has been considered as a local optimizer. The UEGO algorithm executed with WLM to solve the medianoid problem will be called UEGO_med throughout.

A global description of UEGO_med is given in Algorithm 3. The input given parameter nf_1 indicates the additional leader facility, which has to be taken into account apart from the m pre-existing facilities. Additionally, UEGO_med has four more user given parameters: (1) N, the maximum number of function evaluations (f.e.) allowed for the entire optimization process; (2) L, the maximum number of

---

Algorithm 3: Algorithm UEGO_med(nf_1, N, L, M, R_L)

1: Init_species_list
2: Optimize_species(n_1)
3: **for** $i = 2$ to L **do**
4:    Determine R_i, new_i, n_i
5:    Create_species(new_i)   {# *budget_per_species* = new_i/*length(species_list$_i$)*}
6:    Fuse_species(R_i)
7:    Shorten_species_list(M)
8:    Optimize_species(n_i)   { # *budget_per_species* = n_i/M}
9:    Fuse_species(R_i)

---

levels (iterations) of the algorithm; (3) M, which refers to the maximum length of the *species_list*, and (4) R_L, which indicates the minimum radius that a species can have. Furthermore, from these four input parameters, three important values are computed at each level i: the maximum number of f.e. for the creation of new species (new_i), the maximum number of f.e. for the optimization of species (n_i), and the radius assigned to the new species (R_i). The equations linking all these parameters are detailed in [23, 31].

In the following, the different key stages of UEGO_med are described:

- *Init_species_list:* The initial *species_list* is composed of a single species. The value of nf_2 is randomly computed and the corresponding radius is set to R_1.
- *Create_species(create_evals):* In terms of evolutionary computation, this procedure can be interpreted as an algorithm to create offspring. The input parameter *create_evals* indicates the number of function evaluations allowed for the creation procedure at the current level. The most remarkable aspect of this mechanism is that every species in the *species_list* is able to generate a new progeny without participation from the remaining ones. The parameter *create_evals* is internally divided by the current number of existing species (*length(species_list$_i$)*), which means that the budget available per species for the creation of new points is equal to:

$$budget\_per\_species = new_i/length(species\_list_i).$$

For each single species, the creation method proceeds as follows: New random exploratory points are created within the area defined by its radius, and for every pair of those points, a new candidate solution is created at the middle of the *segment* connecting the pair. Then, all the candidate points are evaluated, and the one with the best objective function value replaces the center of the original species in the case that it improves the objective function of the center. Later, the merit of the extreme points to become a new species, is analyzed. Both extreme points are inserted into the *species_list* if their objective function values are better than the one at the corresponding midpoint. Every new inserted species is assigned the current radius value (R_i).

- *Fuse_species(radius)*: This procedure unites species from the *species_list* that are closer than the distance defined by the parameter *radius*. Then, for every pair of species in the list, the Euclidean distance is computed. If such a distance is smaller than the given radius, the species with the lowest fitness are removed. The radius of the species that remains is set equal to the maximum of the radii of the original two species.
- *Shorten_species_list (max_list_length)*: It deletes species to reduce the list length to *max_list_length* value. The species with the smaller radius are deleted first.
- *Optimize_species(opt_evals)*: In this procedure, every species calls a local optimizer once, using the *nf₂* value of the caller species as initial point. If after the execution of the local method a new point with a better objective function is found, then the original *nf₂* is updated. The budget per species for the optimization process, in terms of number of function evaluations, is n_i/M. For the problem at hand, a Weiszfeld-like algorithm has been considered as local optimizer.

4.2.1 Weiszfeld-Like Algorithm WLM

This algorithm is a steepest descent method. The derivatives of the objective function are equated to zero and the next iterate is obtained by implicitly solving these equations. Notice that, here, the derivatives are computed taking the F_l and G_l functions described in Sect. 4.1 into account. Of course, they should be recomputed if any other expression is considered.

If we denote

$$r_i = \sum_{j=1}^{m} u_{ij}, \; t_i = \widehat{w}_i \sum_{j=k+1}^{m} u_{ij},$$

$$H_i(nf_2) = \frac{\partial \Pi_2}{\partial d_i(z_2)} = -\frac{dF_2}{dM_2} \cdot \frac{\alpha_2 \gamma_i t_i g_i'(d_i(z_2))}{(\gamma_i \alpha_2 + r_i g_i(d_i(z_2)))^2} - \frac{d\Phi^i}{dd_i(z_2)},$$

and $d_i(z_2)$ is a distance function such that

$$\frac{\partial d_i(z_2)}{\partial x_2} = x_2 A_{i1}(z_2) - B_{i1}(z_2), \quad \frac{\partial d_i(z_2)}{\partial y_2} = y_2 A_{i2}(z_2) - B_{i2}(z_2), \tag{8}$$

then the Weiszfeld-like algorithm for solving the corresponding problem is described by Algorithm 4 (for more details see [18]).

Algorithm 4: WLM (Weiszfeld-like algorithm)

1: Set iteration counter $ic = 0$
2: Initialize $nf_2^{(0)} = (x_2^{(0)}, y_2^{(0)}, \alpha_2^{(0)})$
3: **while** stopping criteria are not met **do**
4:    Update $nf_2^{(ic+1)} = (x_2^{(ic+1)}, y_2^{(ic+1)}, \alpha_2^{(ic+1)})$
5:    **if** $nf_2^{(ic+1)}$ is unfeasible **then**
6:        $nf_2^{(ic+1)} \in [nf_2^{(ic)}, nf_2^{(ic+1)}] \cap \partial S_2$
7:    $ic = ic + 1$

Values of $x_2^{(ic+1)}$ and $y_2^{(ic+1)}$ in Algorithm 4 are obtained as:

$$x_2^{(ic+1)} = \frac{\sum_{i=1}^{n} H_i(nf_2^{(ic)})B_{i1}(z_2^{(ic)})}{\sum_{i=1}^{n} H_i(nf_2^{(ic)})A_{i1}(z_2^{(ic)})}, \qquad y_2^{(ic+1)} = \frac{\sum_{i=1}^{n} H_i(nf_2^{(ic)})B_{i2}(z_2^{(ic)})}{\sum_{i=1}^{n} H_i(nf_2^{(ic)})A_{i2}(z_2^{(ic)})}$$

and $\alpha_2^{(ic+1)}$ as a solution of the equation:

$$\frac{dF_2}{dM_2} \cdot \sum_{i=1}^{n} \frac{\gamma_i t_i g_i(d_i(z_2^{(ic+1)}))}{(\gamma_i \alpha_2 + r_i g_i(d_i(z_2^{(ic+1)})))^2} - \frac{dG_2}{d\alpha_2} = 0.$$

Two stopping rules are applied in WLM: (1) the algorithm stops if

$$\|(x_2^{(ic-1)}, y_2^{(ic-1)}) - (x_2^{(ic)}, y_2^{(ic)})\|_2 < \epsilon_1 \text{ and } |\alpha_2^{(ic-1)} - \alpha_2^{(ic)}| < \epsilon_2,$$

for given tolerances $\epsilon_1, \epsilon_2 > 0$; and (2) the procedure finishes if a maximum number of iterations ic_{max} is achieved or the number of function evaluations exceeds the budget assigned.

In Step 6 of Algorithm 4, $nf_2^{(ic+1)}$ is set to a point in the segment $[nf_2^{(ic)}, nf_2^{(ic+1)}]$ which is also on the border ∂S_2 of the feasible region S_2.

The l_{2b} distance, given by

$$d_i(z_l) = \sqrt{b_1(x_l - p_{i1})^2 + b_2(y_l - p_{i2})^2},$$

satisfies the conditions in (8). Furthermore, it has proved to be a good distance predicting function (see [17]), and it is therefore a good distance function to be used in competitive location models, as it measures distances (or travel time) as they are perceived by customers on their ways to and from facilities.

4.3   Solving the Centroid Problem

Four heuristics are introduced in [39] for handling the centroid problem, namely, a grid search procedure (GS), an alternating method called AlternatMed and two evolutionary algorithms based on the UEGO_med structure. These two variants, which differ basically in the considered local optimizer, are named UEGO_cent.WLM and UEGO_cent.SASS.

A comprehensive computational study in [39] shows that UEGO_cent.SASS is the algorithm which provides the best results. In fact, in all the considered problems, it is the algorithm giving the best solutions. In view of those results, only the algorithm UEGO_cent.SASS is explained below. For the sake of brevity, only the fundamental differences concerning UEGO_med are mentioned. The interested reader can always consult [39] for a detailed account of the remaining methods.

Species definition: A species is now defined by the vector (nf_1, nf_2, R), where nf_1 refers to the leader point, nf_2 is the solution obtained by UEGO_med when taking the original m existing facilities and nf_1 into account, and R is the radius of the species.

Create_species procedure: This procedure is, in essence, the same as the creation process described in Sect. 4.2. However, some amendments have been made to comply with certain computational requirements.

In this procedure, random trial points for nf_1 are also created within the area defined by the radius of the species. Additionally, similar to what is done in UEGO_med, the midpoint of each pair of solutions is also computed. However, not all candidate solutions are evaluated, but only the most promising ones, i.e., we do not solve the corresponding medianoid problem associated to each new point to obtain the follower's facility. This is done in this way because this procedure is too costly and the number of points to be evaluated is very high. On the contrary, we first analyze the merit of the candidate solutions by computing an approximate objective value. More precisely, the follower's facility associated to the species from which they were generated is used to obtain an approximate fitness for the leader's candidate solutions.

After this process, for every species in the *species_list* we have a sublist of 'candidate' points to generate new species. Notice that in this creation process, the candidate solutions never replace the original species, as happens in UEGO_med. This is because the comparison in terms of fitness may be misleading, since the objective value at the midpoints or at the endpoints of the segments is only an *approximation*.

Furthermore, in order to reduce the large number of candidate points, those 'candidate' points are merged as described in Sect. 4.2 (using the procedure *Fuse_species*). Finally, for each candidate point in this reduced list, its corresponding follower's facility is computed applying UEGO_med, and the objective value for the leader's facility is evaluated. The new species (with the corresponding radius according to the iteration) are inserted in the *species_list*.

Algorithm 5: Algorithm LeaderOpt

1: Let (nf_1, nf_2, R) be the species to be optimized.
2: $opt\_nf_1 = \text{SASS+WLM}(nf_1, nf_2, R)$
3: $opt\_nf_2 = \text{UEGO\_med}(opt\_nf_1)$
4: **if** $opt\_nf_1 = nf_1$ **then**
5:      **if** $\Pi_2(nf_1, nf_2) > \Pi_2(nf_1, opt\_nf_2)$ **then**
6:          $opt\_nf_2 = nf_2$
7:      Update the original species to (nf_1, opt_nf_2, R).
8: **else if** $\Pi_1(opt\_nf_1, opt\_nf_2) > \Pi_1(nf_1, nf_2)$ **then**
9:      Update the original species to (opt_nf_1, opt_nf_2, R)

Optimize_species procedure: For every species in the list, the local optimization process described in Algorithm 5 is applied. In Step 2, the SASS+WLM local search is applied (see [39]). This method tries to obtain a better solution for the leader (nf_1) based on the current choice of the follower (nf_2). To do so, this algorithm uses the stochastic hill climber SASS (see [46]) for updating the leader's facility and WLM for updating the follower's. Notice that the algorithm WLM is used because obtaining the exact new follower's facility every time the leader's facility changes, using UEGO_med, makes the process very time-consuming. Nevertheless, to prevent that the objective value for the leader becomes misleading (overestimated), UEGO_med is used in Step 3 of Algorithm 5. Finally, the species is replaced only in case a better objective function value is obtained (see steps 5–9 of Algorithm 5).

4.4   The Cost of a Myopic Decision

A study is carried out to know how important it is to consider the follower's reaction. To this aim, for fourteen problems, we have calculated the leader's profit by solving the medianoid problem but interchanging the roles of the leader and the follower and only taking the original m facilities into account, i.e., the *reverse medianoid problem*. The corresponding optimal solution will be denoted by $nf_1^{(myop)}$. Then, we have solved the corresponding medianoid problem, taking the existing m facilities and $nf_1^{(myop)}$ into account, using UEGO_med. And finally, we have evaluated $\Pi_1^{(myop)} = \Pi_1(nf_1^{(myop)}, \text{UEGO\_med}(nf_1^{(myop)}))$.

Table 2 shows the obtained results. The first column refers to the setting of the problems solved (for three settings, more than one problem was generated, and the letters a, b, and c at the end of the setting has been added to highlight it). Columns two and three show the values of $nf_1^{(myop)}$ and $\Pi_1^{(myop)}$. The following two columns provide the values of the facility (nf_1^*) and the profit (Π_1^*) obtained with UEGO_cent.SASS. Finally, the loss in profit caused by the myopic decision as compared to the long term decision, in percentage, is shown.

Table 2 Comparison between the myopic and the long term view

(n, n, k)	$nf_1^{(myop)}$			$\Pi_1^{(myop)}$	nf_1^*			Π_1^*	% loss
	x_1	x_2	α_1		x_1	x_2	α_1		
(21,5,2)	2.234	3.352	1.524	226.645	2.981	4.482	2.218	228.394	0.76
(21,5,3)	3.024	6.576	0.536	363.451	2.234	3.352	1.162	379.943	4.34
(50,5,0)a	6.082	2.378	2.230	9.156	6.082	2.378	2.230	9.156	0.00
(50,5,0)b	5.419	6.411	5.000	67.569	5.417	6.906	4.851	94.044	28.15
(50,5,1)	4.452	5.920	3.839	116.424	4.917	5.150	3.418	143.498	18.87
(50,5,2)a	2.264	2.096	2.421	189.113	2.228	2.138	2.122	189.653	0.28
(50,5,2)b	3.573	4.044	2.554	109.514	3.572	4.044	2.549	111.246	1.56
(50,6,3)a	1.122	3.362	3.224	291.052	1.161	4.222	3.663	292.554	0.51
(50,6,3)b	1.733	5.848	3.991	194.486	7.151	3.487	3.123	212.358	8.42
(50,6,3)c	6.851	3.459	4.486	218.890	4.103	3.055	4.255	230.329	4.97
(50,8,4)	5.677	2.830	2.973	198.546	5.893	2.629	2.864	223.983	11.36
(100,2,0)	4.471	4.704	5.000	168.430	4.724	4.591	5.000	169.717	0.76
(100,2,1)	3.379	6.298	5.000	271.951	3.255	6.366	5.000	272.027	0.03
(100,10,0)	2.758	5.119	5.000	40.944	2.758	5.119	5.000	40.944	0.00

As can be seen, the loss is less than 1% for half of the problems, it is over 4% for 6 out of 14 problems, and it exceeds 11% in three of them. This clearly indicates how important anticipating the competitor's reaction is, since the loss that can be produced may be substantial. Furthermore, note that the obtained results are independent of the setting (n, m, k) of the problem. Notice, for example, that the two extreme cases, with 0% loss and 28.15% loss, have the same configuration $(50, 5, 0)$. What is important is the actual distribution of the demand points and the actual locations and qualities of the existing facilities. Notice also that even though $nf_1^{(myop)}$ may be close to nf_1^*, the value of $\Pi_1^{(myop)}$ may be very different from Π_1^*, see problem (50,5,0)b.

4.5 High Performance Computing for the Leader-Follower Problem

UEGO_cent.SASS is a costly algorithm, since the evaluation of the objective function value implies the resolution of a global optimization problem. Its parallelization may allow to reduce the execution time and to increase the size of the problems that can be solved. In [40], a *master-slave* algorithm and four *coarse-grain* methods are presented to parallelize UEGO_cent.SASS. The efficiency of the parallel algorithms is tested through an extensive computational testbed. Results showed that the *master-slave* method outperforms all the *coarse-grain* proposals, i.e. it is able to solve more instances using fewer processing elements and to obtain efficiencies close to or even greater than the ideal one.

Algorithm 6: Algorithm MS

1: Init_species_list
2: Optimize_species(n_1)
3: **for** $i = 2$ to L **do**
4:     Determine R_i, new_i, n_i
5:     Create_species_paral(new_i)
6:     Fuse_species(R_i)
7:     Shorten_species_list(M)
8:     Optimize_species_paral(n_i)
9:     Fuse_species(R_i)

In the following, the main features of the *master-slave* strategy are detailed. Readers interested in delving into the *coarse-grain* methods as well as into the performance comparison among parallel algorithms are referred to [40].

4.5.1   A Master-Slave Strategy (MS)

Broadly speaking, in this parallel strategy, two types of processing elements are considered: the *master* processor, which makes global decisions and delivers data among the slaves, and the *slaves*, which execute different tasks simultaneously.

In our particular master-slave (MS) model (see Algorithm 6), the master processor executes UEGO_ cent.SASS sequentially. The parallelism has been included in new creation and optimization procedures (see Steps 5 and 8 in Algorithm 6). Next, they are briefly described.

- *Create_species_paral*: In this procedure, the master obtains a new offspring of candidate solutions for the leader sequentially. The parallelism comes from the simultaneous resolution of the medianoid problems to evaluate the new leader's trial points. To do so, the master divides the list of candidate solutions by the number of processors P and delivers the resulting sublists among all the processing elements (including itself). Each processing element applies UEGO_med to every received leader's facility to obtain the associated follower's location.

  The master processor does not receive information from the slaves until it has finished its work (first synchronization point). When it does so, it picks up all the follower sublists sent by the slaves, updates the candidate solutions list with such information and includes it in the *species_list$_i$*, with the radius value associated to the current level i.

- *Optimize_species_paral*: In this procedure, the master divides the *species_list$_i$* among all the processing elements (again including itself). Once the sublist has been received, each slave applies the local optimization process SASS+WLM to every leader's facility and executes UEGO_med to obtain the corresponding follower (see [40]). Finally, once the master finishes its work, it starts to receive the new species sublists from the slaves (second synchronization point).

Note that the synchronization points are imposed because the master is working with the whole *species_list_i*, or because it is needed to know the fitness value at the points of the leader before executing the next stage of the optimization procedure.

4.5.2 Improving the Quality of the Solution: A New Creation Procedure

Parallel algorithms can use more computational resources. Then, they can incorporate computationally intensive techniques that help at intensifying the search for more effective solutions. In [40], new alternative procedures to be included in UEGO_cent.SASS are studied. In particular, new creation methods that explore the search space deeper are analysed. After an exhaustive computational study, where several options are examined, it is found that the procedure named *Create_species*$_{21}$ is the best choice, since it maintains a good balance between the quality of the final solution and the execution time and memory resources required by UEGO_cent.SASS.

The idea behind this method is to take advantage of the non-consumed evaluations of the previous level. The budget per species in the *Optimize_species* procedure is $bo_i = n_i/M$. This means that there is a remainder of $n_i - bo_i \cdot length(species\_list_i)$ function evaluations in the optimization process, when the length of the *species_list_i* is not equal to the maximum allowed. Then, these function evaluations can be used to force the creation of more candidate solutions at the next level. Therefore, the budget per species in the level $i + 1$ is:

$$bc_{i+1} = \frac{new_{i+1} + n_i - bo_i \cdot length(species\_list_i)}{length(species\_list_{i+1})}.$$

As a consequence of the previous generation procedure, a huge list of candidate solutions is obtained. To reduce the list length while keeping the most promising solutions, a fusion procedure with the radius set to $2R_i$ is applied.

This new creation procedure makes the sequential UEGO_cent.SASS run out of memory most of the times. Then, to be able to use it, high performance computers are required. In [40], this new proposal is checked with the *master-slave* parallel model, since this algorithm does not modify the behavior of the sequential version, i.e., it considers the same number of function evaluations and acts over the species in the same way as the sequential algorithm. For the studies, the use of two processing elements has been enough to solve all the problems. An exhaustive analysis has proved that the *Creation_species*$_{21}$ method can improve the objective value more than 1% in some instances, which is not a negligible value.

4.5.3 Efficiency Results of MS

In this subsection the behavior of MS is analyzed by solving a representative set of location problems. The settings (n, m, k) employed in this experiment can be

Table 3 Settings of the larger test problems

n	100			150			200		
m	1	2	5	1	3	7	2	5	10
k	0	0, 1	0, 2	0	0, 1	0, 3	0, 1	0, 2	0, 5

Table 4 Efficiency results

n	P	Av(Obj)	Av(T)	Eff(P, Q)
100	2	472.66	2512.24	–
	4	472.66	1218.48	1.03
	8	472.67	580.96	1.08
	16	472.66	271.28	1.06
	32	472.66	152.44	1.03
150	4	646.90	2271.08	–
	8	646.90	1161.28	0.99
	16	646.90	582.28	0.98
	32	646.90	295.71	0.96
200	8	850.70	964.53	–
	16	850.70	474.53	1.02
	32	850.70	238.74	1.01

seen in Table 3. For every setting, five problems are generated. Furthermore, all the instances are solved five times and average values are considered.

Table 4 shows average results (for all the values of m and k) for each value of n and P. In the column labelled $Av(Obj)$, the average objective function value is given, in $Av(T)$ the average computational time and in the last column $Eff(P, Q)$, efficiency values are given.

Results reveal how costly solving the centroid problem is. As can be seen in Table 4, the higher the number of demand points of the problem at hand, the larger the minimum number of processing elements required to solve it. Nevertheless, the performance of the parallel algorithm is good, i.e. its efficiency is larger than the ideal one for problems with 100 and 200 demand points, and very close to ideal for problems with $n = 150$.

5   A Model with Costs and Variable Demand

5.1   The Model

The model considered in this section, introduced in [42], extends the previous model by relaxing the assumption that the demand is fixed. On the contrary, an endogenous (variable) demand is contemplated so that it varies depending on several factors. In real problems, for example, consumer expenditures on services or products that are offered by the facilities may increase depending on different reasons related to the location of the new facility. So, opening new outlets may increase the overall utility of the product. Also, the 'marketing presence' of a product may be increased with the marketing expenditures resulting from the new facilities. Another thing that can happen is that some consumers who did not patronize any of the facilities may now be induced to do so. The quality of the facilities may also modify consumer expenditures because a better service usually leads to more sales. The fact that the demand is endogenous is commonly disregarded in literature, usually due to the difficulty of the problems to be solved (see [41]).

The demand at a demand point p_i is now assumed to be a function of $U_i(nf_1, nf_2) = u_{i,nf_1} + u_{i,nf_2} + \sum_{j=1}^{m} u_{i,j}$, in the form

$$w_i(U_i(nf_1, nf_2)) = w_i^{\min} + incr_i \cdot e_i(U_i(nf_1, nf_2)),$$

where $incr_i = w_i^{\max} - w_i^{\min}$, and $w_i^{\max}$ (resp. $w_i^{\min}$) denotes the maximum (resp. minimum) possible demand at p_i. Function $e_i(U_i(nf_1, nf_2))$ can be interpreted as the share of the maximum possible increment that a customer decides to spend given a location scenario.

The objective functions Π_2 for the follower problem and Π_1 for the leader one, are formulated as in Sect. 4.1 (see (6) and (7), respectively), although the market share function expressions (M_l) contain the variable demand function $w_i(U_i(nf_1, nf_2))$ instead of the constant $\widehat{w}_i$:

$$M_2(nf_1, nf_2) = \sum_{i=1}^{n} w_i(U_i(nf_1, nf_2)) \frac{u_{i,nf_2} + \sum_{j=k+1}^{m} u_{i,j}}{u_{i,nf_1} + u_{i,nf_2} + \sum_{j=1}^{m} u_{i,j}},$$

$$M_1(nf_1, nf_2) = \sum_{i=1}^{n} w_i(U_i(nf_1, nf_2)) \frac{u_{i,nf_1} + \sum_{j=1}^{k} u_{i,j}}{u_{i,nf_1} + u_{i,nf_2} + \sum_{j=1}^{m} u_{i,j}}.$$

The operating costs also are modified to include the variable demand in the $\Phi_l^i(d_i(z_l))$ functions, so that now

$$\Phi_l^i(d_i(z_l)) = Aver_{A_i}(w_i(U_i(nf_1, nf_2)))/((d_i(z_l))^{\phi_l^{i0}} + \phi_l^{i1}).$$

$Aver_{A_i}(w_i(U_i(nf_1, nf_2)))$ stands for the average value of $w_i(U_i(nf_1, nf_2))$ over the feasible set and can be thought of as an estimation of the demand at p_i by a fixed number (see [41] for more details about how to compute this average). In [42] *linear expenditures* is considered, i.e., $w_i^{\min} = 0$, $w_i(U_i(nf_1, nf_2)) = w_i^{\max} \cdot e_{i_1}(U_i(nf_1, nf_2))$, where $e_{i_1}(U_i(nf_1, nf_2)) = q_i U_i(nf_1, nf_2)$, with q_i a given constant such that $q_i \leq 1/U_i^{\max}$, where $U_i^{\max}$ is the maximum utility that could be observed by a customer at i.

Certainly, other functions could be defined depending on the real problem considered, and for each real application the most appropriate F_l and G_l functions should be discovered. In [49] a pseudo-real application to the case of the location of supermarkets in the Autonomous Region of Murcia, in Southern Spain, can be found. Although in that paper the demand was assumed to be exogenous (fixed) and no reaction from the competitor was expected, the parameters and functions have the same meaning as those in this section.

It must be emphasized that although the objective function of the follower's problem with exogenous demand is multimodal, it tends to be smoother than the one of the follower's problem with endogenous demand, which has much more local optima and whose landscape is much steeper. Consequently, the complexity of the centroid problem is greatly increased due to the endogenous demand assumption.

5.1.1   A Real Example

In order to show the difficulty of the problem at hand, and its differences with the exogenous demand case, in [42] the quasi-real example introduced in [49] dealing with the location of supermarkets in an area around the city of Murcia was solved. There are five supermarkets in the area: three from a first chain, 'E', and two from another chain, 'C'. Two problems have been considered: the first one assumes that the leader belongs to chain 'E' and the second one assumes that it belongs to chain 'C'. Each problem was solved both considering fixed and variable demand. The numerical results are shown in Table 5. The interested reader can find a detailed description of the example with some illustrative figures in [42].

As can be seen, when the leader belongs to chain 'E', in the exogenous demand case, the optimal location for the leader is near the city of Alcantarilla ($x_1 = 3.303, y_1 = 6, 433$), with a quality of 0.5. At that location, the market share captured by the new leader's facility is $m_1 = 2.112$, which coincides with the 5.94% of the total market share. Taking into consideration all its facilities, chain 'E' obtains 53.22% of the market, and a profit $\Pi_1 = 593.352$. The location for the follower's facility is near the city of Molina ($x_1 = 3.259, y_1 = 4.285$), with a quality of 3.696, where it captures 20.04% of the total market share. However, the results are rather different for the endogenous case, where the leader's optimal location is in the suburb of Puente Tocinos ($x_1 = 5.407, y_1 = 5.798$), in Murcia city, with a quality of 0.961. The market share captured by the facility is 0.419, which is only 5.94% of the total one. The whole chain obtains 43.68% of the market and a smaller profit $\Pi_1 = 73.454$. The location for the follower's facility is near the suburb of San Benito ($x_1 = 5.190, y_1 = 6.276$), in Murcia city, with a quality of 0.571, where it only captures 3.875% of the total market share.

For the second problem, where it is assumed that chain 'C' is the leader, then, in the exogenous demand case, the optimal location for the leader is near the city of Orihuela, with a quality of 3.277, where the facility gets 17.57% of the total market share. The location for the follower's facility is near the city of Alcantarilla, with a quality of 0.5, where it captures 6.15% of the total market share. However, the leader's optimal location in the endogenous demand case is near the suburb of San Benito, in Murcia city, with a quality of 1.042 and only captures 6.52% of the total

Table 5 Examples

Demand	nf_1	M_1	m_1	Π_1	nf_2	M_2	m_2	Π_2
Leader: chain E								
Exogenous	(3.303, 6.433, 0.500)	18.915	2.112	593.352	(3.259, 4.285, 3.696)	16.625	7.123	461.776
Endogenous	(5.407, 5.798, 0.961)	2.807	0.419	73.454	(5.190, 6.276, 0.571)	3.618	0.249	101.563
Leader: chain C								
Exogenous	(8.487, 3.026, 3.277)	15.961	6.247	442.122	(3.274, 6.441, 0.500)	19.579	2.187	614.652
Endogenous	(5.368, 6.166, 1.042)	3.822	0.453	106.320	(5.298, 6.228, 0.571)	2.6378	0.2489	70.227

market share. The location for the follower's facility is near the suburb of San Benito too, with a quality of 0.571, where it captures 3.88% of the total market share.

These two examples indicate how important it is to consider endogenous demand. As can be seen, depending on whether endogenous or exogenous demand is considered, the maximum profit for a chain is obtained at different locations and with different qualities. Additionally, it is interesting to remark that even the percentage of market share captured by the chains may change to the point that the chain obtaining more profit may be the competitor's one.

5.2   Solving the Centroid Problem

Considering the algorithms proposed for solving the centroid problem with exogenous demand (see Sect. 4.3), the following three algorithms are implemented to solve the centroid problem with endogenous demand [42]: a grid search procedure, a multistart method named MSH, and an evolutionary algorithm named TLUEGO. MSH and TLUEGO require the use of a local optimizer. In particular, a local optimizer based on SASS and WLM has been designed. In fact, two variants of the local optimizer have been implemented, leading to two versions of MSH and TLUEGO. Next we describe the corresponding algorithms.

5.2.1   The Local Optimizer SASS+WLMv

In [39], after studying several strategies, a local procedure SASS+WLMv, similar to SASS+WLM in Sect. 4.3 is proposed. The main differences between this local algorithm and SASS+WLM are:

- The Weiszfeld-like algorithm used now for updating the follower's facility is WLMv, a variant of WLM to take the variability of the demand into account (see [41]). Similar to what was considered for WLM (see Sect. 4.2.1), WLMv stops when either two consecutive iterations are closer than the tolerance $\epsilon_1 = \epsilon_2 = 0.0001$, or when a maximum number of $ic_{max} = 400$ iterations is reached.
- Due to the high increment in the complexity of the problem when using endogenous demand, the WLMv algorithm is not as reliable as the corresponding method WLM for the fixed demand case. Consequently, due to the cumulative error, a large number of consecutive iterations in SASS could give rise to the leader achieving overestimated solutions. To deal with this drawback, the number of consecutive iterations in SASS+WLMv has been reduced to only 15. In addition, in order to compensate the possible error obtained using WLMv, after every 15 iterations, the medianoid problem is solved accurately using a reliable global optimizer. Two global optimizers have been considered: iB&B [18] or UEGO_med (see Sect. 4.2), resulting in two versions of the local optimizer.

5.2.2 TLUEGO: A Two-Level Evolutionary Global Optimization Algorithm

The evolutionary algorithm TLUEGO is rather similar to the UEGO_cent.SASS algorithm introduced in Sect. 4.3 for the fixed demand case. The main differences are the following:

- *Create_species procedure:* In the same way that for UEGO_cent.SASS, after the *creation* procedure it is very important to precisely evaluate the fitness of the new species. In this problem, two alternative algorithms to compute a reliable follower solution have been implemented: iB&B or UEGO_med.

- *Optimize_species procedure:* The *local optimizer* algorithm used in TLUEGO is SASS+WLMv. There is another difference: this local optimizer is executed twice in order to have more chances of obtaining a better point. The input parameter value of σ_{ub} passed to SASS+WLMv is always (the two times it is called) the radius associated to the calling species. Therefore, the scope of the local optimizer coincides with the region covered by the species. As it has been mentioned in 5.2.1, the execution of SASS+WLMv implies that a reliable optimization algorithm, iB&B or UEGO_med, is run at the end of the algorithm (Step 9 in Algorithm 7). As a result, the inclusion of iB&B or UEGO_med in TLUEGO derives two algorithms for solving the centroid problem, TLUEGO_BB and TLUEGO_UE, respectively. The reader is referred to [42] for a more detailed description of these procedures.

Algorithm 7: Algorithm SASS+WLMv$(nf_1, nf_2, iter_{max}(= 15), \sigma_{ub})$

1: Initialize SASS parameters. Set $iter = 1, nf_1^{opt} = nf_1, \Pi_1^{opt} = \Pi_1(nf_1, nf_2)$.

2: **while** $iter \leq iter_{max}$ **do**

3:     Update SASS parameters considering the previous successes at improving the objective function value of the leader.

4:     Generate a location for the leader $nf_1^{(iter)}$ within the updated radius.

5:     Solve the corresponding medianoid problem using WLMv and let $nf_2^{(iter)}$ denote the solution obtained.

6:     **if** $\Pi_1(nf_1^{(iter)}, nf_2^{(iter)}) > \Pi_1^{opt}$ **then**

7:         set $nf_1^{opt} = nf_1^{(iter)}$ and $\Pi_1^{opt} = \Pi_1(nf_1^{(iter)}, nf_2^{(iter)})$.

8:     $iter = iter + 1$.

9: Compute the corresponding follower nf_2^{opt} for nf_1^{opt} using either iB&B or UEGO_med.

10: **if** $\Pi_1(nf_1^{opt}, nf_2^{opt}) > \Pi_1(nf_1, nf_2)$ **then**

11:     return (nf_1^{opt}, nf_2^{opt})

12: **else**

13:     Return (nf_1, nf_2).

5.2.3   MSH: A Multistart Heuristic Algorithm

The MSH algorithm consists of randomly generating *MaxStartPoints* feasible candidate solutions for the leader and then applying a local optimizer to each one in order to improve it to an optimized leader solution. The final solution provided by the algorithm will be obtained by selecting the solution with best objective function value.

For this problem with exogenous demand, the considered local optimizer has been SASS+WLMv (see Algorithm 7). In order to provide a better balance between exploitation and exploration of the search space, this method has also been executed twice as in TLUEGO, but with different values for σ_{ub} because the multistart heuristic does not have a cooling process for the radius. In the first call, a value of $\sigma_{ub} = 2.083895$ (the one corresponding to level 10 in TLUEGO) was considered. This value was chosen because then the initial random candidate solutions in the multistart strategy can cover the whole searching space, and at the same time, they can search on an area small enough so that the local procedure can find a good local optimum. In the second call, a value of $\sigma_{ub} = 0.162375$ (level 23 in TLUEGO) was used to improve the quality of the local optima obtained with the first call. These σ_{ub} values were selected after doing some preliminary studies, in which eight problems of different sizes were solved trying different strategies for the heuristic algorithm.

As in TLUEGO, two versions of the MSH method have been implemented: MSH_BB and MSH_UE. They differ in whether iB&B or UEGO_med is used as a method of computing the follower nf_2^{opt} in Step 9 of Algorithm 7.

5.2.4   Computational Studies

To study the performance of the algorithms, a set of 24 problems has been generated varying the number n of demand points, the number m of existing facilities and the number k of those facilities belonging to the leader's chain. The actual settings (n, m, k) employed are detailed in Table 6. For each setting, the problem has been generated by randomly choosing its parameters within given intervals. In all the problems, $S_1 = S_2 = ([0, 10], [0, 10])$ and $\alpha_1, \alpha_2 \in [0.5, 5]$.

For every heuristic algorithm, each problem has been solved ten times and average values have been computed. However, the heuristic GS has only been run once and the results obtained in that run (no average results) are given. All results for all the problems are shown in [42]. In this section only some average results for $n = 15$ and $n = 50$ are shown in Table 7. In the column labeled '*Time*', the average

Table 6 Settings of the test problems

n	15			25			50		
m	2	5	10	2	5	10	2	5	10
k	0,1	0,1,2	0,2,4	0,1	0,1,2	0,2,4	0,1	0,1,2	0,2,4

Table 7 Results for the problems with $n = 15$ and $n = 50$

(n)	Algorithm	Time	Max dist	Objective function			
				Min	Av	Max	Dev
15	TLUEGO_BB	226	0.015	15.478	15.478	15.479	0.000
	TLUEGO_UE	891	0.009	15.478	15.478	15.479	0.001
	MSH_BB	258	1.164	15.350	15.413	15.453	0.038
	MSH_UE	1091	0.516	15.290	15.409	15.469	0.067
	GS	490,338	–	–	15.445	–	–
50	TLUEGO_BB	9470	0.186	39.866	39.960	40.065	0.081
	TLUEGO_UE	8259	0.185	39.912	40.072	40.174	0.102
	MSH_BB	11,090	2.855	25.597	31.329	37.722	4.508
	MSH_UE	9911	2.769	23.769	33.088	38.084	5.346
	GS	3,003,794	–	–	37.280	–	–

TLUEGO_BB ($\epsilon_1 = \epsilon_2 = 0.0001$), TLUEGO_UE, MSH_BB and MSH_UE and GS

time in the ten runs (in seconds) of each problem is shown; the '*MaxDist*' column indicates the maximum Euclidean distance (for the three variables (x_1, y_1, α_1)) between every pair of solutions provided by the algorithm in different runs, which gives an idea of how far these solutions can be; in the following three columns, the minimum, the average and the maximum objective value are computed. Finally, in the '*Dev*' column, the standard deviation is shown. As can be seen in these tables, two versions of TLUEGO and MSH algorithms have been executed. It is worth mentioning that the number of times that MSH_BB (resp. MSH_UE) was allowed to repeat its basic local optimizer was chosen so that the CPU time employed by MSH_BB (resp. MSH_UE) was, on average (when considering all the problems with the same value of n), similar to the CPU time employed by TLUEGO_BB (resp. TLUEGO_UE) or a bit higher. In particular, for the problems with 15 and 50 demand points, the number of starting points were 150 and 250, respectively.

Analyzing the results, it can be seen that the method used to reliably solve the medianoid problem does not seem to have an influence on the quality of the final solution, i.e., TLUEGO and MSH behave similarly, regardless whether iB&B or UEGO_med is employed. This is due to the reliability of UEGO (in spite of its metaheuristic nature). The iB&B technique is faster than UEGO_med for small size problems ($n = 15$), which directly reduces the execution time of both TLUEGO and MSH. Specifically, the use of iB&B reduces the computing time of TLUEGO_BB by 74.6% as compared to TLUEGO_UE. A similar behavior in computing time can be seen in MSH when iB&B is used instead of UEGO_med. Nevertheless, for medium size problems (with $n = 50$ demand points), TLUEGO_UE and MSH_UE reduce the computing time as compared to TLUEGO_BB and MSH_BB, by 12.79% and 10.63%, respectively. These results are also consistent with the ones showed in [37], where it was observed that the increase of requirements for iB&B with the size of the problem was greater than for UEGO_med.

Focusing now on the strategies proposed to solve the current centroid problem, it can be stated that TLUEGO (in both versions) is the algorithm achieving the best results. Their average objective function values are always higher than the ones provided by both MSH and GS. It is also remarkable that the minimum objective function value found by TLUEGO in the ten runs is always better than the average values obtained by both MSH and GS (see columns '*Min*' and '*Av*'). Additionally, TLUEGO is the most robust algorithm in the sense that it usually attains the same solution in all the runs, whereas MSH is more erratic, and can provide different solutions in each run (see the values of '*MaxDist*' and '*Dev*').

5.3   Influence of the Fuse Process in the Creation Procedure

Taking into account the main structure of TLUEGO, based on UEGO_cent.SASS algorithm, it can be seen that in the creation procedure, for every species in the list, a set of possible new solutions is computed, fused and evaluated with the objective of finding new promising species, and therefore increasing the species-list. This creation process is applied independently to each species as no relation among species exists.

Taking into consideration that the evaluation of a single species in TLUEGO requires intensive computational effort, since it implies the execution of another expensive optimization algorithm (UEGO_med or iB&B) to obtain the optimal location of the follower (by solving the corresponding medianoid problem), TLUEGO had to be designed to maintain a small-size species-list. This was done by including a '*fuse*' process just after the creation of candidate solutions and before the evaluation of the resulting ones.

However, it is known that working with larger species-list sizes helps to explore the search space deeply and consequently to obtain better solutions. With this aim, in this section, new creation procedures are proposed, where the fuse process is relaxed in part by modifying the threshold distance to apply the fusion of two species. Now two species will be fused if the distance between their centers is smaller than the new thresholds R_t, $R_t/2$ or 0 instead of $2R_t$. In what follows, only TLUEGO_UE will be used, since it can solve larger instances. It will simply be denoted by TLUEGO. For the analysis at hand, only medium size problems have been considered, i.e. $n = 50, 100$ (the actual settings can be seen in Table 8).

Considering that each run of TLUEGO may provide a different solution, each problem has been solved ten times and average values have been computed. Table 9 shows the average results obtained by the algorithms considering all the

Table 8 Settings of the test problems

n	50			100		
m	2	5	10	2	5	10
k	0,1	0,1,2	0,2,4	0,1	0,1,2	0,2,4

Table 9 Effectiveness evaluation of the fuse process in TLUEGO (sequential algorithm) for problems with $n = 100$ and $n = 50$ demand points

n	Threshold	Time	MaxDist	Π_1	Dev	Dif Π_1	DifSol
50	$2R_t$	10,993	0.520	148.316	0.578	–	–
	R_t	17,689	0.307	149.616	0.177	0.782	1.812
	$R_t/2$	18,686	0.129	150.296	0.113	1.235	2.364
	0	22,898	0.135	151.002	0.064	1.794	2.940
100	$2R_t$	32,029	0.755	177.364	1.992	–	–
	R_t	52,125	0.146	183.341	0.490	3.260	4.221
	$R_t/2$	56,932	0.133	185.710	0.272	4.562	5.998
	0	65,470	0.056	186.551	0.058	5.033	7.027

configurations for the problems with $n = 50$ and $n = 100$, respectively. In [42] a complete set of tables with detailed results for each configuration can be found. The first column gives the size of the problem. The second one indicates the threshold value used in the fuse process. In the third column, the average time in the ten runs (in seconds) is computed. The *MaxDist* column provides the maximum Euclidean distance [for the three variables (x_1, y_1, α_1)] between any pair of solutions provided by the algorithm in the ten runs, which gives an idea of how far the solutions computed by the algorithm in different runs can be. The average objective function value (column Π_1) in the ten runs and the corresponding standard deviation (column *Dev*) are given next. Column *Dif* Π_1 shows the relative improvement in the objective function value between the solution obtained by the algorithms when a threshold different from $2R_t$ is used as compared to the result obtained when using $2R_t$. The final column shows the relative difference between the solutions.

As can be seen, the CPU time increases as the threshold decreases, and when this is set to 0, the time is more than double as compared to the $2R_t$ case. The algorithm also becomes more robust (see the decrease in columns *Dev*), in the sense that the objective function value at different runs are more similar. In addition, analysing column Π_1 it can be deduced that the quality of the solution also becomes better. Regarding the relative improvement in the objective function value, it can be seen that for the problems with $n = 50$ demand points is moderate, with an average of 1.794%. However when the threshold is set to 0, for the problems with $n = 100$ it attains a significant 5.033%. This clearly shows that the smaller the threshold, the better the solutions are. Unfortunately this is at the cost of increasing the CPU time and the memory requirements.

5.4 High Performance Computing

Due to the high computational cost of TLUEGO, which is even higher than that of UEGO_cent.SASS, a parallelization of the algorithm is required, especially if real problems, with more demand points than the studied in the previous section must be solved. In [1], three programming paradigms for the parallelization of TLUEGO are designed. More specifically, a pure message passing paradigm, a pure shared memory programming model and a hybrid one which combines message passing with shared memory are implemented and their efficiency and effectiveness are analyzed and compared. Results showed that both pure message passing and pure shared memory paradigms have almost the same performance, while the hybrid one shows less efficiency though it can exploit all computational resources of the parallel architecture.

Considering that TLUEGO structure is similar to UEGO_cent.SASS, the message passing algorithm is based on a *master-slave strategy* like the one described in Sect. 4.5. For this reason only the main features of pure shared memory strategy are detailed here. Readers interested in a deep description of the three strategies as well as in the performance comparison among them are referred to [1].

5.4.1 Shared Memory Programming for TLUEGO: SMP_TLUEGO

For the implementation of this parallel strategy, OpenMP has been selected, since it is a portable and scalable model, and gives programmers a simple and flexible interface for developing parallel applications.

Concerning the parallel model, it can be considered a *pseudo* master-slave technique, similar to the MS described in Sect. 4.5. OpenMP includes mechanisms to distribute the species list among the different processors without the existence of a master processor. Therefore there does not exist a master processor which globally controls the algorithm and manages the species list. This task can be done in parallel by all the processors. However, the existence of a kind of pseudomaster processor to be in charge of applying the *Selection* procedure and updating the species list that will be accessible to all processors, is still necessary. Accordingly, the parallelism is applied to the evaluation of the new candidate solutions in the *Creation* and *Optimization* procedures. Consequently, new creation and optimization procedures have also been designed. They are briefly described next.

The parallel algorithm developed considers that the species-list is stored in shared memory. When the *Create_species_paral* is executed, each processor picks up a new single species and evaluates it. Once a processor has finished this task, it collects another species. This cyclical process finished when all the new offspring are evaluated. Notice that mutual exclusion is not needed because each processor accesses different memory areas.

The *Optimize_species_paral* procedure maintains a similar structure to the previous method *Create_species_paral*. But instead of only evaluating the species, it applies the local search procedure. Considering that the number of function evaluations required to optimize a single species, and therefore, the computational load assumed by each processor, may be quite different, this strategy of selecting the species one by one helps to balance the computational burden and to reduce the waiting time of the processors.

Table 10 Settings of the test problems

n	50			100		
m	2	15	25	2	15	25
k	0,1	0,5,10	0,7,15	0,1	0,5,10	0,7,15

Table 11 Efficiency results for SMP_TLUEGO

P	n	Time	Eff(P)	n	Time	Eff(P)
1	100	65,470	–	500	565,358	–
2		32,878	1.00		283,707	1.00
4		16,928	0.97		143,416	0.99
8		8703	0.94		73,065	0.97

5.4.2 Efficiency Results of SMP_TLUEGO

In this subsection the behavior of SMP_TLUEGO is analyzed by solving a set of 24 problems whose settings can be found in Table 10. For every setting one problem was generated. Additionally, all the instances are solved ten times and average values are considered.

Table 11 shows, for the problems with $n = 100$ and $n = 500$ demand points, the average computing time (in secs.) and the mean efficiency Eff(P) obtained. As can be seen, SMP_TLUEGO has either optimal or near-optimal efficiency for up to $P = 8$ processors. For a given n the efficiency values slightly decrease as the number of processors P increases. Notice, however, that the algorithm is *scalable*, as it shows a better performance (see Eff(P) columns) when the problem size increases, i.e. the efficiency improves with higher n values.

6 Solving the Models with Costs Exactly

In this section we propose an exact solution method for the problems described in sections 4 and 5, i.e. when operational costs are taken into account. As already mentioned, the B&B method described in Sect. 3 works only when no costs are present, that is, the zero-sum property holds for the objective functions of the leader and follower. The method we propose to solve these harder problems exactly is a generalization of the algorithm presented in [48]. In that paper almost the same problem is solved exactly on networks, although with fixed qualities. Here, we propose a modification of this method to be able to solve the problem on the plane having the quality as additional variables for the new facilities.

In [48] a B&B method is used to solve the leader problem, while in an embedded way another B&B was used to refine the follower. The main difference between this method and Algorithms 2 and 1 is that the follower problem has to be solved for a set of leader placements instead of for a leader point. This is much more challenging, and it may even be impossible if the aim is to solve the problem with a small accuracy. Therefore, instead of solving the follower problem in the inner B&B to optimality, its searching set is only refined, and the solution (set of sets) is stored together with the leader set. The method proposed next differs from that in [48] mainly in the searching space and the solution sets, that instead of being segments of edges of the network, they are now 3-dimensional boxes (vector of intervals) in $\mathbb{R}^3$.

6.1 Overcoming the Difficulty of the Lack of the Zero-Sum Property

In Sect. 3 we have already seen that when the objective function is the market share (no costs are present), and the qualities of the facilities are given parameters, the problem can be solved efficiently by a B&B method. The key point there is the zero-sum property of the objective functions: minimizing the objective of the leader, one directly maximizes the objective of the follower and vice-versa. What makes the method very efficient is that although (reverse) medianoid problems have to be solved to obtain bounds, the other new facility is always fixed to a point. This is no longer the case when costs are taken into account. It may even happen that changing the location of the follower increase both the leader and the follower objective. Therefore the result of Lemma 1 cannot be used directly, and so a new trick is needed to overcome this difficulty.

When operational costs are present, for the bound calculations of the leader, all possible locations (and qualities) of the follower have to be considered. On the one hand, until the follower is not enclosed tightly in a set of boxes, it might mean that the obtained bounds are very loose. On the other hand, until the leader box is not small enough, it is not possible to enclose the follower tightly. Thus, what is needed is a good and possibly cheap bound calculation procedure in order to overcome the above problem. One promising approach is to use *interval bounds*, as done in [48].

6.2 Interval Arithmetic Bounds

We propose to use *Interval Arithmetic* to obtain lower and upper bounds of the objective functions automatically when one or both facilities are in boxes. The main idea of Interval Analysis is to change all real arithmetic operators and elementary real functions to their interval versions. As a result, an interval containing all possible results from points from the input intervals is obtained, maybe with some overestimation. See [21] for details of interval analysis in global optimization.

Let us denote intervals with capital letters, e.g. $X = [\underline{x}, \overline{x}]$, where $\underline{x} \le \overline{x}$ are the lower and upper bounds of X, respectively.

For a given box NF_l containing a new facility nf_l, an interval U_{i,nf_l} containing the utility of any point within NF_l can be computed as

$$U_{i,nf_l} = [\underline{u_{i,nf_l}}, \overline{u_{i,nf_l}}] = [\underline{\alpha_l}/g_i(\overline{d_i(Z_l)}), \overline{\alpha_l}/g_i(\underline{d_i(Z_l)})]$$

where

$$\underline{d_i(Z_l)} = \sqrt{(\max\{\underline{x_l} - p_{i1}, p_{i1} - \overline{x_l}, 0\})^2 + (\max\{\underline{y_l} - p_{i2}, p_{i2} - \overline{y_l}, 0\})^2},$$

$$\overline{d_i(Z_l)} = \sqrt{\max\{(\underline{x_l} - p_{i1})^2, (p_{i1} - \overline{x_l})^2\} + \max\{(\underline{y_l} - p_{i2})^2, (p_{i2} - \overline{y_l})^2\}}.$$

Given a fixed box (or a point) $\widetilde{NF_2}$ for the follower, an upper bound of Π_1 at the box NF_1 can be calculated with interval arithmetic as

$$UB(\Pi_1(NF_1, \widetilde{NF_2})) = c \cdot UB(M_1(NF_1, \widetilde{NF_2})) - LB(G_1(NF_1)),$$

where the upper bound of the market share is given by the formula

$$UB(M_1(NF_1, \widetilde{NF_2})) = \sum_{i=1}^{n} \widehat{w}_i \frac{\overline{u_{i,nf_l}} + \sum_{j=1}^{k} u_{ij}}{\underline{u_{i,nf_1}} + \underline{u_{i,nf_2}} + \sum_{j=1}^{m} u_{ij}},$$

when the demand is fixed, and

$$UB(M_1(NF_1, \widetilde{NF_2})) = \sum_{i=1}^{n} w_i^{\max} q_i(\overline{u_{i,nf_l}} + \sum_{j=1}^{k} u_{ij}),$$

when the demand is endogenous but linear as introduced in Sect. 5.

The lower bound $LB(G_1(NF_1))$ of the operational cost function G_1, when it has the form

$$G_1(nf_1) = \sum_{i=1}^{n} w_i/((d_i(z_1))^{\phi_1^{i0}} + \phi_1^{i1}) + \exp(\alpha_1/\xi_1^0 + \xi_1^1) - \exp(\xi_1^1)$$

(where w_i stands for $\widehat{w}_i$ when the demand is fixed, and for $w_i(U_i(nf_1, nf_2))$ when the demand varies) can be computed as

$$LB(G_1(NF_1)) = \sum_{i=1}^{n} \frac{\widehat{w}_i}{\overline{d_i(Z_1)}^{\phi_1^{i0}} + \phi_1^{i1}} + \exp(\underline{\alpha_1}/\xi_1^0 + \xi_1^1) - \exp(\xi_1^1)$$

when the demand is fixed, and as

$$LB(G_1(NF_1)) = \sum_{i=1}^{n} \frac{w_i^{\min}}{d_i(Z_1)^{\phi_1^{i0}} + \phi_1^{i1}} + \exp(\underline{\alpha_1}/\xi_1^0 + \xi_1^1) - \exp(\xi_1^1)$$

when it varies.

Of course, if an upper bound for the leader's profit is required when the follower is in a set of boxes $\mathbb{NF}_2$, it can be obtained as

$$UB(\Pi_1(NF_1, \mathbb{NF}_2)) = c \cdot \max_{NF_2 \in \mathbb{NF}_2} UB(M_1(NF_1, NF_2)) - LB(G_1(NF_1)).$$

The interval arithmetic lower bound of the profit can be obtained by interchanging upper bounds and lower bounds in the above formulae. The bounds for the follower are straightforward by the rules above.

One can see that even those computations might be time-consuming for obtaining an upper or a lower bound. However, notice that in the fixed demand case, we can still use the zero-sum property of the market share for its bound calculations, so that if bounds for the follower's market share are known, they can be used directly for the leader's bounds on the market share and vice-versa.

6.3   Solution Method

A B&B method is designed to solve the leader's problem, and consequently the follower's problem as well. The main goal of the method is for every subproblem to simultaneously tighten the set containing the global optimizer of the leader and the set that contains all the global optimizers for the follower problem.

Without loss of generality, it is assumed that the feasible set of both the leader and the follower is a box. We define subproblems of the leader as boxes. For a given box of the leader, the follower's possible position can be in many places, and until the leader is not enclosed tightly, the follower can only be bounded to a set of boxes. Therefore, for every box of the leader we need to store the subboxes that may contain the global optimal solutions of the follower. Hence, a partial solution or subproblem of the leader refers to a box containing the leader and the set of boxes that contain the corresponding solution of the follower problem.

An inner B&B method tightens the boxes of the follower, and a main (outer) B&B method tightens the boxes of the leader. Thus, lower and upper bounds for the leader's (follower's) profit are needed when the follower (leader) is enclosed in a box. For the calculation of the lower and upper bounds of the follower in a given box NF_2, its corresponding single leader's box NF_1 is taken into account. These lower and upper bounds are $LB(\Pi_2(NF_1, \widehat{nf}_2))$ and $UB(\Pi_2(NF_1, NF_2))$, respectively, where $\widehat{nf}_2 \in NF_2$ is a feasible solution within the follower's box. For the calculation of the bounds for a leader's box NF_1, every box of the follower corresponding to it has to be considered, i.e. $LB(\Pi_1(\widehat{nf}_1, \mathbb{NF}_2))$ and $UB(\Pi_1(NF_1, \mathbb{NF}_2))$, where $\widehat{nf}_1$ is a feasible solution in the leader's box and $\mathbb{NF}_2 \ni NF_2$ the set of the corresponding boxes of the follower.

6.3.1 Inner B&B

Both the leader's and their corresponding follower's boxes need to be refined for the algorithm to converge. The inner B&B takes care of the refinement of the follower's boxes.

The termination criterion of the inner B&B is to have the size of each follower's box at least as small as the corresponding leader's box. The algorithm returns the modified list of the boxes of the follower. The selection rule chooses the largest box, while the branching rule bisects the box perpendicularly to the coordinate direction of maximum width.

Given a leader box, this method is applied to the set of follower boxes associated to it, until the corresponding follower's sub-boxes have a size smaller than or equal to that of the leader's box. Each time a new leader box is created, the inner B&B is run until its follower's boxes are refined.

6.3.2 Outer B&B

The outer B&B refines the leader's boxes and calls the inner B&B method for each new box of the leader. Recall that a subproblem of the leader is a box with the corresponding set of boxes for the follower. Thus, the initial subproblem is the starting box of the leader, and the starting box of the follower. However it might be more efficient to make a pre-division at the very beginning, as the first lower and upper bounds obtained by the algorithm are usually useless, but computing them needs time.

The output is a set of boxes containing any global optimizer, and the interval containing their objective values contains the global optimum of the problem. The selection rule selects the leader box with the highest upper bound of the leader's profit, while the branching rule bisects the leader's box perpendicularly to the coordinate direction of maximum width and leaves the follower's boxes unchanged but duplicated for the new boxes of the leader. The algorithm stops when the interval containing the objective values of all leader's boxes gets smaller than a prescribed tolerance or the size of all the boxes becomes smaller than another tolerance parameter.

6.4 Algorithm

The pseudocode of the inner and outer B&B algorithms are given in Algorithm 8. For the sake of simplicity let us denote the objective function as Π (Π_1 for the outer and Π_2 for the inner B&B).

Algorithm 8: The inner and outer B&B methods

 1: **Input:** Λ, GLB for the inner B&B
 2:     $\Lambda = \{S\}$, $GLB = -\infty$ for the outer B&B
 3: Remove all NF^i from Λ with $UB^i < GLB$
 4: **while** $\Lambda \neq \emptyset$ **do**
 5:     Select NF from Λ
 6:     Bisect NF into NF^1 and NF^2
 7:     **for** $i := 1$ **to** 2 **do**
 8:         Determine an upper bound UB^i on NF^i
 9:         **if** not $UB^i < GLB$ **then**
10:             Compute a lower bound LB^i of Π at midpoint(NF^i)
11:             **if** $LB^i > GLB$ **then**
12:                 $GLB := LB^i$, $BestPoint := $ midpoint(NF^i)
13:                 Remove all NF^j from Λ with $UB^j < GLB$
14:             **if** not TerminationCriterion(NF^i) **then**
15:                 **if** outer **then**
16:                     Call the inner B&B on the set of follower boxes of NF^i
17:                 $\Lambda := \Lambda \cup \{NF^i\}$
18:             **else**
19:                 $\Gamma := \Gamma \cup \{NF^i\}$
20: **Output:** Γ, *BestPoint*

In line 3 we remove each box known not to contain any global optimizer from list Λ. The main cycle of the general B&B method is listed from line 4 to line 19. The main difference of the outer B&B from the inner B&B is the call of the inner method added in lines 15 and 16. In fact, the additional differences between the inner and outer procedures are hidden in the bound calculations, as well as in the selection and termination rules.

The output of Algorithm 8 is the set of boxes which could not be eliminated and thus contain any global optimizer, and the point at which the best lower bound was achieved.

The proposed method should be tested on a set of test problems to know the size of the problems that it can solve, for both exogenous and endogenous demand. However, this is not the aim of this section, but to show that an exact algorithm can be designed even if operational costs are considered, the qualities are variables of the model and the demand is endogenous.

7   Conclusions and Future Research

Despite its inherent difficulty, facility location leader-follower (or Stackelberg) problems can be addressed when the location space considered is the plane, at least in its simple case, when only one new facility is going to be located by the leader and the follower. Exact (interval) branch-and-bound methods can be put to work for solving small instances, whereas evolutionary algorithms can handle large instances. If so required, parallel implementations of the algorithms can help to solve larger instances and with more accuracy.

Dealing with problems where more than one facility is to be located by the leader and/or the follower seems to still be a challenge when the location space is the plane. An extension which deserves to be explored is to allow the existing facilities to modify their quality, or even close some of them. Studying the problems with other patronizing behavior of customers is another line of future research. From the computational point of view, the design of high performance computing approaches for the exact branch-and-bound algorithms is also worth exploring.

Acknowledgements This research has been supported by grants from the Spanish Ministry of Economy and Competitiveness (MTM2015-70260-P, and TIN2015-66680-C2-1-R), the Hungarian National Research, Development and Innovation Office—NKFIH (OTKA grant PD115554), Fundación Séneca (The Agency of Science and Technology of the Region of Murcia, 19241/PI/14), Junta de Andalucía (P12-TIC301), in part financed by the European Regional Development Fund (ERDF). Juana López Redondo is a fellow of the Spanish 'Ramón y Cajal' contract program.

References

1. Arrondo, A.G., Redondo, J.L., Fernández, J., Ortigosa, P.M.: Solving a leader-follower facility problem via parallel evolutionary approaches. J. Supercomput. **70**(2), 600–611 (2014)
2. Bhadury, J., Eiselt, H.A., Jaramillo, J.H.: An alternating heuristic for medianoid and centroid problems in the plane. Comput. Oper. Res. **30**(4), 553–565 (2003)
3. Biesinger, B., Hu, B., Raidl, G.: Models and algorithms for competitive facility location problems with different customer behavior. Ann. Math. Artif. Intell. **76**(1), 93–119 (2016)
4. Daskin, M.S.: Network and Discrete Location: Models, Algorithms and Applications. Wiley, New York (1995)
5. Dempe, S.: Foundations of Bilevel Programming. Springer, New York (2002)
6. Dorta-González, P., Santos-Peñate, D.R., Suárez-Vega, R.: Spatial competition in networks under delivered pricing. Pap. Reg. Sci. **84**, 271–280 (2005)
7. Drezner, Z.: Competitive location strategies for two facilities. Reg. Sci. Urban Econ. **12**(4), 485–493 (1982)
8. Drezner, T.: Locating a single new facility among existing unequally attractive facilities. J. Reg. Sci. **34**(2), 237–252 (1994)
9. Drezner, T.: Optimal continuous location of a retail facility, facility attractiveness, and market share: an interactive model. J. Retail. **70**(1), 49–64 (1994)
10. Drezner, T., Drezner, Z.: Replacing continuous demand with discrete demand in a competitive location model. Nav. Res. Logist. **44**, 81–95 (1997)
11. Drezner, T., Drezner, Z.: Facility location in anticipation of future competition. Locat. Sci. **6**(1), 155–173 (1998)
12. Drezner, T., Drezner, Z.: Retail facility location under changing market conditions. IMA J. Manag. Math. **13**(4), 283–302 (2002)
13. Drezner, T., Drezner, Z., Kalczynski, P.: Strategic competitive location: improving existing and establishing new facilities. J. Oper. Res. Soc. **63**(12), 1720–1730 (2012)
14. Drezner, T., Drezner, Z., Kalczynski, P.: A leader–follower model for discrete competitive facility location. Comput. Oper. Res. **64**, 51–59 (2015)
15. Eiselt, H.A., Laporte, G.: Sequential location problems. Eur. J. Oper. Res. **96**(2), 217–231 (1996)
16. Eiselt, H.A., Laporte, G.,Thisse, J.F.: Competitive location models: a framework and bibliography. Transp. Sci. **27**(1), 44–54 (1993)

17. Fernández, J., Fernández, P., Pelegrín, B.: Estimating actual distances by norm functions: a comparison between the $l_{k,p,\theta}$-norm and the $l_{b_1,b_2,\theta}$-norm and a study about the selection of the data set. Comput. Oper. Res. **29**(6), 609–623 (2002)
18. Fernández, J., Pelegrín, B., Plastria, F., Tóth, B.: Solving a Huff-like competitive location and design model for profit maximization in the plane. Eur. J. Oper. Res. **179**(3), 1274–1287 (2007)
19. Fernández, J., Salhi, S., Tóth, B.G.: Location equilibria for a continuous competitive facility location problem under delivered pricing. Comput. Oper. Res. **41**(1), 185–195 (2014)
20. Hakimi, S.L.: On locating new facilities in a competitive environment. Eur. J. Oper. Res. **12**(1), 29–35 (1983)
21. Hansen, E., Walster, G.W.: Global Optimization Using Interval Analysis. Marcel Dekker, New York (2004). Second revised and expanded edition
22. Huff, D.L.: A programmed solution for approximating an optimum retail location. Land Econ. **42**(3), 293–303 (1966)
23. Jelásity, M.: The shape of evolutionary search: discovering and representing search space structure. Ph.D. thesis, Leiden University (2001)
24. Jelásity, M., Ortigosa, P.M., García, I.: UEGO, an abstract clustering technique for multimodal global optimization. J. Heuristics **7**(3), 215–233 (2001)
25. Küçükaydin, H., Aras, N., Altinel, I.K.: Competitive facility location problem with attractiveness adjustment of the follower: a bilevel programming model and its solution. Eur. J. Oper. Res. **208**(3), 206–220 (2011)
26. Küçükaydin, H., Aras, N., Altinel, I.K.: A leader-follower game in competitive facility location. Comput. Oper. Res. **39**(2), 437–448 (2012)
27. Lederer, P.J., Hurter, A.P.: Competition of firms: discriminatory pricing and location. Econometrica **54**(3), 623–640 (1986)
28. McGarvey, R.G., Cavalier, T.M.: Constrained location of competitive facilities in the plane. Comput. Oper. Res. **32**, 359–378 (2005)
29. Miller, T.C., Friez, T.L., Tobin, R.L.: Equilibrium Facility Location on Networks. Springer, New York (1996)
30. Mirchandani, P.B., Francis, R.L. (eds.): Discrete Location Theory. Wiley, New York (1990)
31. Ortigosa, P.M., García, I., Jelásity, M.: Reliability and performance of UEGO, a clustering-based global optimizer. J. Glob. Optim. **19**(3), 265–289 (2001)
32. Plastria, F.: GBSSS, the generalized big square small square method for planar single facility location. Eur. J. Oper. Res. **62**, 163–174 (1992)
33. Plastria, F.: Avoiding cannibalization and/or competitor reaction in planar single facility location. J. Oper. Res. Soc. Jpn. **48**, 148–157 (2005)
34. Plastria, F., Carrizosa, E.: Optimal location and design of a competitive facility. Math. Program. **100**(2), 247–265 (2004)
35. Redondo, J.L., Ortigosa, P.M., García, I., Fernández, J.J.: Image registration in electron microscopy. A stochastic optimization approach. In: Proceedings of the International Conference on Image Analysis and Recognition, ICIAR 2004. Lecture Notes in Computer Science, vol. 3212(II), pp. 141–149. Springer, Berlin/Heidelberg (2004)
36. Redondo, J.L., Fernández, J., García, I., Ortigosa, P.M.: Parallel algorithms for continuous competitive location problems. Optim. Methods Softw. **23**(5), 779–791 (2008)
37. Redondo, J.L., Fernández, J., García, I., Ortigosa, P.M.: A robust and efficient global optimization algorithm for planar competitive location problems. Ann. Oper. Res. **167**(1), 87–106 (2009)
38. Redondo, J.L., Fernández, J., García, I., Ortigosa, P.M.: Solving the multiple competitive facilities location and design problem on the plane. Evol. Comput. **17**(1), 21–53 (2009)
39. Redondo, J.L., Fernández, J., García, I., Ortigosa, P.M.: Heuristics for the facility location and design $(1|1)$-centroid problem on the plane. Comput. Optim. Appl. **45**(1), 111–141 (2010)
40. Redondo, J.L., Fernández, J., García, I., Ortigosa, P.M.: Solving the facility location and design $(1|1)$-centroid problem via parallel algorithms. J. Supercomput. **58**(3), 420–428 (2011)
41. Redondo, J.L., Fernández, J., Arrondo, A.G., García, I., Ortigosa, P.M.: Fixed or variable demand? Does it matter when locating a facility? Omega **40**(1), 9–20 (2012)

42. Redondo, J.L., Fernández, J., Arrondo, A.G., García, I., Ortigosa, P.M.: A two-level evolutionary algorithm for solving the facility location and design (1|1)-centroid problem on the plane with variable demand. J. Glob. Optim. **56**(3), 983–1005 (2013)
43. Saidani, N., Chu, F., Chen, H.: Competitive facility location and design with reactions of competitors already in the market. Eur. J. Oper. Res. **219**(1), 9–17 (2012)
44. Sáiz, M.E., Hendrix, E.M.T., Fernández, J., Pelegrín, B.: On a branch-and-bound approach for a Huff-like Stackelberg location problem. OR Spectrum **31**, 679–705 (2009)
45. Serra, D., ReVelle, C.: Competitive location in discrete space. In: Facility Location: A Survey of Applications and Methods, pp. 367–386. Springer, New York (1995)
46. Solis, F.J., Wets, R.J.B.: Minimization by random search techniques. Math. Oper. Res. **6**(1), 19–30 (1981)
47. Tóth, B., Fernández, J.: Interval Methods for Single and Bi-objective Optimization Problems - Applied to Competitive Facility Location Problems. Lambert Academic, Saarbrücken (2010)
48. Tóth, B.G., Kovács, K.: Solving a Huff-like Stackelberg location problem on networks. J. Glob. Optim. **64**(2), 233–257 (2016)
49. Tóth, B., Plastria, F., Fernández, J., Pelegrín, B.: On the impact of spatial pattern, aggregation, and model parameters in planar Huff-like competitive location and design problems. OR Spectrum **31**(1), 601–627 (2009)

A Game Theoretic Approach to an Emergency Units Location Problem

Vito Fragnelli, Stefano Gagliardo, and Fabio Gastaldi

1 Introduction

Emergency management represents a hard task in several situations. In fact, it can be viewed under many different lights and involves a large number of parameters. Consequently, the complexity of the problem allows and requires many different experiences and expertise and the synergic use of different methods and approaches in order to reach an efficient result. We refer to the problem of locating units in the area controlled by an emergency service as the *emergency units location problem* (*EULP*). In this paper, we introduce a new class of games to deal with it.

Location problems are broadly studied in operations research. Among the wide literature, we address to the books [9] and [10] for an analysis of the theory and possible applications, and to the book [29] for a common theory of location models (continuous, discrete and network location problems). Moreover, we refer to three survey papers [11, 18, 31].

Also EULPs receive great attention in the field of location analysis, applying several disciplines and techniques. In a deterministic environment, the pivotal work is [43], who refer to a set covering problem, refined first in [5] adding a constraint on the available units. Schilling et al. [37] goes on accounting different types of units, while [40] introduces travel-time constraints.

One of the first probabilistic models is [7], that considers a request only if all capable units are not engaged with other interventions; [13] applies the method to the city of Bangkok; [15] uses a tabu search simulation; [35] introduces variations

V. Fragnelli (✉) • F. Gastaldi
Università del Piemonte Orientale, Viale T. Michel, 11, 15121 Alessandria, Italy
e-mail: vito.fragnelli@uniupo.it; fabio.gastaldi@uniupo.it

S. Gagliardo
Università degli Studi di Genova, Via Dodecaneso 35, 16146 Genova, Italy
e-mail: gagliardo@dima.unige.it

© Springer International Publishing AG 2017
L. Mallozzi et al. (eds.), *Spatial Interaction Models*, Springer Optimization and Its Applications 118, DOI 10.1007/978-3-319-52654-6_8

in travel speed. Two other probabilistic models are [36] and [1]: the first aims to position a certain number of units in order to maximize the population satisfied within a time window with a fixed reliability; the second accounts for the probability of system failure.

The works just introduced use linear programming [1, 36] and simulation [15] to find the solution of the considered problem.

In the last years, GIS (Geographical Information System) integrates with simulation and gives rise to decision support software and tools; for instance, [3] and [8] integrate GIS, GPS (Global Positioning System) and GSM (Global System for Mobile communication), offering a solution to the problem of ambulance management and emergency accident handling in the prefecture of Attica in Greece. A survey about location methods can be found in [4], while for a survey about applications of GIS to location problems we address to [28].

In the last 30 years, the interest in location analysis gives birth to some game theoretic papers that mainly deal with the cost sharing aspect. In a cooperative setting, some of the first papers introducing cost allocation games arising from location problems are [17], dealing with single facility location problems in tree graphs, [22], generalizing location games on graphs, [41], considering coverage models on graphs, and [6], studying games arising from p-facilities problems in graphs. Puerto et al. [32] introduces a new class of games linked to continuous single facility location problems, where the location for a facility has to be found in order to minimize the transportation cost for the users, which depends on their distances from the facility; the authors give some sufficient conditions in order that a game in this class has a non-empty core and define an allocation rule which is in the core for two classes of location games (Weber location problems and minimax location problems). The model in [32] is extended in [26] designing a continuous single facility location problem in which the fixed cost depends on the region where the new facility is located and finding two sufficient conditions to have a non-empty core for the cost allocation game. Goemans and Skutella [14] finds fair cost allocations among the customers of a service in several situations and show strongly connections with linear programming relaxations. Pal and Tardos[30], Leonardi and Shäfer [24], Xu and Du [44], and Immorlica et al. [21] address the problems of choosing a subset of providers and a subset of users which will be part of a service network and of sharing the building costs of the network itself. Finally, [33, 34] studies the core and the polynomial representations of it for new classes of cooperative games related to facility location models defined on metric spaces.

In a non-cooperative setting, after [23, 42] introduces the multifacility location game in the context of supermodular games. In [20] possible collusions are considered. Mallozzi [25] starts from the idea that real-world transportation costs are seldom linear with respect to the distance to extend the idea of [42] to a more general context. Mazalov and Sakaguchi [27] analyzes the Hotelling's duopoly model (see [19]) on the plane, in which two firms are located in different points inside a circle and the customers are distributed in it according to a density function. A survey on game theoretic models is provided in [12].

1.1 Utility in an Emergency Environment

Referring to emergency management, it is necessary to distinguish among different types of emergencies, for instance medical, environmental and urban, each one involving particular features, which may be coped in different ways that require different means and tools.

The concept of emergency is strongly related to the idea of urgency and then to the idea of fast intervention. The ideal time necessary to solve an emergency situation varies: for example, medical interventions should be carried out in minutes, environmental ones, e.g. fires or floods, may allow longer time and rescue missions after an earthquake can be organized also some days (or weeks) after the event.

In this work we are interested in those particular situations in which, after a given period, there is a sudden and large reduction of the usefulness of the intervention. In particular, we refer to a situation in which the utility is *0–1 type*, i.e. it is equal to 1 until a given distance (measured as ground distance, time distance, Euclidean distance, etc.), and then goes down to zero (Fig. 1a). A classical real-world example is represented by the ambulances that, from a statistical point of view, are required, for the most serious and urgent situations (the so-called *red* and *yellow codes*), to reach the emergency place within a fixed maximum time. In Italy, for example, the time allowed is 8 min for the calls coming from the city and 20 min for the ones coming from the province, including the time for answering the call (2 min, on average). Statistically, emergency management does not consider as performed the missions that are completed after the fixed time threshold. However, if an ambulance arrives later, the intervention is anyhow completed (e.g., an injured person is transported to hospital). This means that in a forecasting model, only the area that is reachable respecting the threshold is considered as covered by an ambulance and that the utility of the service from the point of view of the emergency management is 1 within this area and 0 otherwise.

This situation is different from the standard location problem, where the utility of a service located in a given place decreases when the distance increases (Fig. 1b). In Fig. 1c we represent a real-life oriented utility function that is a good approximation of the one in Fig. 1a. For instance, we may refer to a fire plane intervention. In this case, the distance from water reservoir and the fuel tank capacity have to be taken into account. We may assume that the utility is constant until a certain distance,

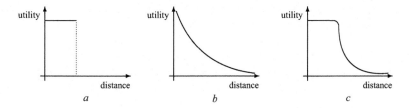

Fig. 1 Possible utility functions for location problems

that allows for the number of flights necessary to extinguish the fire, without refuelling. From that point on, the utility quickly decreases because refuelling becomes necessary, making the time to extinguish the fire longer and the risk higher.

Despite of this, we choose the situation represented in Fig. 1a as the utility function used by emergency managers in their performance analysis.

Finally, we want to stress that the utility functions in Fig. 1a, c can be good approximations of the utilities of some location problems which arise from other situations. For example, we may think to the problem of locating antennas for mobile phones or internet services, where there exists a distance within which the signal has its maximum power, and then it suddenly decreases. Another interesting situation is represented by the transportation of perishable goods through appropriate vehicles that can preserve the quality of the goods only for a certain time.

1.2   Game Theory and EULP

The main idea that has driven us in our game theoretic approach to this problem is that the different candidate locations for hosting an emergency unit interact among them. In fact, the choice of deploying an emergency unit cannot take into account only the characteristics of a candidate location, i.e. the extension of the area that can be covered within the maximum time allowed (or with the maximum utility), the probability of a call in that area, etc., but should also consider where the other emergency units are located. In other words, we look for maximizing the utility associated to the selected set of locations, rather than for a set of locations with maximal utility; in particular, we account the *marginal contribution* of an ambulance to each possible set of units located in the other candidate zones of the area, i.e. we consider what a further candidate location may add to the service when an ambulance is located there. The average marginal contribution may be considered a measure of the relevance of the candidate locations. We are going to do that solving a suitable cooperative game for *ranking by relevance* the whole set of the candidate locations. In view of this, we may not account the number of available ambulances.

We may stress that the problem under analysis is a *centralized decision situation* in which the emergency management decides where to locate vehicles. Cooperative games have been widely used to deal with situations in which interacting agents realize that they may improve their payoffs by cooperating; in the EULP case, cooperation ends in supplying the best possible service to a set of users. This is the reason why we decide to consider the EULP in a cooperative setting. A non-cooperative approach can be reasonable if we think that the inhabitants of a zone prefer to have an emergency unit close to their houses and that sometimes "political" problems can arise: if an association supplies or pays for a vehicle, it wants to decide where to locate it (probably close to itself). In spite of this, the solution of a non-cooperative approach does not take into account the global welfare and may be inefficient from a social point of view, and this is another reason why we prefer to deal with a cooperative setting.

To the best of our knowledge, our model is the first, in which the cooperative game theoretic approach is used to support the location of the facilities, disregarding cost allocation problems.

In the following, we devote Sect. 2 to introduce the EULP; Sect. 3 recalls some basic game theoretical concepts and introduce the model; in Sect. 4 we propose the Shapley value as the most suitable solution for the class of games introduced in the previous section; Sect. 5 provides a simple and computationally efficient algorithm for solving the EULP; in Sect. 6 we describe an application to the real-world situation of 118 of Milan; Sect. 7 concludes.

2   EULP

In this section, we formally introduce the emergency units location problem. In particular, let us consider an area divided into zones and let us use the following notations:

- $M = \{1, \ldots, m\}$ is the set of zones of the area;
- $N = \{1, \ldots, n\} \subseteq M$ is the subset of zones which are candidate locations for an emergency unit;
- $C = (c_{ij}) \in \mathbb{R}^{n \times m}$ is the *coverage matrix*, s.t. $c_{ij} = 1$ if an emergency unit located in i covers zone j, $c_{ij} = 0$ otherwise;
- $w \in \mathbb{R}^m$ the vector of the demands of the zones of the area.

The (M, N, C, w)-*EULP* is the problem of ranking the n candidate locations in the area described by the previous parameters in order to satisfy the demand in the best possible way, where the utility of the service is modelled as in Fig. 1a.

Example 1 Consider the area in Fig. 2, where each box represents a zone and the value in each of them is the demand of the zone; suppose that each zone is a candidate location and that an ambulance located in a zone is able to cover it and the two adjacent ones (those immediately to the left and to the right, if any).

In this case we have:

- $M = N = \{a, b, c, d, e, f\}$;
- C is given by the following table:

a	b	c	d	e	f
2.46	2.28	0.91	1.52	3.34	2.31

Fig. 2 A simple EULP

	a	b	c	d	e	f
a	1	1	0	0	0	0
b	1	1	1	0	0	0
c	0	1	1	1	0	0
d	0	0	1	1	1	0
e	0	0	0	1	1	1
f	0	0	0	0	1	1

- $w = (2.46, 2.28, 0.91, 1.52, 3.34, 2.31)$.

To make clear the importance of considering the interaction among the possible locations, observe that, in order to cover the whole area, it is sufficient to locate two units in zones b and e; allocating the units in the most demanding zones the area is covered when four ambulances are located in zones e, a, f and b; accounting the locations with the highest aggregated demands[1] that are 4.74, 5.65, 4.71, 5.77, 7.17, 5.65, respectively, the area is covered when three ambulances are located in zones e, d and b, preferring this last with respect to zone f that has equal aggregated demand.

A final remark is devoted to the meaning of the *demand* of a zone. In general, it may represent different quantities, for instance the expected number of calls originated from each zone or its spatial extension; the choice will orient the model towards *efficiency* in the former case or *equity* in the latter; other choices are possible.

3   Game Theoretical Model

In this section, we recall some notions and notations of game theory and formally introduce the game theoretical model for the EULP.

3.1   TU-Games and the Shapley Value

A *cooperative game with transferable utility* or *TU-game* is a pair (N, v), where $N = \{1, 2, \ldots, n\}$ denotes the finite set of *players* and $v : 2^N \to \mathbb{R}$ is the *characteristic function*, with $v(\varnothing) = 0$. A group of players $S \subseteq N$ is called a *coalition* and $v(S)$ is the *worth* of the coalition, i.e. what the players in S may obtain independently from the other players. N is called the *grand coalition*. Often, a TU-game (N, v) is identified with the corresponding characteristic function v.

---

[1] By aggregated demand we mean the sum of the demands coming from all the zones covered by a location.

Denoting by $\mathscr{G}_N$ the class of TU-games with player set N, a *(point-valued) solution* is a function $\psi : \mathscr{G}_N \to \mathbb{R}^n$. One of the most well-known solutions to a TU-game is the *Shapley value*, introduced in [39]. It is based on the concept of *marginal contribution*: given a coalition S, the marginal contribution to it of player i is the value that player i adds to S entering it. The Shapley value assigns to each player his average marginal contribution over all the possible permutations of players.

Definition 1 (Shapley Value) Given a TU-game (N, v), the *Shapley value* ϕ assigns to player $i \in N$

$$\phi_i(v) = \frac{1}{n!} \sum_{\pi \in \Pi} [v(P(\pi; i) \cup \{i\}) - v(P(\pi; i))] \,, \tag{1}$$

where Π is the set of all the permutations of the players and $P(\pi; i)$ is the set of players that precede i in permutation π.

Shapley characterized his value as the unique solution ϕ which satisfies the following axioms:

A1. **Efficiency**: for each game (N, v), $\sum_{i \in N} \phi_i(v) = v(N)$;
A2. **Symmetry**: if two players i and j are *symmetric* for a game (N, v), i.e. $v(S \cup \{i\}) = v(S \cup \{j\})$ for each $S \subseteq N \setminus \{i, j\}$, then $\phi_i(v) = \phi_j(v)$;
A3. **Null Player**: if i is a *null player* for a game (N, v), i.e. $v(S \cup \{i\}) = v(S)$ for each $S \subseteq N \setminus \{i\}$, then $\phi_i(v) = 0$;
A4. **Additivity**: given two games u and v with the same set of players N, let the *sum game* $(u + v)$ be the game with the same set of players N and $(u+v)(S) = u(S) + v(S)$ for each $S \subseteq N$; then $\phi_i(u + v) = \phi_i(u) + \phi_i(v)$ for each $i \in N$.

The main problem in dealing with the Shapley value is its high computational complexity: according to (1), it is necessary to consider $n!$ orderings of the players that, in general cases, could be dozens, and for the EULP also hundreds. Moreover, the definition of the characteristic function can be really complex (with n players we need to define the worth of 2^n coalitions).

3.2 Coverage Games

Given an EULP, we introduce a new class of TU-games related to it, namely the *coverage games*, denoted by $\mathscr{C}$.

Definition 2 (Coverage Games) The *coverage game* is the TU-game (N, v) defined by

$$v(S) = \sum_{j \in A_S} w_j \quad \forall S \subseteq N \,,$$

where $A_S = \{j \in M \mid \exists i \in S \text{ s.t. } c_{ij} = 1\}$, i.e. the set of zones which are covered by at least one emergency unit located in S, when each zone in S hosts one emergency unit.

The main idea of the coverage games is to evaluate the demand covered by a coalition of possible locations. In fact, looking at Definition 2, the value of a coalition S in the coverage game is the sum of the demands of the zones that are covered locating one emergency unit in each location of S.

As we said in the Introduction, the definition of the coverage game related to an EULP does not consider the number of available units; ranking by relevance the possible locations makes our approach adaptable to a variable number of units to activate. Finally, we stress that we do not consider that the ambulances may not satisfy the whole demand, for instance due to the high number of calls or to their time distribution (in an efficiency-oriented case).

4 Shapley Value for the Coverage Games

Looking for a good solution to the EULP, we propose to use the Shapley value of the coverage games. As we already said in Sect. 1.2, a good reason to do that is the concept of marginality: it is important to take into account not only the demand of a zone or the aggregated demand that a candidate location can cover, but mainly the contribution that an ambulance located there can add to the other locations. As we already said, our aim is ranking by relevance the candidate locations accounting their marginal contributions; in view of this the Shapley value represents a very good solution. Then, the available ambulances are deployed according to the ordering of relevance of the candidate locations.

Also the Banzhaf value (see [2]) takes into account marginal contributions, but the Shapley value is also *efficient*, allowing sharing among the locations the whole demand that can be potentially covered by all of them. Moreover, it has pretty good fairness properties with respect to our problem. To choose a "good" solution, in fact, we identify two suitable *fairness criteria* it should satisfy, called *coverage indifference* and *demand indifference*; we may notice that these properties are related more to the problem than to the game, allowing improving the fairness of the solution of the location problem.

In particular, the coverage indifference looks at the situation from the point of view of the users and requires to give the same importance to the units that cover a zone allowing for equally sharing the demand of the zone among them; in a sense, any of those units has the same probability to satisfy a call coming from the considered zone. This property is suitable to the EULP as the utility function we use (Fig. 1a) implies that the most important requirement is to satisfy the demand within the fixed time threshold, independently from the unit which does it and from the actual time required. The demand indifference looks at the situation from the point of view of the emergency service provider and gives the same importance to each demand, wherever it comes from and whatever the required intervention is. Also this property is suitable to the EULP as all emergency calls (yellow and red codes) have the same importance for the emergency service.

It is interesting to notice that other usual fairness criteria, such as *monotonicity* and *equal treatment of equals*, are not so important in the situation at hand. For example, looking at the situation in Fig. 2, the Shapley value of the corresponding coverage game is $(1.99, 2.29, 1.57, 1.92, 2.78, 2.27)$, so that deploying two units in the second and the fifth zones we obtain the optimal solution. This choice does not satisfy the two mentioned criteria: the second zone has an aggregated demand of 5.65 and receives one unit, while the fourth zone receives nothing even if it has an aggregated demand of 5.77 (non-monotonicity); the second and the sixth zones have both an aggregated demand of 5.65, but one receives a unit and the other does not (they are not equally treated). This is essentially due to the fact that monotonicity and equal treatment of equals focus on the features of each location.

4.1 Coverage and Demand Indifference

Before introducing the two properties of coverage indifference and demand indifference, we need to define a sub-class of coverage games, the *jth zone sub-games*, in which uniquely zone j has a positive demand, i.e we put down to zero the demands of all the zones but j.

Definition 3 (*j*th Zone Sub-Game) Let v be a coverage game. Given $j \in M$, the *jth zone sub-game* of v is the coverage game v^j defined, for each $S \subseteq N$, by

$$v^j(S) = \begin{cases} w_j & \text{if } j \in A_S \\ 0 & \text{otherwise} \end{cases}.$$

The following result proves that the coverage game is the sum of all its zone sub-games.

Lemma 1 Let $v \in \mathscr{C}$ be any coverage game. Then, for every $S \subseteq N$,

$$v(S) = \sum_{j \in M} v^j(S). \tag{2}$$

Proof By Definitions 2 and 3,

$$v(S) = \sum_{j \in A_S} w_j = \sum_{j \in A_S} v^j(S) = \sum_{j \in M} v^j(S),$$

where the last equality holds because $v^j(S) = 0$ if $j \in M \setminus A_S$. □

Now, we can formally introduce the two fairness criteria.

Definition 4 (Coverage Indifference: CI) A solution ψ satisfies *coverage indifference* if for each $v \in \mathscr{C}$ and each jth zone sub-game v^j of v, if $j \in A_{\{i\}} \cap A_{\{l\}}$, $i, l \in N$, then

$$\psi_i(v^j) = \psi_l(v^j) .$$

Definition 5 (Demand Indifference: DI) A solution ψ satisfies *demand indifference* if for each $v \in \mathscr{C}$, $i \in N$,

$$\psi_i(v) = \sum_{j \in M} \psi_i(v^j) .$$

We can notice that DI is very similar to the additivity property A4. However, DI is a different condition as we require the additivity to be satisfied only with respect to the zone sub-games. Moreover, it is easy to observe that

$$\psi_i(v^j) = \begin{cases} \dfrac{w_j}{\sum_{l=1}^{n} c_{lj}} & \text{if } j \in A_{\{i\}} \\ 0 & \text{otherwise} \end{cases} \quad i \in N, \tag{3}$$

satisfies CI. We may notice that (3) recalls the other Shapley axioms (efficiency, symmetry, null player), but from a "local" point of view, in the sense that it does not concern the coverage game, but its zone sub-games; in particular, it does not imply neither A1., nor A2., nor A3. for the coverage game, but it does only together with DI.

The following proposition shows that the Shapley value satisfies both properties.

Proposition 1 *The Shapley value of a coverage game satisfies CI and DI.*

Proof At first, we prove that ϕ satisfies CI. Let be $j \in M$ and let us consider the jth zone sub-game v^j. Then:

- if $i \in N$ is s.t. $j \notin A_{\{i\}}$, then i is a null player in the game v^j and by A2. $\phi_i(v^j) = 0$;
- if $i, l \in N$ are s.t. $j \in A_{\{i\}} \cap A_{\{l\}}$, then i and l are symmetric players in the game v^j and by A1. $\phi_i(v^j) = \phi_l(v^j)$;

Let be $\alpha = \phi_i(v^j)$ for all i s.t. $j \in A_{\{i\}}$, i.e. for all i s.t. $c_{ij} = 1$ (if such an i exists). Then, by the efficiency of the Shapley value:

$$w_j = v^j(N) = \sum_{i \in N} \phi_i(v^j) = \sum_{i:c_{ij}=1} \alpha = \alpha \sum_{i=1}^{n} c_{ij} \implies \alpha = \frac{w_j}{\sum_{i=1}^{n} c_{ij}} .$$

Then,

$$\phi_i(v^j) = \begin{cases} \dfrac{w_j}{\sum_{l=1}^{n} c_{lj}} & \text{if } j \in A_{\{i\}} \\ 0 & \text{otherwise} \end{cases} ,$$

that is CI.

DI immediately follows from (2) and A4. □

As a consequence of DI and of (3), the following proposition holds.

Proposition 2 *The Shapley value of a coverage game is given by*

$$\phi_i(v) = \sum_{j \in A_{\{i\}}} \frac{w_j}{\sum_{l \in N} c_{lj}}. \tag{4}$$

5 Algorithm for the Shapley Value

In this section, we show how (4) leads to a computationally efficient algorithm for implementing the solution of the corresponding coverage game.

The algorithm simply requires the construction of a $n \times m$ matrix D, called the *division matrix*, where, for each $i \in N$, $j \in M$,

$$d_{ij} = \begin{cases} \dfrac{w_j}{\sum_{l \in N} c_{lj}} & \text{if } c_{ij} = 1 \\ 0 & \text{otherwise} \end{cases}.$$

By (4), the Shapley value of the coverage game for player $i \in N$ can be obtained simply summing up the values in the ith row of D. The computational complexity of this algorithm is then polynomial in n and m, differently from the one of formula (1) that is exponential in n. Moreover, we may stress that the algorithm does not require defining the 2^n values of the characteristic function of the coverage game, further reducing the complexity of the implementation of the solution.

Example 2 Let us consider the following EULP: the area depicted in Fig. 3 is made up of 18 zones, three of which, namely **d**, **j** and **m**, are candidate locations to host an ambulance; the thicker lines represent the coverage of the candidate locations and the numbers in the boxes the demands of the corresponding zones. This leads to the following coverage matrix C and demand vector w:

$$C = \begin{pmatrix} 1\,1\,1\,1\,1\,0\,0\,1\,1\,1\,0\,0\,0\,0\,0\,0\,0\,0 \\ 0\,0\,0\,1\,1\,1\,0\,0\,1\,1\,1\,0\,0\,1\,1\,1\,0\,0 \\ 0\,0\,0\,0\,0\,0\,1\,1\,1\,0\,0\,1\,1\,1\,0\,0\,1\,1 \end{pmatrix} \begin{matrix} \longleftrightarrow \mathbf{d} \\ \longleftrightarrow \mathbf{j} \\ \longleftrightarrow \mathbf{m} \end{matrix}$$

$$w = (2, 3, 4, 6, 8, 1, 3, 4, 3, 8, 1, 2, 3, 4, 5, 3, 2, 1)$$

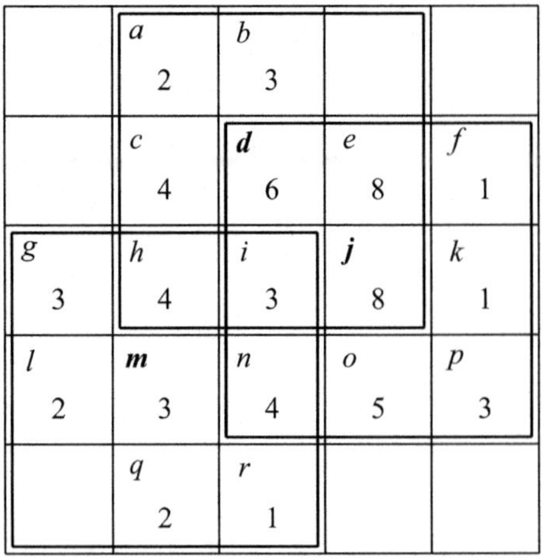

Fig. 3 A simple example of EULP

Following the algorithm previously described, we obtain the following division matrix and Shapley value:

$$D = \begin{pmatrix} 2\,3\,4\,3\,4\,0\,0\,2\,1\,4\,0\,0\,0\,0\,0\,0\,0\,0 \\ 0\,0\,0\,3\,4\,1\,0\,0\,1\,4\,1\,0\,0\,2\,5\,3\,0\,0 \\ 0\,0\,0\,0\,0\,0\,3\,2\,1\,0\,0\,2\,3\,2\,0\,0\,2\,1 \end{pmatrix} \begin{array}{l} \longleftrightarrow \phi_d(v) = 23 \\ \longleftrightarrow \phi_j(v) = 24 \\ \longleftrightarrow \phi_m(v) = 16 \end{array}$$

which provides the ranking $\mathbf{j} \succ \mathbf{d} \succ \mathbf{m}$ among the three locations.

5.1 Computational Experiments

Table 1 summarizes the results of some random computational experiments for testing the performance of our algorithm: the first column gives the numbers m of zones; the second column is the random-generated number $n \leq m$ of possible locations, with a uniform integer distribution on the interval $[m/5, m/3]$; the third column gives the average computational time (calculated on 30 runs per each pair (m, n) and given in milliseconds) for computing the Shapley value of the corresponding coverage game.[2]

As we may notice, the algorithm is able to ranking a large number of candidate locations in few milliseconds.

---

[2] The algorithm ran on a Desktop PC with an Intel Core i5-2400 Processor, 3.10 GHz, with 8 GB RAM.

Table 1 Average computational time of the algorithm on random experiments

m	n	Time
25	7	0.242
100	30	0.364
225	75	0.719
400	113	1.122
625	181	2.753
900	224	5.266
1225	295	10.115
1600	338	16.579

6 Example of Milan

The 118 is a public institution that manages the ambulances in the area of each Italian province. A project named DECEMbRIA (DECisioni in EMergenza sanitaRIA, i.e. decisions in medical emergency) started in 2006, involving 118 of Milan, University of Milan, Polytechnic of Milan and University of Eastern Piedmont, with the objective of providing an emergency decisions support system at strategic, operative and tactical level to the 118 of Milan. A part of the project deals with the ambulances location problem. The 118 of Milan controls the area of the city and of the province of Milan, for a total surface of about $1580 \, km^2$, populated by three million people; on average, the dispatch center receives 1580 calls per day, resulting in 670 missions activated per day. The number of ambulances available is 53 for the whole area, 25 in the city and 28 in the province. This gives an idea of the dimensions that the ambulances location problem may assume and of the importance of providing good (and fast) solutions to it.

The performance of our algorithms has been tested at first on a common benchmark provided by the 118 of Milan, represented by a hypothetical city made up of 1127 zones, all of which are possible locations, in which 53 ambulances have to be deployed. This benchmark was used by the researchers of DECEMbRIA project for testing the results obtained with different approaches. It is worthwhile to remark that our algorithm finds a solution to the problem in few seconds, ranking in the first 564 positions 46 of the 53 locations found by the approximated mathematical programming approach.

In the following, we present in details the results given by the application of the coverage game to the extra-urban area of Milan. The situation corresponds to the EULP where:

- M is the set of the 117 municipalities in the province of Milan (excluding the city itself);
- N is the subset of 65 municipalities in which an emergency unit can be located;

- C is obtained assigning 1 to those entries of the coverage matrix which correspond to an average travelling time smaller than the time threshold of 18 min, and 0 to the others[3];
- w is defined for 13 different scenarios, as described in the following.

We may remark that in the extra-urban area locations are fixed, as ambulances can be hosted only in emergency service sites (such as hospitals, Red Cross sites, and so on). Figure 4 represents the province of Milan, where the black circles show the 65 possible locations.

Speaking about the demands, the first 12 out of the 13 scenarios look at the problem from an *efficiency* point of view; their identification is based on the results obtained by the University of Milano team; each scenario is a segment of the year where the daily call distribution is similar, so that the treatment could be the same (see Table 2). The demand of each zone is the average number of calls per day received by the 118. The 13th scenario uses an *equity* point of view and the demands of the zones are given by their spatial extensions.

Table 3 summarizes the results we obtained. The first column is the list of the candidate locations in which an ambulance may be located.[4] Towns in capital letters are those in which the 118 located an ambulance. The following 12 columns refer to the efficiency approach applied to the scenarios introduced above, while the last one (marked with a E) refers to the equity approach: an "X" in one of these columns means that the candidate location of the corresponding row by the corresponding approach is ranked among the first 28, that is the number of ambulances available in the province of Milan.

The algorithms are implemented in Matlab and the results are obtained in a very short time: the computational time (see footnote 2) of the different approaches and scenarios varies from 3.691 to 4.842 ms. This is one of the most important features of our approach, as its low computational complexity allows studying many different scenarios and situations very quickly, using several trials, with different parameters and splitting the day in very refined time periods.

We compared the coverage of our approach with that of the real deployment of the ambulances, obtaining a reduction of about 2%, that we consider a good performance, in view of the limited amount of time required for computing our solution. We considered also a hypothetical situation with a tighter threshold for the intervention time, from 18 to 13 min, preserving the number of 28 ambulances; also with this hypothesis our method covers more than 90% of the area in all the cases.

---

[3]The travelling time have been obtained thanks to the information provided by the 118 of Milan on the real average travelling time of an ambulance from each of the possible locations to each of the municipalities in the area; the time threshold of 18 min is determined by the 20 min allowed to reach the location of an event, minus the 2 min spent (on average) to answer the call.

[4]We omitted the candidate locations which an ambulance is never assigned to: Bellinzago, Bollate, Cassano, Ceriano, Cislago, Cuggiono, Inzago, Lainate, Novate, Pozzuolo, Solaro, Uboldo.

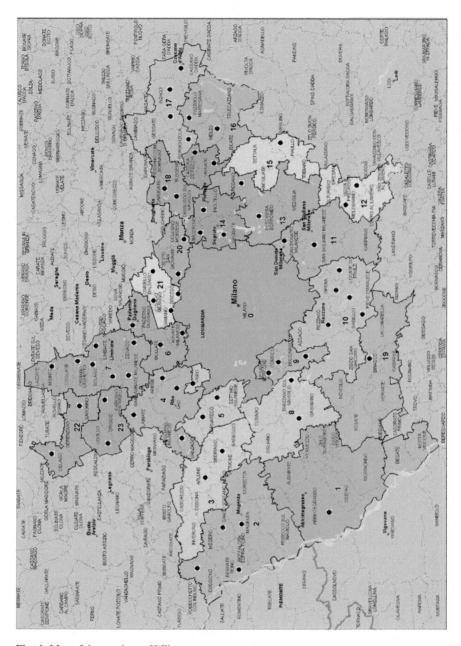

Fig. 4 Map of the province of Milan

Table 2 The analyzed scenarios (first 12 cases)

Winter 1	Winter	From 00 a.m. to 08 a.m.
Winter 2	Winter	From 08 a.m. to 02 p.m.
Winter 3	Winter	From 02 p.m. to 12 p.m.
SA_{mf} 1	Spring–Autumn	From Monday to Friday, from 00 a.m. to 07 a.m.
SA_{mf} 2	Spring–Autumn	From Monday to Friday, from 07 a.m. to 13 a.m.
SA_{mf} 3	Spring–Autumn	From Monday to Friday, from 01 p.m. to 12 p.m.
SA_{ss} 1	Spring–Autumn	Saturday and Sunday, from 00 a.m. to 07 a.m.
SA_{ss} 2	Spring–Autumn	Saturday and Sunday, from 07 a.m. to 13 a.m.
SA_{ss} 3	Spring–Autumn	Saturday and Sunday, from 01 p.m. to 12 p.m.
Summer 1	Summer	From 00 a.m. to 07 a.m.
Summer 2	Summer	From 07 a.m. to 02 p.m.
Summer 3	Summer	From 02 p.m. to 12 p.m.

Table 3 Results of the different approaches on the province of Milan

Town	Winter			SA_{mf}			SA_{ss}			Summer			E
	1	2	3	1	2	3	1	2	3	1	2	3	
ABBIATEGRASSO	X	X	X	X	X	X	X	X	X	X	X	X	X
Arese	X	X	X	X	X	X	X	X	X	X	X	X	
ARLUNO	X	X	X	X	X	X	X	X	X	X	X	X	X
BASIGLIO													X
Bernate													X
BINASCO	X	X	X	X		X	X	X			X	X	X
BRESSO	X	X	X	X	X	X	X	X	X	X	X	X	
BUCCINASCO													
CARONNO													
CARUGATE													
Cassina		X					X			X	X	X	
Cernusco	X	X	X	X	X	X	X	X	X	X	X	X	
CESANO													
Cesate							X						
CINISELLO	X		X	X	X	X	X	X	X	X	X	X	
COLOGNO	X	X	X	X	X	X	X	X	X	X	X	X	
Cormano	X	X	X	X	X	X	X	X	X	X	X	X	
Cornaredo	X		X	X	X	X		X	X				X
CORSICO		X	X	X	X	X		X	X	X	X	X	X
CUSANO	X	X	X	X	X	X	X	X	X	X	X	X	
Gaggiano	X	X	X	X	X	X	X	X	X	X	X	X	X
Garbagnate	X	X	X	X	X	X	X	X	X	X	X	X	
GORGONZOLA													X
Limbiate	X						X			X			
Locate		X	X	X	X	X	X				X	X	

(continued)

Table 3 ((Continued))

Town	Winter			SA_{mf}			SA_{ss}			Summer			E
	1	2	3	1	2	3	1	2	3	1	2	3	
MAGENTA													X
Marcallo													X
MELEGNANO	X	X	X	X	X	X	X	X	X	X	X	X	X
MELZO							X			X			X
MISINTO													
OPERA													
Paderno	X	X	X	X	X	X	X	X	X	X	X	X	
PAULLO													X
Pero	X	X	X	X	X	X	X	X	X	X	X	X	X
PESCHIERA													
Pieve			X										X
PIOLTELLO	X	X		X	X	X	X	X	X	X	X	X	X
RHO	X						X						
Rodano													X
Rozzano	X	X	X	X	X	X	X	X	X	X	X	X	X
SAN DONATO	X	X	X	X	X	X	X	X	X	X	X	X	X
San Giuliano	X	X	X	X	X	X	X	X	X	X	X	X	X
SARONNO													
Sedriano	X	X	X	X	X			X	X	X	X	X	X
Segrate		X	X		X	X	X	X	X	X	X	X	
Senago	X	X	X	X	X	X	X	X	X	X	X	X	
Sesto	X	X	X	X	X	X	X	X	X	X	X	X	
SETTIMO	X	X	X	X	X	X	X	X	X	X	X	X	X
TREZZANO				X						X			
Vanzago	X			X			X	X					X
Vignate							X						X
Vimodrone	X	X	X	X	X	X	X	X	X	X	X	X	
Vizzolo													X

7 Concluding Remarks

In the paper we presented a cooperative game theoretical model for a particular set of location problems, where the utility of the service is maximal within a given distance, and then goes down to zero. An algorithm for computing the Shapley value, based on two suitable fairness criteria is provided and an application to the real-world situation of the emergency service of Milan is described.

Note that there exist other important point-valued solutions, e.g. the *nucleolus* (see [38]), which is based on the idea of minimizing the maximum dissatisfaction of each coalition. However, in view of what we said in the paper, the Shapley value, in our opinion, turns out to be the most suitable solution to the problem.

We may observe that our approach can give a solution which is not the best one, i.e. the distribution of a given number of ambulances so that the covered area is maximum (in terms of the demands of the covered zones). As it is well-known, the value of a coalition $S \neq N$ has no relationship with the sum of the Shapley values of its members, i.e. $v(S) \neq \sum_{i \in S} \phi_i(v)$, but using the coalition of maximal worth with a fixed cardinality, corresponding to the number of available ambulances, requires to compute the whole characteristic function.

Looking at Table 3, we may notice that eight of the locations chosen by the 118 (Buccinasco, Caronno, Carugate, Cesano, Misinto, Opera, Peschiera and Saronno) are never selected and a little more than one third of the remaining, precisely just 12 (Abbiategrasso, Arluno, Binasco, Bresso, Cinisello, Cologno, Corsico, Cusano, Melegnano, Pioltello, San Donato and Settimo) are among the most relevant locations. Moreover, 12 locations that the 118 has not chosen (Arese, Cernusco, Cormano, Gaggiano, Garbagnate, Paderno, Pero, Rozzano, San Giuliano, Senago, Sesto and Vimodrone) are selected in all the scenarios.

Another interesting point is that the algorithm may be used as a first step in approaching those cases in which a large number of candidate locations is involved, because of its low computational complexity. We may think about a problem with several hundreds of candidate locations. We can use our approach to restrict the size of the problem, finding a subset of the locations with a high Shapley value. Then, a new problem involving only this subset of candidate locations can be built and solved, possibly exactly, with other approaches, such as mathematical programming, stochastic optimization, and so on. We can use the results obtained on the common benchmark mentioned in Sect. 6 to make things clearer. As we said, 46 of the 53 locations found by an approximated mathematical programming approach are placed in the first half of the ranking of the corresponding coverage game. Therefore, our approach is able to transform the original allocation problem in a new one, where the number of possible locations is about one half. This dramatically reduces the complexity of the problem and may allow finding a better solution to the problem through mathematical programming (or similar methods), reducing the time needed. This way of acting may allow obtaining an exact solution to an approximated problem (as some of the best locations may be lost), instead of an approximated solution to the given problem, taking some advantages from the computational point of view, which could be really important if we need to find solutions for different scenarios in a short time.

Finally, we may stress that the Shapley approach can give more than one optimal set of locations and the approximated problem may restrict the choice; even if sometimes this may cause the exclusion of all the optimal sets previously found, when this does not happen the approximated problem may quickly determine one of the optimal solutions in the approximated problem.

Possible extensions and further researches are in the direction of allowing more than one ambulance in each location in order to satisfy the whole demand; also the service time of the ambulances, from the beginning to the end of the mission, may deserve a deeper analysis; moreover, the *interaction index* (see [16]) instead of the Shapley value may be used to find a solution to the coverage game, considering the marginal contributions given by coalitions of more than one player.

Moreover, the situation described in Fig. 1c can be modelled considering also the zones that are not covered within the time threshold, but giving them a weight in the interval [0, 1] depending on the distance from the candidate location, and adjusting the implementation of the Shapley value according to the weights.

Finally, some experiments have been done also to exploit the issue of *multi-covering*: due to the definition of the coverage game, if two ambulances largely overlap on a part of the considered area with large demand, they both can be chosen by the Shapley value despite of a completely non-served area with small demand. This feature can be exploited defining a new game, that we may call *multicoverage game*, where only the multi-covered zones are considered in the definition of the characteristic function; algorithms similar to those described before allow fast calculation of the Shapley value also for multicoverage games. Finally, the results of the two approaches may be also combined to have a solution which profits from both the issues of covering and multi-covering.

Acknowledgements The authors gratefully acknowledge the financial support of DECEMbRIA project.

References

1. Ball, M.O., Lin, F.L.: A reliability model applied to emergency service vehicle location. Oper. Res. **41**(1), 18–36 (1993)
2. Banzhaf, J.: Weighted voting doesn't work: a mathematical analysis. Rutgers Law Rev. **10**, 317–343 (1965)
3. Branas, C.C., MacKenzie, E.J., ReVelle, C.S.: A trauma resource allocation model for ambulances and hospitals. Health Serv. Res. **35**, 489–507 (2000)
4. Brotcorne, L., Laporte, G., Semet, F.: Ambulance location and relocation models. Eur. J. Oper. Res. **147**, 451–463 (2003)
5. Church, R.L., ReVelle, C.S.: The maximal covering location problem. Pap. Reg. Sci. Assoc. **32**, 101–118 (1974)
6. Curiel, I.: Cooperative Game Theory and Applications. Kluwer Academic, Dordrecht (1997)
7. Daskin, M.S.: A maximum expected covering location model: formulation, properties and heuristic solution. Transp. Sci. **17**(1), 48–70 (1983)
8. Derekenaris, G., Garofalakis, J., Makris, C., Prentzas, J., Sioutas, S., Tsakalidis, A.: Integrating GIS, GPS and GSM technologies for the effective management of ambulances. Comput. Environ. Urban Syst. **25**, 267–278 (2001)
9. Drezner, Z.: Facility Location: A Survey of Applications and Methods. Springer, Berlin (1995)
10. Drezner, Z., Hamacher, H.W.: Facility Location: Application and Theory. Springer, Berlin (2001)
11. Farahani, R.Z., SteadieSeifi, M., Asgari, N.: Multiple criteria facility location problems: a survey. Appl. Math. Model. **34**, 1689–1709 (2010)
12. Fragnelli, V., Gagliardo, S.: Open problems in cooperative location games. Int. Game Theory Rev. **15**, 1340015.1–1340015.13 (2013)

13. Fujiwara, O., Makjamroen, T., Gupta, K.K.: Ambulance deployment analysis: a case study of Bangkok. Eur. J. Oper. Res. **31**(1), 9–18 (1987)
14. Goemans, M.X., Skutella, M.: Cooperative facility location games. J. Algorithms **50**, 194–214 (2004)
15. Goldberg, J., Dietrich, R., Chen, J.M., Mitwasi, M.G.: A simulation model for evaluating a set of emergency vehicle base locations: development, validation and usage. Socio Econ. Plan. Sci. **24**, 124–141 (1990)
16. Grabisch, M., Roubens, M.: An axiomatic approach to the concept of interaction among players in cooperative games. Int. J. Game Theory **28**(4), 547–565 (1999)
17. Granot, D.: The role of cost allocation in locational models. Oper. Res. **35**, 234–248 (1982)
18. Hale, T.S., Moberg, C.R.: Location science research: a survey. Ann. Oper. Res. **123**, 21–35 (2003)
19. Hotelling, H.: Stability in competition. Econ. J. **39**, 41–57 (1929)
20. Huck, S., Knoblauch, V., Müller, W.: On the profitability of collusion in location games. J. Urban Econ. **54**, 499–510 (2003)
21. Immorlica, N., Mahdian, M., Vahab, S.M.: Limitations of cross-monotonic cost-sharing schemes. ACM Trans. Algorithms **4**(2), Article 24 (2008)
22. Knoblauch, V.: Generalizing location games to a graph. J. Ind. Econ. **39**, 683–688 (1991)
23. Knoblauch, V.: A pure strategy Nash equilibrium for a three-firm location game on a two dimensional set. Locat. Sci. **4**, 247–250 (1997)
24. Leonardi, S., Shäfer, G.: Cross-monotonic cost sharing methods for connected facilities location games. Theor. Comput. Sci. **326**, 431–442 (2004)
25. Mallozzi, L.: Noncooperative facility location games. Oper. Res. Lett. **35**, 151–154 (2007)
26. Mallozzi, L.: Cooperative games in facility location situations with regional fixed costs. Optim. Lett. **5**, 173–181 (2011)
27. Mazalov, V., Sakaguchi, M.: Location game on the plane. Int. Game Theory Rev. **5**, 13–25 (2003)
28. McLafferty, S.L.: GIS and health care. Annu. Rev. Public Health **24**, 25–42 (2003)
29. Nickel, S., Puerto, J.: Location Theory: A Unified Approach. Springer, Berlin (2005)
30. Pal, M., Tardos, E.: Group strategyproof mechanisms via primal-dual algorithms. In: Proceedings of the 44th Annual IEEE Symposium on Foundations of Computer Science, pp. 584–593 (2003)
31. Plastria, F.: Static competitive facility location: an overview of optimisation approaches. Eur. J. Oper. Res. **129**, 461–470 (2001)
32. Puerto, J., Garcia-Jurado, I., Fernandez, F.R.: On the core of a class of location games. Math. Methods Oper. Res. **54**, 373–385 (2001)
33. Puerto, J., Tamir, A., Perea, F.: A cooperative location game based on the 1-center location problem. Eur. J. Oper. Res. **214**, 317–330 (2011)
34. Puerto, J., Tamir, A., Perea, F.: Cooperative location games based on the minimum spanning Steiner subgraph problem. Discrete Appl. Math. **160**, 970–979 (2012)
35. Repede, J.F., Bernardo, J.J.: Developing and validating a decision support system for locating emergency medical vehicles in Louisville, Kentucky. Eur. J. Oper. Res. **75**, 567–581 (1994)
36. ReVelle, C.S., Hogan, K.: The maximum reliability location problem and α-reliable p-center problem: derivatives of the probabilistic location set covering problem. Ann. Oper. Res. **18**(1), 155–173 (1989)
37. Schilling, D.A., Elzinga, D.J., Cohon, J.L., Church, R.L., ReVelle, C.S.: The TEAM-FLEET models for simultaneous facility and equipment sitting. Transp. Sci. **13**, 163–175 (1979)
38. Schmeidler, D.: The nucleolus of a characteristic function game. SIAM J. Appl. Math. **17**, 1163–1170 (1969)
39. Shapley, L.S.: A value for n-person games. In: Kuhn, H.W., Tucker, A.W. (eds.) Contributions to the Theory of Games, vol. 2, pp. 307–317. Princeton University Press, Princeton (1953)
40. Taillard, E.D., Badeau, P., Gendreau, M., Guertin, F., Potvin, J.Y.: A tabu search heuristic for the vehicle routing problem with soft time windows. Transp. Sci. **31**, 170–186 (1997)

41. Tamir, A.: On the core of cost allocation games defined on locational problems. Transp. Sci. **27**, 81–86 (1992)
42. Topkis, D.: Supermodularity and Complementarity. Princeton University Press, Princeton (1998)
43. Toregas, C., Swain, R., ReVelle, C., Bergman, L.: The location of emergency service facilities. Oper. Res. **19**(6), 1363–1373 (1971)
44. Xu, D., Du, D.: The k-level facility location game. Oper. Res. Lett. **34**, 421–426 (2006)

An Equilibrium-Econometric Analysis of Rental Housing Markets with Indivisibilities

Mamoru Kaneko and Tamon Ito

1 Introduction

1.1 General Idea

We develop a theory of an equilibrium-econometric analysis of rental housing markets, and test it with some data from the Tokyo area. Our theory has the following salient features:

(*i*) An econometric method is developed through an (market) equilibrium theory.

(*ii*) Both *economic agents* and an *econometric analyzer* are facing statistical components in the economy. We show that an equilibrium theory without statistical components is regarded as an idealization, and that its structure is estimated by our equilibrium-econometric analysis.

(*iii*) We define the measure of discrepancy between the prediction by our theory and the best statistical estimator, and show that the prediction is quite satisfactory in the example of the Tokyo area.

Feature (*ii*) tells why and how we can use an equilibrium theory for an econometric analysis of (*i*). Feature (*iii*) is a requirement from the econometric point of view. Here, focusing on these features, we discuss our motivations and backgrounds.

M. Kaneko (✉)
Waseda University, Tokyo 186-8050, Japan
e-mail: mkanekoepi@waseda.jp

T. Ito
Saganoseki Hospital, Oita 879-2201, Japan
e-mail: tm-ito@sekiaikai.jp

© Springer International Publishing AG 2017
L. Mallozzi et al. (eds.), *Spatial Interaction Models*, Springer Optimization
and Its Applications 118, DOI 10.1007/978-3-319-52654-6_9

One fundamental question arises in an application of an equilibrium theory to real economic problems with an econometric method: what is the source for errors in the econometric analysis? This may be answered in the same way as classical statistics: the source is attributed to partial observations. In many economic problems, this answer is applied not only to the economic analyzer but also to economic agents. Both face non-unique (perturbed) rents of goods. Error terms represent the effects of variables not included in available information to either economic agents or the econometric analyzer.

We look at a rental housing market in Tokyo. In the Tokyo area, the rental housing market is held, day by day, in a highly decentralized manner, i.e., many households (demanders) and many landlords (suppliers) look for better opportunities.[1] Various weekly magazines, daily newspapers, and internet services for listing apartments for rental prices (rents) are available as media for information transmission of supplied units together with rents from suppliers to demanders.[2] With the help of those media, rental housing markets function well, even though rents are not uniform over the "same" category of apartment units. We will call these media *housing magazines*.

Housing magazines give *concise and coarse* date about each listed apartment unit, following a fixed number of criteria, rents, size, location, age, geography, etc. This information is far from the description of its full characteristics. This is because the number of weekly listed units is large; e.g., 100–1200 listed around one railway station, and an weekly issue may exceed 500 pages.

The data of rents show that they are heterogeneous over the "same" category of apartment units. The market can still be regarded as "perfectly competitive" in that each has many competitors. These may appear contradictory, but can be reconciled; households and landlords look at summary statistics, and behave as if they are facing uniform rents. Taking this interpretation into account, the econometric analyzer may make estimation of a structure of the market. These are the two faces of our theory.

We call the attributes listed in the magazine as *systematic* components and the others as *non-systematic* factors. The systematic ones are described as a market model $\mathbb{E}$, which is assumed to be an equilibrium theory without perturbations, and the non-systematic factors are summarized by error terms ϵ. The listed rents in housing magazines are given as $p(\mathbb{E}) + \epsilon$, where the rent vector $p(\mathbb{E})$ is determined by $\mathbb{E}$. Both economic agents and econometric analyzer observe the rents $p(\mathbb{E}) + \epsilon$, but they have different purposes. The economic agents use them for their behavioral choices, while the econometric analyzer does for the estimation of the systematic components of $\mathbb{E}$. Those structures are depicted in Fig. 1. In the left box, the

---

[1]In the city of Tokyo (about 12 millions of residents), the percentage of households renting apartments is about 55% in 2005, and in the entire Japan, the percentage is about 37%.

[2]There are many decentralized real estate agents. In our analysis, we do not explicitly count real-estate companies. But we should remember that behind the market description, many real-estate companies are included.

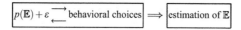

Fig. 1 $\mathbb{E}(\epsilon) = (\mathbb{E}; \epsilon)$

rents and behavioral choices are simultaneously determined as an equilibrium. The determined, yet perturbed, rents are used by the analyzer. We study each of those, and then synthesize them.

We adopt the theory of *assignment markets* for the systematic part $\mathbb{E}$, which was initiated by Böhm-Bawerk [19] and developed by von Neumann-Morgenstern [20] and Shapley-Shubik [16]. In this theory, housings are treated as indivisible commodities, which significantly differs from the urban economics literature of bid-rent theory from Alonso [1].[3] In particular, we adopt a theoretical model given by Kaneko [7] in which income effects are allowed. In the model, apartments units are classified into a finite number, T, of *categories,* and are traded for rents measured by the composite commodity other than housing services.

1.2   Specific Developments

Let us discuss specific developments of our theory. First, the *systematic* part of the housing market is summarized as:

$$\mathbb{E} = (M, u, I; N, C), \tag{1}$$

where M is the set of households, u their utility functions, I the income distribution for households, and N the set of landlords, C the cost functions for landlords. The details of (1) and the market equilibrium theory are given in Sect. 2.

In the systematic part, the rents are uniform over each category of apartments. However, the rents listed in housing magazines are not uniform over each category. Those non-uniform rents are resulted by *non-systematic factors* other than the components listed in (1). The effects of non-systematic factors are summarized by one random variable ϵ_k for each category k. That is, the apartment rent for a unit d in category k is determined by $p_k + \epsilon_{kd}$, where p_k is the competitive rent for category k and ϵ_{kd} is an independent random variable identical to ϵ_k. This ϵ_{kd} represents properties of unit d such as its specific location in addition to the systematic components in $\mathbb{E}$. The rent p_k is *latent* in that only $p_k + \epsilon_{kd}$ is observed in housing magazines. The market model with housing magazines is denoted by $\mathbb{E}(\epsilon) = (\mathbb{E}; \epsilon)$.

---

[3] See van der Laan et al. [17] and its references for recent papers for the literature of assignment markets, and see Arnott [2] for a recent survey on the urban economics literature from Alonso [1].

As described above, the housing market model $\mathbb{E}(\epsilon)$ has two faces: it is purely the trading place with media for information transmissions; and it is a target of an econometric study. In both faces, housing magazines serve information about rents to households/landlords and to the econometric analyzer. Here, we emphasize that these two faces are asymmetric.

An economic agent pursues his utility or profit in the market, rather than to understand the market structure. If he looks at the average of the rents of randomly taken 10 apartment units from one category, its variance becomes $1/10$ of the original distribution. Thus, the uniform rent assumption for each category seems to be an approximation. This interpretation will be expressed by the *convergence* theorem (see Theorem 3). Once this is obtained, we can use a housing market model $\mathbb{E}$ without errors as representing a market structure.

The econometric analyzer estimates the components in $\mathbb{E}$. Let Γ be some class of market models $\mathbb{E}$ so that each $\mathbb{E}$ in Γ has a *competitive rent vector* $p(\mathbb{E})$. He minimizes the total sum of square residuals $T_R(P_D, p(\mathbb{E}))$ from the observed data P_D to $p(\mathbb{E})$ by choosing $\mathbb{E}$ in Γ. This will be formulated in Sect. 4.

Here, we consider two specific choice problems:

A: a measure η of discrepancy between the data and predicted rent vector;
B: a candidate set of market models Γ.

For A, the discrepancy measure η is defined in terms of $T_R(P_D, p(\mathbb{E}))$ in Sect. 4, to describe how much the estimated result deviates from the optimal estimates. In our application to the data in Tokyo, the value of the measure will be shown to be 1.025–1.032, i.e., 2.5–3.2% of the optimal estimates, by specifying certain classes of market models with homogeneous utility functions.

As an application, we examine the law of diminishing marginal utility for the household. It holds strictly with respect to, particularly, the consumption other than the housing services.

For B, we consider two classes of market models. We show the *Ex Post Rationalization Theorem* in Sect. 6 that we make the value of the discrepancy measure exactly 1 by choosing a certain set Γ of market models. However, this has no prediction power in that only after observations, we adjust a model to fit to the data, because this candidate set Γ has enough freedom. This means that a too general candidate set is meaningless for an econometric analysis. As a kind of opposite, we consider the standard linear regression in our equilibrium-econometric analysis. When the households have the common linear utility functions with respect to attributes of housing and consumption, our econometric analysis becomes linear regression, which is "too specific" in that income effects cannot be taken into account. The choice of an appropriate candidate set is subtle.

This chapter is organized as follows: In Sect. 2, the market equilibrium theory of Kaneko [7] is described together with the example from the Tokyo area. In Sect. 3, a market equilibrium theory with perturbed rents is discussed. In Sect. 4, statistical/econometric treatments are developed as well as a definition of the measure for discrepancy is defined. In Sect. 5, we apply those concepts to a data set from the Tokyo metropolitan area. In Sect. 6, we consider two classes of utility functions. Section 7 gives conclusions and concluding remarks.

2   Equilibrium Theory of Rental Housing Markets

In Sect. 2.1, we describe the market structure $\mathbb{E}$ of (1), and state the existence results of a competitive equilibrium in $\mathbb{E}$ due to Kaneko [7]. In Sect. 2.2, we describe a rental housing market in the Tokyo area.

2.1   Basic Theory: The Assignment Market

The target situation is summarized as $\mathbb{E} = (M, u, I; N, C)$, where

M1: $M = \{1, \ldots, m\}$—the set of *households*, and each $i \in M$ has a utility function u_i and an income $I_i > 0$ measured by the composite commodity other than housing services;

M2: $N = \{1, \ldots, T\}$—the set of *landlords* and each $k \in N$ has a cost function C_k.

Each $i \in M$ looks for (at most) one unit of an apartment, and each $k \in N$ supplies some units of apartments to the market. The apartments are classified into *categories* $1, \ldots, T$. These categories of apartments are interpreted as potentially supplied. Multiple units in one category of apartments may be at the market. When no confusion is expected, we use the term "apartment" for either one unit or a category of apartments.

Each household $i \in M$ chooses a consumption bundle from the consumption set $X := \{\mathbf{0}, \mathbf{e}^1, \ldots, \mathbf{e}^T\} \times \mathbf{R}_+$, where $\mathbf{e}^k$ is the unit T-vector with its k-th component 1 for $k = 1, \ldots, T$ and $\mathbf{R}_+$ is the set of nonnegative real numbers. We may write $\mathbf{e}^0$ for $\mathbf{0}$, meaning that he decides to rent no apartment. A typical element $(\mathbf{e}^k, m_i)$ means that household i rents one unit from the k-th category and enjoys the consumption $m_i = I_i - p_k$ after paying the rent p_k for $\mathbf{e}^k$ from his *income* $I_i > 0$.

The *initial endowment* of each household $i \in M$ is given as $(\mathbf{0}, I_i)$ with $I_i > 0$. His *utility function* $u_i : X \to \mathbf{R}$ is assumed to satisfy:

Assumption A (Continuity and Monotonicity)   For each $x_i \in \{\mathbf{0}, \mathbf{e}^1, \ldots, \mathbf{e}^T\}$, $u_i(x_i, m_i)$ is a continuous and strictly monotone function of m_i and $u_i(\mathbf{0}, I_i) > u_i(\mathbf{e}^k, 0)$ for $k = 1, \ldots, T$.

The last inequality, $u_i(\mathbf{0}, I_i) > u_i(\mathbf{e}^k, 0)$, means that going out of the market is preferred to renting an apartment by paying all his income.

Remark 1   The emphasis of the model $\mathbb{E}$ is on the households and their behavior, rather than on the landlords. We simplify the descriptions of landlords: As long as competitive equilibrium is concerned, we can *assume without loss of generality* that only one landlord k provides all the apartments of category k (cf., Sai [15]). Still, he is a price-taker.

By this remark, we assume that the set of landlords is given as $N = \{1, \ldots, T\}$, where only one landlord k provides the apartments of category k ($k = 1, \ldots, T$).

Each landlord k has a cost function $C_k(y_k) : \mathbf{Z}_+^* \to \mathbf{R}_+$ with $C_k(0) = 0 < C_k(1)$, where $\mathbf{Z}_+^* = \{0, 1, \dots, z^*\}$ and z^* is an integer greater than the number of households m. The cost of providing y_k units is $C_k(y_k)$. No fixed costs are required when no units are provided to the market.[4] The finiteness of $\mathbf{Z}_+^*$ will be used only in Theorem 3.

We impose the following on the cost functions:

Assumption B (Convexity) For each landlord $k \in N$,
$$C_k(y_k + 1) - C_k(y_k) \leq C_k(y_k + 2) - C_k(y_k + 1) \text{ for all } y_k \in \mathbf{Z}_+^* \text{ with } y_k \leq z^* - 2.$$

This means that the marginal cost of providing an additional unit is increasing.

We write the set of all economic models $\mathbb{E} = (M, u, I; N, C)$ satisfying Assumptions A and B by Γ_0.

Now, we define the concept of a competitive equilibrium in $\mathbb{E} = (M, u, I; N, C)$. Let (p, x, y) be a triple of $p \in \mathbf{R}_+^T$, $x \in \{\mathbf{0}, \mathbf{e}^1, \dots, \mathbf{e}^T\}^m$ and $y \in (\mathbf{Z}_+^*)^T$. We say that (p, x, y) is a *competitive equilibrium* in $\mathbb{E}$ iff

UM (Utility Maximization Under the Budget Constraint): for all $i \in M$, $I_i - px_i \geq 0$; and $u_i(x_i, I_i - px_i) \geq u_i(x_i', I_i - px_i')$ for all $x_i' \in \{\mathbf{0}, \mathbf{e}^1, \dots, \mathbf{e}^T\}$ with $I_i - px_i' \geq 0$;

PM (Profit Maximization): for all $k \in N$, $p_k y_k - C_k(y_k) \geq p_k y_k' - C_k(y_k')$ for all $y_k' \in \mathbf{Z}_+^*$;

BDS (Balance of the Total Demand and Supply): $\sum_{i \in M} x_i = \sum_{k=1}^T y_k \mathbf{e}^k$.

Note $px_i := \sum_{k=1}^T p_k x_{ik}$. These conditions constitute the standard notion of competitive equilibrium. Here, each agent maximizes his utility (or profits) as if he can observe all rents $p_1, \dots, p_T$, and then the total demand and supply balance.

The above housing market model is a special case of Kaneko [7], where the existence of a competitive equilibrium is proved.

Theorem 1 (Existence) *In each $\mathbb{E} = (M, u, I; N, C)$ in Γ_0, there is a competitive equilibrium (p, x, y).*

A competitive equilibrium may not be unique, but we choose a particular competitive rent vector, we say that p is a *competitive rent vector* iff (p, x, y) is a competitive equilibrium for some x and y, and that $p = (p_1, \dots, p_T)$ is a *maximum competitive rent vector* iff $p \geq p'$ for any competitive rent vector p'. By definition, a maximum competitive rent vector would be unique if it ever exists. We have the existence of a maximum competitive rent vector in $\mathbb{E} = (M, u, I; N, C)$. This fact has been known in slightly different models since the pioneering work of Shapley-Shubik [16] and Gale-Shapley [4]. Also, see Miyake [13].

Theorem 2 (Existence of a Maximum Competitive Rent Vector) *There is a maximum competitive rent vector in each $\mathbb{E} = (M, u, I; N, C)$ in Γ_0.*

---

[4]The cost functions here should not be interpreted as measuring costs for building new apartments. In our rental housing market, the apartment units are already built and fixed. Therefore, $C_j(y_j)$ is the valuation of apartment units y_j below which he is not willing to rent y_j unit for the contract period. This will be clearer in the numerical example in Sect. 2.2.

We can define also a minimal competitive rent vector, but here we focus on the maximum one.

2.2 Application to a Rental Housing Market in Tokyo (1)

Consider the JR (Japan Railway) Chuo line from Tokyo station in the west direction along which residential areas are spread out. See Fig. 2. The line has 30 stations from Tokyo to Takao station, which is almost on the west boundary of the Tokyo great metropolitan area. Here, we consider only a submarket: we take six stations and three types of sizes for apartments. We explain how we formulate this market as a market model $\mathbb{E} = (M, u, I; N, C)$.

Look at Table 1. The first column shows the time distance from Tokyo to each station, i.e., 18, 23, 31, 52, 64, and 70 min. It is assumed that people commute to Tokyo station (office area) from their apartments. The first raw designates the sizes of apartments, and the three intervals are represented by the medians, 15, 35, and 55 m^2. Thus, the apartments are classified into $T = 6 \times 3 = 18$ categories.

We assume that the households have the common *base utility function* $U^0(t, s, m_i)$, from which the utility function $u_i(x_i, m_i)$ in the previous sense follows: it is given as

$$U^0(t, s, m_i) = -2.2t + 4.0s + 100\sqrt{m_i}, \qquad (2)$$

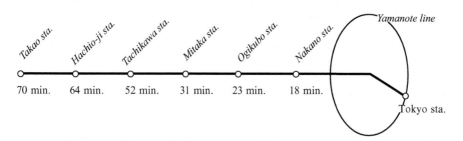

Fig. 2 Chuo line

Table 1 Basic data

Time(min) : size(m²)	<25			25–45			45–65		
	k	h_k	w_k	k	h_k	w_k	k	h_k	w_k
18: Nakano	11	20.4	1176	5	100.4	761	1	180.4	269
23:Ogikubo	12	9.4	1153	6	89.4	739	2	169.4	367
31:Mitaka	14	−8.2	716	8	71.8	571	3	151.8	267
52:Tachikawa	16	−54.4	460	10	25.6	283	4	105.6	260
64:Hachio-ji	17	−80.8	1095	13	−0.8	346	7	79.2	184
70:Takao	18	−94.0	103	15	−14.0	105	9	66.0	102

where t is the time distance, 18, 23, 31, 52, 64, or 70 min, s is the size 15, 35, or 55 m^2, and m_i is the consumption after paying the rent. A pair (t, s) determines a category. By calculating the first part $-2.2t + 4.0s$ of $U^0(t, s, c)$, we obtain h_k for the corresponding cell of Table 1. These h_k's give the ordering over the 18 categories: For example, $-2.2t + 4.0s$ takes the largest value at $(t, s) = (18, 55)$; we label $k = 1$ to the category of $(t, s) = (18, 55)$. Similarly, it takes the 7-th value at $(t, s) = (64, 55)$, and thus $k = 7$. We have the correspondence $\lambda_0(t, s) = k$ from (t, s)'s to k's. We call λ_0 the *category function*.

Now, we define the utility function $u : X = \{e^0, e^1, \ldots, e^{18}\} \times \mathbf{R}_+ \to \mathbf{R}$ by

$$u(e^k, m_i) = h_k + 100\sqrt{m_i}, \tag{3}$$

where $\lambda_0(t, s) = k$ and $h_k = -2.2t + 4.0s$ for $k \geq 1$ and h_0 is chosen so that $h_0 + 100\sqrt{I_m} > h_1$. The derived utility function in (3) satisfies Assumption A. The concavity of $100\sqrt{m_i}$ expresses the law of diminishing marginal utility of consumption.

The third entry w_k of category k in Table 1 is the number of units listed for sale in housing magazines; particularly, *the Yahoo Real Estate* (15, June 2005). The largest number of supplied units is $w_{11} = 1176$ for the smallest apartments in the Nakano area, and the smallest number is $w_9 = 102$ for the largest apartments in the Takao area. The total number of apartment units on the market is $\sum_{k=1}^{18} w_k = 8957$. These large numbers will be important for statistical treatments in subsequent sections.

We *assume* that the same number, $m = 8957$, of households come to the market to look for apartments and they rent all the units.

To determine a competitive equilibrium, we separate between the cost functions for $k = 1, \ldots, T - 1$ and $k = T$. For $k = 1, \ldots, T - 1$, we define the cost function $C_k(y_k)$ as:

$$C_k(y_k) = \begin{cases} c_k y_k & \text{if } y_k \leq w_k \\ \text{``large''} & \text{if } y_k > w_k, \end{cases} \tag{4}$$

where $c_k > 0$ for $k = 1, \ldots, T - 1$ and "large" is a number greater than I_1. Thus, only the supplied units are in the scope of cost functions. For $k = T$, we assume that more units are waiting for the market. Let w_T^0 be an integer with $w_T^0 > w_T$. We define $C_T(y_T)$ by (4) with $c_T > 0$ and substitution of w_T^0 for w_k. Hence, the market rent for an apartment in category T must be c_T. This satisfies Assumption B.

For calculation of the maximum competitive rent vector, we take $c_{18} = 48.0$ and $c_1, \ldots, c_{17}$ are "small" in the sense that all the w_k units are supplied at the competitive rents for $k = 1, \ldots, 17$. The cost 48,000 yen is about the average rents of the smallest category in Takao around in 2005.

Table 2 Calculated and average rents

Time(min) : size(m²)	<25				25–45				45–65			
	k	p_k	$\bar{p}_k$	s_k	k	p_k	$\bar{p}_k$	s_k	k	p_k	$\bar{p}_k$	s_k
18:Nakano	11	78.5	74.4	12.7	5	113.9	112.5	23.8	1	154.8	162.7	26.7
23:Ogikubo	12	74.3	75.8	13.6	6	108.6	107.0	23.1	2	149.0	146.2	20.9
31:Mitaka	14	68.7	68.9	9.8	8	110.6	102.1	21.2	3	140.0	143.1	21.6
52:Tachikawa	16	56.4	59.8	11.0	10	80.7	78.1	12.5	4	116.6	116.0	16.5
64:Hachio-ji	17	50.0	51.5	7.5	13	71.0	73.3	11.3	7	104.0	103.5	17.9
70:Takao	18	48.0	46.4	5.9	15	67.2	65.1	9.6	9	98.1	86.1	11.3

Finally, we assume that the (monthly) income distribution $I = (I_1, \ldots, I_{8957})$ over $M = \{1, \ldots, 8957\}$ is uniform from 100,000 yen to 850,000 yen. Hence, $I_{8957} = 100{,}000$ and $I_1 = 850{,}000$. In fact, this uniform distribution is just for the purpose of calculation, and can be changed into other distributions.[5]

Under the above specification of $\mathbb{E} = (M, u, I; N, C)$, we can calculate the maximum competitive rent vector $p = (p_1, \ldots, p_{18})$, which is given in Table 2. The average rents $\bar{p} = (\bar{p}_1, \ldots, \bar{p}_{18})$ as well as the standard deviations $(s_1, \ldots, s_T)$ from *the Yahoo Real Estate* are given. Figure 3 depicts the average rents $\bar{p} = (\bar{p}_1, \ldots, \bar{p}_{18})$ from the data of as well as $p = (p_1, \ldots, p_{18})$.

In Sect. 4.1, we will define the discrepancy measure in order to consider how much the calculated rent vector $p = (p_1, \ldots, p_T)$ fits the data from housing magazines. For the present data, the value is about 1.032, i.e., the discrepancy is 3.2%.

3 Rental Housing Markets with Housing Magazines

In a competitive equilibrium in $\mathbb{E} = (M, u, I; N, C)$, all the apartment units in each category are uniformly priced, but in reality, prices are not uniform. This non-uniformity represents the effects of non-systematic factors. Here, we modify a housing market model by taking non-systematic factors into account. We show that the market model $\mathbb{E}$ can still be used as an analytic tool for the markets with non-systematic factors.

---

[5]At this stage, the result is not sensitive with the uniform distribution assumption, i.e., if we change it to a truncated normal distribution, the calculated rents are not much changed. However, in the later calculation in Sect. 5, a change of this assumption seems to affect the result.

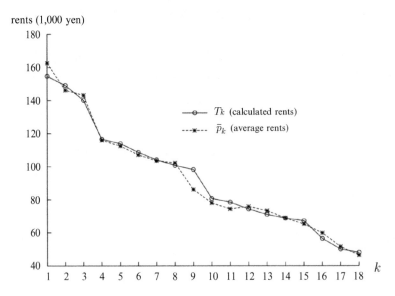

Fig. 3 Calculated and average rents and average prices

Fig. 4 weekly markets

3.1 Time Structure of the Rental Housing Market

Since our approach is a static equilibrium theory, we do not need time indices. However, it would be easier first to describe the economy with the time structure for the consideration of decision making with housing magazines. We use time indices only for this explanation.

The market is recurrent and is described using the "week" due to Hicks [5] in Fig. 4. In week t, market $\mathbb{E}^t(\varepsilon^t) = (\mathbb{E}^t; \epsilon^t)$ has, in addition to the systematic part $\mathbb{E}^t = (M^t, u^t, I^t; N^t, C^t)$, a perturbation term $\epsilon^t = (\epsilon^t_1, \ldots, \epsilon^t_T)$ as the summary of non-systematic factors.

Interactions between information from housing magazines and decision making by households/landlords have a complex temporal structure. For logical clarity, here we simplify the story in the following manner. Before going to the market of week t, households M^t look at housing magazines of week $t-1$ and decide which category they go to. Landlords N^t decide to a supply quantity, also looking at the same housing magazine (recall Remark 1). Decision making by each household is only about category choice; and decision making by each landlord is only about supply quantity choice. Then, households M^t and landlords N^t go to the market of week t, and there they trade apartment units (no explicit decision making is considered here in our theory) and disappear from the market.

In $\mathbb{E}^t(\epsilon^t)$, the rental prices are realized with error term ϵ^t. This ϵ^t is a T-vector of independent random variables which perturb the market rents p_k^t for apartments in category $k = 1, \ldots, T$ to $p_k^t + \epsilon_k^t$. However, when apartment unit d in category k is provided, the error term applied to the unit d is ϵ_{kd}^t. Here, it is assumed that ϵ_{kd}^t is independently and identically distributed as ϵ_k^t over those units in category k.

To distinguish between random variables and their realizations, we prepare the underlying probability space $(\Omega, \mathscr{F}, \mu)$ which all the random variables in this paper follow.

In week $t - 1$, apartment units of categories $1, \ldots, T$ were brought to the housing market. Let $D_1^{t-1}, \ldots, D_T^{t-1}$ are the (finite nonempty) sets of those units. Each unit d in D_k^{t-1} is listed in housing magazines with its realized rent $p_k^{t-1} + \epsilon_{kd}^{t-1}(\omega_o^{t-1})$, where ω_o^{t-1} is the realized value of the state of nature. The entire *housing magazine of week* $t - 1$ is described as

$$\{p_1^{t-1} + \epsilon_{1d}^{t-1}(\omega_o^{t-1}) : d \in D_1^{t-1}\} \quad \cdots \quad \{p_T^{t-1} + \epsilon_{Td}^{t-1}(\omega_o^{t-1}) : d \in D_T^{t-1}\}. \tag{5}$$

Each household $i \in M^t$ looks at the housing magazines (5) of week $t - 1$, and then forms an estimate of the rent distribution:

$$P_k^{i,t} = p_k^{t-1} + \epsilon_k^{i,t} \text{ for each } k = 1, \ldots, T. \tag{6}$$

In general, $P_k^{i,t}(\omega^t) = p_k^{t-1} + \epsilon_k^{i,t}(\omega^t)$ is a random variable for each k; possibly, it may be degenerated such as $P_k^{i,t}(\omega^t) = p_k^{t-1} + \epsilon_{kd}^{t-1}(\omega_o^{t-1})$ given by the observation of a particular unit d. Household i makes a choice of a category by looking at his rent estimator in (6). That is, he maximizes the expected utility (subject to the budget constraint) relative to this rent expectation.

Each landlord k ($k = 1, \ldots, T$) decides the supply quantity of apartment units in category k based on his estimate $p_k^{t-1} + \epsilon_k^{k,t}$ of the rents of apartments in category k.

3.2 Equilibrium with Subjective Estimates

Assuming that the market is stationary and $P_k^{i,t}$, as a random variable, is independent of week t, we drop the superscript t from $\mathbb{E}^t(\epsilon^t)$ and $P_k^{i,t}$. The economy where the households and landlords have their estimators are denoted by $\mathbb{E}(\epsilon; \epsilon^{M \cup N}) = (\mathbb{E}(\epsilon); \epsilon^{M \cup N})$, where $\epsilon^{M \cup N} = (\{\epsilon^i\}_{i \in M}, \{\epsilon^k\}_{k \in N})$. Thus, each household $i \in M$ has his own subjective estimate $P_k^i = p_k + \epsilon_k^i$ in each $k = 1, \ldots, T$, and each landlord $k \in N$ has the rent estimate $P_k^k = p_k + \epsilon_k^k$. We assume that these rent estimates do not take negative values:

$$P_k^i(\omega) \geq 0 \text{ and } P_k^k(\omega) \geq 0 \text{ for all } \omega \in \Omega. \tag{7}$$

Since the realization ω_o^{t-1} of the previous week differs typically from the realization ω_o^t of the present week, we should distinguish between the realization of the previous week and the present one. We still use the symbol ω_o^{t-1} for the previous week, and the symbol ω without the time index for the present week.

We give two examples for such subjective rent estimates.

Example 1 (Average Rents)   Looking at the housing magazine (5), household i (landlord k) takes some samples of rents from category k. Let L_i be the samples taken. Then, if he uses the average of the observed rents, he has a single-value for estimate:

$$P_k^i(\omega) = \sum_{d \in L_k} (p_k + \epsilon_{kd}(\omega_o^{t-1}))/ |L_k| , \tag{8}$$

which is independent of ω. On the other hand, if he is more careful and take some uncertainty about rents into account, he could have the random average P_k^i:

$$P_k^i(\omega) = \sum_{d \in L_k} (p_k + \epsilon_{kd}(\omega))/ |L_k| . \tag{9}$$

This exact form must be very rare. The point of this example is: The number of samples $|L_k|$ is typically small such as $5 \sim 25$. In the second case, since ϵ_{kd} are independent random variables identical to ϵ_k for $d \in D_k$, the expected value is $E(P_k^i) = p_k + E(\epsilon_k)$ and its variance is $E(P_k^i - E(P_k^i))^2 = E(\epsilon_k - E(\epsilon_k))^2/ |L_k|$. Thus, the variance is reciprocal to the number of samples.

The above examples suggest that the economic agents may take rents with smaller variances than the actual variances of $\epsilon = (\epsilon_1, \ldots, \epsilon_T)$. If household i (landlord k) very carefully scrutinizes the housing magazines by drawing a histogram. Since the number of units listed in the magazine is quite large, it is close to the true $P_k^i = p_k + \epsilon_k^i = p_k + \epsilon_k$. Since, however, the magazine is quite large and not well-organized, it is costly to extract the distribution $p_k + \epsilon_k(\cdot)$. Instead, often, the information publicly used is the average rent of samples; the examples of (8) and (9) are better fitting to reality.

In the model $\mathbb{E}(\epsilon; \epsilon^{MUN})$, the concept of a competitive equilibrium is adjusted by incorporating each agent's rent estimation. We, first, take this estimation into account in utility maximization for each household, and then we formulate a landlord's profit maximization.

To capture the budget constraint for household i with estimation $P^i = (P_1^i, \ldots, P_T^i)$, we define the following utility function: for $x_i \in \{0, e^1, \ldots, e^T\}$ and $\omega \in \Omega$,

$$U_i(x_i, I_i - P^i(\omega) \cdot x_i) = \begin{cases} u_i(x_i, I_i - P^i(\omega) \cdot x_i) & \text{if } 0 \leq I_i - P^i(\omega) \cdot x_i \\ \\ u_i(0, I_i) & \text{otherwise.} \end{cases} \tag{10}$$

In the second case, his budget is violated; so, no trade occurs. In general, this utility function $U_i(x_i, I_i - P^i(\cdot) \cdot x_i)$ is a random variable. We define the expected utility before going to a category:

$$EU_i(x_i, I_i - P^i \cdot x_i) = \int_{\omega \in \Omega} U_i(x_i, I_i - P^i(\omega) \cdot x_i) d\mu(\omega). \tag{11}$$

He chooses a category by maximizing this expected utility function over $\{0, e^1, \ldots, e^T\}$.

We assume that each landlord k has a *risk-neutral utility function*. Then his expected utility is calculated as the expected payoff:

$$E(y_k P_k^k - C_k(y_k)) = y_k E(P_k^k) - C_k(y_k). \tag{12}$$

If $E(\epsilon_k^k) = 0$, i.e., $E(P_k^k) = p_k$, (12) becomes simply the profit function $y_k p_k - C_k(y_k)$. However, we treat landlords in the same way as households in that he may construct his rent estimate P_k^k without assuming $E(\epsilon_k^k) = 0$.

In the housing market $\mathbb{E}(\epsilon; \epsilon^{M \cup N})$, a competitive equilibrium is simply defined by substituting the objective functions (11) and (12) for the utility functions and profit functions in UM and PM. Nevertheless, we need to take two approximations: a γ-competitive equilibrium and a convergent sequence of rent estimates.

Let γ be a nonnegative real number. We call (p, x, y) is a *γ-competitive equilibrium* $\mathbb{E}(\epsilon; \epsilon^{M \cup N})$ when the following two conditions and the BDS condition, $\sum_{i \in M} x_i = \sum_{k=1}^{T} y_k e^k$, hold:

γ -Expected Utility Maximization: for all household $i \in M$,

$$EU_i(x_i, I_i - P^i \cdot x_i) + \gamma \geq EU_i(x_i', I - P^i \cdot x_i') \text{ for all } x_i' \in \{0, e^1, \ldots, e^T\}.$$

γ -Expected Profit Maximization: for all landlord $k = 1, \ldots, T$,

$$E(P_k^k y_k - C_k(y_k)) + \gamma \geq E(P_k^k y_k' - C_k(y_k')) \text{ for all } y_k' \in \mathbf{Z}_+^*.$$

The other notion is that $\epsilon^{M \cup N}$ is "small perturbations". To describe this, we introduce the convergence of the vectors of estimators $\epsilon^{M \cup N}$. We say that an error sequence $\{\epsilon^{M \cup N, \nu} : \nu = 1, \ldots\} = \{(\{\epsilon^{i,\nu}\}_{i \in M}, \{\epsilon^{k,\nu}\}_{k \in N}) : \nu = 1, \ldots\}$ is *convergent to 0 in probability* iff for any $\delta > 0$,

$$\mu(\{\omega : \max_{j \in M \cup N} \|\epsilon^{j,\nu}(\omega)\| < \delta\}) \to 1 \text{ as } \nu \to +\infty, \tag{13}$$

where $\|\cdot\|$ is the max-norm $\|(y_1, \ldots, y_T)\| = \max_{1 \leq t \leq T} |y_t|$. This mean that when ν is large enough, the estimation $\epsilon^{j,\nu}(\omega)$ is distributed closely to $\mathbf{0}$.

We have the following theorem. The proof will be given in the appendix.

Theorem 3 (Convergence to $\mathbb{E}$) *Suppose that the sequence of estimation errors* $\{\epsilon^{M\cup N,\nu} : \nu = 1,\ldots\}$ *is convergent to 0 in probability.*

(1): If (p,x,y) be a competitive equilibrium in $\mathbb{E}$, then for any $\gamma > 0$, there is a ν_o such that for any $\nu \geq \nu_o$, (p,x,y) is a γ-competitive equilibrium in $\mathbb{E}(\epsilon; \epsilon^{M\cup N,\nu})$.

(2): Suppose that a triple (p,x,y) satisfies $px^i < I_i$ for all $i \in M$. Then, the converse of (1) holds.

When the rent expectation for landlord $k \in N$ satisfies $E(\epsilon^k) = 0$, his expected profit is simply given as the profit function, and so we need to consider neither the convergent sequence nor the γ-modification for landlord k. Also, if a competitive equilibrium (p,x,y) is strict in the sense that a household (landlord) maximizes his utility at a unique choice, then we do not need the γ-modification for the household (landlord).

The convergence condition is interpreted as meaning that the subjective rent expectation of each household (landlord) has a small variance. Theorem 3 states that when each household i (landlord k) has his rent expectation ϵ^i (or ϵ^k) with a small variance, his utility maximization (or profit maximization) in the idealized market $\mathbb{E}$ is preserved approximately in the market $\mathbb{E}(\epsilon; \epsilon^{M\cup N,\nu})$ for large ν, and *vice versa*. Thus, the competitive equilibrium in $\mathbb{E}(\epsilon; \epsilon^{M\cup N,\nu})$ can well be represented by one in $\mathbb{E}$.

4 Statistical Analysis of Rental Housing Markets

We turn our attention to estimation of the structures of the rental housing market from the data given in housing magazines. In Sect. 4.1 we develop various concepts to connect the data with possible market models and to evaluate such a connection. In Sect. 4.2 we specify a class of market models for estimation.

4.1 Estimation of the Market Structure

We denote, by $\mathbb{E}^o(\epsilon^o) = (\mathbb{E}^o; \epsilon^o)$, the true market, to distinguish between $\mathbb{E}^o(\epsilon^o)$ and an estimated $\mathbb{E}$. We call $\mathbb{E}^o = (M^o, u^o, I^o; N^o, C^o)$ the *latent* true market structure. We assume that this $\mathbb{E}^o$ satisfies Assumptions A and B of Sect. 2, i.e., $\mathbb{E}^o \in \Gamma_0$. The maximum competitive rent vector $p^o = (p_1^o, \ldots, p_T^o)$ of $\mathbb{E}^o$ is called the *latent market rent vector*. Let D_k^o be a nonempty set of apartment units listed in category $k = 1, \ldots, T$. Once the perturbation term ϵ^o is realized at $\omega \in \Omega$ for each $d \in D_k^o, k = 1, \ldots, T$, we have the housing magazines $\{P_{1d}^o(\omega) : d \in D_1^o\}, \ldots, \{P_{Td}^o(\omega) : d \in D_T^o\}$. Here, ω is not fixed to be a specific ω_o. The listed rent for each unit $d \in D_k^o, k = 1, \ldots, T$ is given as: $P_{kd}^o(\omega) = p_k^o + \epsilon_{kd}^o(\omega)$.

We estimate components of $\mathbb{E}^o$ from the housing magazines; in this paper, specifically, we estimate the utility functions of households.

Let $p = (p_1, \ldots, p_T)$ be a rent vector in $\mathbb{R}^T$, which is intended to be an estimated one. Then, the *total sum of square residuals* $T_R(P_D^o(\omega), p)$ is given as

$$T_R(P_D^o(\omega), p) = \sum_{k=1}^{T} \sum_{d \in D_k^o} (P_{kd}^o(\omega) - p_k)^2. \tag{14}$$

This is the distance between the data and estimated rent vector.

Let Γ be a subset of Γ_0 where T is fixed and $M = \{1, \ldots, m\}$ is determined by $m = \sum_{k=1}^{T} |D_k^o|$. Our problem is to choose $\mathbb{E} = (M, u, I; N, C)$ to minimize $T_R(P_D^o(\omega), p(\mathbb{E}))$ in Γ, where $p(\mathbb{E})$ is the maximum competitive rent vector in $\mathbb{E}$. We write our problem explicitly:

Definition 1 (Γ-MSE)   We choose a model $\mathbb{E}$ from Γ to minimize $T_R(P_D^o(\omega), p(\mathbb{E}))$ subject to the condition:

$(*)$: $(p(\mathbb{E}), x, y)$ is a maximal competitive equilibrium in $\mathbb{E}$ for some (x, y) with $y_k = |D_k^o|$ for $k = 1, \ldots, T$.

The additional condition $y_k = |D_k^o|$ for $k = 1, \ldots, T$ requires $(p(\mathbb{E}), x, y)$ to be compatible with the number of apartment units listed in the housing magazines.

If the latent true structure $\mathbb{E}^o$ belongs to Γ, it is a candidate for the solution of the Γ-MSE. However, we do not know whether or not $\mathbb{E}^o$ belongs to Γ. A simple idea is to choose a large class for Γ to guarantee that $\mathbb{E}^o$ could be in Γ. In fact, this idea does not work well in Sect. 6.1, we discuss the negative result for this; we should somehow look at a narrower class for Γ.

As the benchmark, we consider the *average rent estimator*: given the housing magazines $P_D^o = \{P_{kd}^o : d \in D_k^o \text{ and } k = 1, \ldots, T\}$, we define $\overline{P}^o = (\overline{P}_1^o, \ldots, \overline{P}_T^o)$ by

$$\overline{P}_k^o(\omega) = \frac{\sum_{d \in D_k^o} P_{kd}^o(\omega)}{|D_k^o|} \quad \text{for each } \omega \in \Omega \text{ and } k = 1, \ldots, T. \tag{15}$$

This is the best estimator of the latent market rents $p^o = (p_1^o, \ldots, p_T^o)$. Each realization $\overline{P}^o(\omega)$ $(\omega \in \Omega)$ is the unique minimizer of $T_R(P_D^o(\omega), p)$ with no constraints. When $E(\epsilon_k^o) = 0$, $\overline{P}_k^o$ is an unbiased estimator of p_k^o.

Lemma 1 *(1) For each $\omega \in \Omega$, $T_R(P_D^o(\omega), \overline{P}^o(\omega)) \leq T_R(P_D^o(\omega), p)$ for any $p = (p_1, \ldots, p_T) \in \mathbf{R}^T$.*
(2) When $E(\epsilon_k^o) = 0$, $\overline{P}_k^o$ is an unbiased estimator of p_k^o, i.e., $E(\overline{P}_k^o) = p_k^o$.

Proof (1) Let $\omega \in \Omega$ be fixed. Since $T_R(P_D^o(\omega), p)$ is a strictly convex function of $p = (p_1, \ldots, p_T) \in \mathbf{R}^T$, the necessary and sufficient condition for p to be the minimizer of $T_R(P_D^o(\omega), p)$ is given as $\partial T_E(P_D(\omega), p)/\partial p_k = 0$ for all $k = 1, \ldots, T$. Only the average $\overline{P}^o(\omega) = (\overline{P}_1^o(\omega), \ldots, \overline{P}_T^o(\omega))$ satisfies this condition.

(2) Since ϵ_{kd}^o is identical to ϵ_k^o for all $d \in D_k^o$ and $E(\epsilon_k^o) = 0$, we have $E(\epsilon_{kd}^o) = 0$ for all $d \in D_k^o$. Hence $E(\overline{P}_k^o) = \sum_{d \in D_k^o} E(P_{kd}^o) / |D_k^o| = \sum_{d \in D_k^o} (p_k^o + E(\epsilon_{kd}^o)) / |D_k^o| = p_k^o$. $\qquad\square$

The estimator $\overline{P}^o$ enjoys various desired properties such as consistency (i.e., convergence to the latent market rent vector p^o in probability as $\min_k |D_k|$ tends to infinity) and efficiency in the sense of Cramer-Rao. For these, see van der Vaart [18].

We have the decomposition of the total sum of square residuals, which corresponds to the well-known decomposition property in the regression model (cf., Wooldridge [21]). This will be a base for our further analysis.

Lemma 2 (Decomposition) *For each $\omega \in \Omega$,*

$$T_R(P_D^o(\omega), p) = T_R(P_D^o(\omega), \overline{P}^o(\omega)) + \sum_{k=1}^T |D_k^o| (\overline{P}_k^o(\omega) - p_k)^2. \tag{16}$$

Proof The term $\sum_{d \in D_k^o} (P_{kd}^o(\omega) - p_k)^2$ of $T_R(P_D^o(\omega), p)$ for each k is transformed to:

$$\sum_{d \in D_k^o} (P_{kd}^o(\omega) - \overline{P}_k^o(\omega) + \overline{P}_k^o(\omega) - p_k)^2 = \sum_{d \in D_k^o} (P_{kd}^o(\omega) - \overline{P}_k^o(\omega))^2$$

$$+ \sum_{d \in D_k^o} 2(P_{kd}^o(\omega) - \overline{P}_k^o(\omega)) \cdot (\overline{P}_k^o(\omega) - p_k) + \sum_{d \in D_k^o} (\overline{P}_k^o(\omega) - p_k)^2.$$

The second term of the last expression vanishes by (15). The third is written as $|D_k^o| (\overline{P}_k^o(\omega) - p_k)^2$. We have (16) by summing these over $k = 1, \ldots, T$. $\qquad\square$

The first term of (16) is the residual between the data and the averages (optimal estimates) of rents. The second is the total sum of the differences between the average $\overline{P}^o(\omega)$ and p, and this is newly generated by the estimates $p = (p_1, \ldots, p_T)$. We call the ratio

$$\eta(p)(\omega) = \frac{T_R(P_D^o(\omega), p)}{T_R(P_D^o(\omega), \overline{P}^o(\omega))} = 1 + \frac{\sum_{k=1}^T |D_k^o| (\overline{P}_k^o(\omega) - p_k)^2}{T_R(P_D^o(\omega), \overline{P}^o(\omega))} \tag{17}$$

the *discrepancy measure* of p from of $\overline{P}^o(\omega)$. The second is the *theoretical discrepancy*, relative to the smallest total sum of residuals from $\overline{P}^o(\omega)$. In the example of Sect. 2.2, $\eta \doteq 1.032$ (denoted by η^0), i.e., the theoretical discrepancy is only 3.2%.[6]

---

[6]Incidentally, in the present context, the *coefficient of determination* is defined as $\sum_{k=1}^T |D_k^o| (\overline{P}_k^o(\omega) - \overline{\overline{P}}^o(\omega))^2 / T_V(P_D^o(\omega), \overline{\overline{P}}^o(\omega))$, where $\overline{\overline{P}}^o(\omega)$ is the entire average of P_D^o. It indicates how much the systematic factors explain the observed rental prices. In the above example, the coefficient is approximately 0.757.

4.2   Subclass Γ_{sep} of Γ_0

We estimate the utility functions $u^o = (u_1^o, \ldots, u_m^o)$ in $\mathbb{E}^o = (M^o, u^o, I^o; N^o, C^o)$ from the rents listed in the housing magazines, assuming that the other components in $\mathbb{E}^o = (M^o, u^o, I^o; N^o, C^o)$ are given from the other information included in the housing magazines. For example, the set of households M^o is given as $\{1, \ldots, m\}$, where m is the cardinality of the data set $D^o = \cup_{k=1}^T D_k^o$.

The set of market models Γ_{sep} consists of $\mathbb{E} = (M, u, I; N, C)$ satisfying the following three conditions:

S1: The incomes of households are ordered as $I_1 \geq \ldots \geq I_m > 0$.
S2: Every household in M has the same utility function $u_1 = \ldots = u_m$ expressed as

$$u(\mathbf{e}^k, m_i) = h_k + g(m_i) \quad \text{for all } (\mathbf{e}^k, m_i) \in X, \tag{18}$$

where $h_0, h_1, \ldots, h_T$ are given real numbers with $h_k > h_0$ for $k = 1, \ldots, T$ and $g : \mathbf{R}_+ \to \mathbf{R}$ is an increasing and continuous concave function with $g(m_i) \to +\infty$ as $m_i \to +\infty$ and $h_0 + g(I_i) > h_k + g(0)$ for $k = 1, \ldots, T$.
S3: Each landlord $k = 1, \ldots, T$ has a cost function of the form (4).

In S1, the households are ordered by their incomes. Condition S2 has two parts: Every household has the same utility function; and the utility function is expressed in the separable form. The former part is interpreted as requiring the households to have the same location of their offices. The latter still allows the law of diminishing marginal utility over consumption, i.e., $g(m_i)$ may be strictly concave. Condition S3 is for simplification: Our theory emphasizes on the households' side.

The set Γ_{sep} may be regarded as very narrow from the viewpoint of mathematical economics in that the households have the same utility functions and the landlords' cost functions are also very specific. However, we will show in Sect. 6.1 that the class Γ_{sep} is still too large in that the estimated model has no prediction power. Thus, we will consider a narrower class for Γ.

A method of calculating a maximum competitive equilibrium (p, x, y) in $\mathbb{E} = (M, u, I; N, C)$ was given in Kaneko [8] and Kaneko et al. [10]. This method is used to implement our econometrics. Here, we describe this method without a proof.

Consider a rent vector $p = (p_1, \ldots, p_T)$ with $p_1 \geq \ldots \geq p_T > 0$. This is obtained by renaming $1, \ldots, T$. Then, we regard the units in category 1 as the best, and will suppose that the richest households $1, \ldots, |D_1^o|$ rent them. Similarly, the units in category 2 are the second best and the second richest households $|D_1^o| + 1, \ldots, |D_1^o| + |D_2^o|$ rent them. In general, defining

$$G(k) = \sum_{t=1}^k |D_t^o| \quad \text{for all } k = 1, \ldots, T, \tag{19}$$

we suppose the households $G(k-1)+1, \ldots, G(k)$ rent units in category k. We focus on the boundary households $G(1), G(2), \ldots, G(T-1)$ and their incomes $I_{G(1)}, I_{G(2)}, \ldots, I_{G(T-1)}$.

We have the following lemma due to Kaneko [8] and Kaneko et al. [10]. Our econometric calculation is based on this lemma.

Lemma 3 (Rent Equations) *Consider a vector* $(p_1, \ldots, p_T)$ *with* $p_1 \geq \ldots \geq p_T > 0$. *Let* $\mathbb{E} = (M, u, I; N, C) \in \Gamma_{sep}$ *satisfying*

(1): $p_k \leq I_{G(k)}$ *for all* $k = 1, \ldots, T-1$;
(2): $c_k \leq p_k$ *and* $w_k = \left| D_k^o \right|$ *for all* $k = 1, \ldots, T$, *where* c_k *is the marginal cost given in (4). Suppose also that* $(p_1, \ldots, p_T)$ *satisfies*

$$
\boxed{
\begin{aligned}
h_{T-1} + g(I_{G(T-1)} - p_{T-1}) &= h_T + g(I_{G(T-1)} - p_T) \\
h_{T-2} + g(I_{G(T-2)} - p_{T-2}) &= h_{T-1} + g(I_{G(T-2)} - p_{T-1}) \\
&\cdots \\
h_1 + g(I_{G(1)} - p_1) &= h_2 + g(I_{G(1)} - p_2).
\end{aligned}
}
\tag{20}
$$

Then, there is an allocation (x, y) *such that* (p, x, y) *is a maximum competitive equilibrium in* $\mathbb{E}$ *with* $y_k = \left| D_k^o \right|$ *for all* $k = 1, \ldots, T$.

In (20), the boundary household $G(T-1)$ compares his utility $h_{T-1} + g(I_{G(T-1)} - p_{T-1})$ from staying in category $T-1$ with the utility $h_T + g(I_{G(T-1)} - p_T)$ obtained by switching to category T. Also, the household $G(T-2)$ makes a parallel comparison, between staying in category $T-2$ and moving to category $T-1$ and so on. The logic of this argument is essentially the same as Ricardo's [14] differential rents. The rent $p_T = c_T$ in the worst category T is regarded as the land rent-cost of farm lands, which corresponds to Ricardo's absolute rent.

5 Application to the Market in Tokyo (2)

Here, we apply our equilibrium-econometric analysis to the rental housing market in Tokyo described in Sect. 2.2. First, we give a simple heuristic discussion on our application, and then give a more systematic study of it.

5.1 Heuristic Discussion

For a study of a specific target, we consider a more concrete class for Γ than the class Γ_{sep} given in Sect. 4.2. In Sect. 2.2, we used a specific form of the base utility function $U^0(t, s, m_i) = -2.2t + 4.0s + 100\sqrt{m_i}$ and obtained the resulting value of the discrepancy measure, $\eta^0 = 1.032$. Perhaps, we should explain how we have found it and how good it is relative to others.

Let us compare several other base utility functions with (2):

$U^1(t, s, m_i) = -t + s + 100\sqrt{m_i}$	$\eta^1 = 3.259$
$U^2(t, s, m_i) = -2t + 255\sqrt{s + 1000} + 100\sqrt{m_i}$	$\eta^2 = 1.036$
$U^3(t, s, m_i) = -74t + 165s + 100m_i$	$\eta^3 = 1.124.$

$$(21)$$

With U^1, the discrepancy measure η takes large value 3.259. Thus, the total sum of square residuals from the estimated rents is more than the three-times of that from the average rents. With U^2, the value of η is already almost as small as 1.032 given by (2). With U^3, it is larger than this value, but U^3 is entirely linear. The law of diminishing marginal utility does not hold.

The case of U^1 tells that if coefficients are arbitrarily chosen, the discrepancy value could be large. On the other hand, U^0 is chosen by minimizing the discrepancy measure η by changing the coefficients of t and s in the class of base utility functions:

$$\mathscr{U}(1, 1, \tfrac{1}{2}) := \{U(t, s, m_i) = -\alpha_1 t + \alpha_2 s + 100\sqrt{m_i} : \alpha_1, \alpha_2 \in R\}, \qquad (22)$$

where $1, 1$ and $\tfrac{1}{2}$ are the exponents of t, s and m_i. The coefficient 100 of the third term is chosen to make the values of α_1, α_2 clearly visible. Both U^0 and U^1 belong to this class. Then, $U^0(t, s, m_i)$ is obtained by minimizing the discrepancy measure η in this class. This is not the exact solution but is calculated using a method of grid-search by a computer.

Consider our computation procedure more concretely. Suppose that $U \in \mathscr{U}(1, 1, \tfrac{1}{2})$ is given. For each $(t, s) \in \{18, 23, 31, 52, 64, 70\} \times \{15, 35, 55\}$, we have the value $-\alpha_1 t + \alpha_2 s$, which gives the ranking, $1, \ldots, 18$ over $\{18, 23, 31, 52, 64, 70\} \times \{15, 35, 55\}$. Recall that this is described by the category function λ_0. The k-th category has $h_k = \alpha_1 t + \alpha_2 s$ and $\lambda_0(k) = (t, s)$. This method is the same as in Sect. 2.2. Hence, U determines

$$u(e^k, m_i) = h_k + 100\sqrt{m_i} \text{ for } k = 0, 1, \ldots, T. \qquad (23)$$

Thus, each $U \in \mathscr{U}(1, 1, \tfrac{1}{2})$ determines $\mathbb{E} \in \Gamma_{sep}$.

Now, we consider the subclass $\Gamma(1, 1, \tfrac{1}{2})$ of Γ_{sep} defined by

$$\{(M, u, I; N, C) \in \Gamma_{sep} : u \text{ is determined by some } U \in \mathscr{U}(1, 1, \tfrac{1}{2})\}. \qquad (24)$$

Then, we apply the $\Gamma(1, 1, \tfrac{1}{2})$-MSE problem to the data in Sect. 2.2, and find an approximate solution (α_1, α_2) for it.

An approximate solution will be obtained by the following process.

Step 1: we assume that each of α_1 and α_2 takes a (integer) value from some intervals, say, $[1, 100]$. Then, we have $100^2 = 10^4$ combinations of (α_1, α_2).

Step 2:   for each combination (α_1, α_2), we find a maximum competitive rent vector p compatible with the data set $P_D^o(\omega)$ and we have the value η of discrepancy measure. The algorithm to find a maximum competitive rent vector given by Lemma 3 is used to find the rent vector.

Step 3:   we find a combination (α_1, α_2) with the minimum value of η among 10^4 combinations of (α_1, α_2).

If a solution is on the boundary, we calibrate the intervals, and if not, we repeat these steps by choosing a smaller intervals with finer grids. Hence, the computation to obtain the minimum value of η is not exact: it may be a local optimum as well as an approximation.

By the above simulation method, we have found the utility function $U^0(t, s, m_i)$ of (2) in the class $\Gamma(1, 1, \frac{1}{2})$ with $\eta^0 = 1.032$.

The base utility function U^2 is obtained by minimization in the class $\mathscr{U}(1, \frac{1}{2} \circledast \beta_2, \frac{1}{2})$:

$$\{U(t, s, m_i) = -\alpha_1 t + \alpha_2 \sqrt{s + \beta_2} + 100\sqrt{m_i} : \alpha_1, \alpha_2, \beta_2 \in R\}. \qquad (25)$$

In fact, when β_2 is increased, the optimal value of η is decreasing (we calculated η up to $\beta_2 = 400{,}000$), but it does not reach $\eta^0 = 1.032$. Since β_2 is getting large, the second term is getting closer to the linear function. Therefore, we interpret this result as meaning that the base utility function $U^0(t, s, m_i) = -2.2t + 4.0s + 100\sqrt{m_i}$ of (2) would be the limit function.

The utility function $U^3(t, s, m_i)$ is obtained by minimizing the value η in the class $\mathscr{U}(1, 1, 1)$:

$$\{U(t, s, m_i) = -\alpha_1 t + \alpha_2 s + 100 m_i : \alpha_1, \alpha_2 \in R\}. \qquad (26)$$

That is, the utility functions are entirely linear. The estimation in this class is only interested in seeing the relationship between our Γ-MSE problem and the standard linear regression. This will be discussed in Sect. 6.2.

5.2   Law of Diminishing Marginal Utility

In the above classes of base utility functions, $U^0(t, s, m_i)$ gave the best value to the discrepancy measure. The law of diminishing marginal utility holds strictly only for the consumption term m_i, but not for the other variables, the commuting time-distance t and size of an apartment s. One possible test of this observation is to broaden the class of base utility functions. Here, we will give this test.

Consider the following class $\mathscr{U}(\pi_1 \circledast \beta_1, \pi_2 \circledast \beta_2, \pi_3 \circledast \beta_3)$:

$$U(t, s, m_i) = \alpha_1(\beta_1 - t)^{\pi_1} + \alpha_2(s + \beta_2)^{\pi_2} + 100(m_i + \beta_3)^{\pi_3}, \qquad (27)$$

where $\alpha_1, \alpha_2, \beta_1, \beta_2, \beta_3, \pi_1, \pi_2, \pi_3$ are all real numbers. The introduction of β_1 is natural, since the commuting time-distance t has a limit. The parameters β_2 and β_3 will be interpreted after stating the calculation result. The parameters π_1, π_2, π_3 are related to the law of diminishing marginal utility. When they are close to 1, the law is regarded as not holding, and when they are far away from 1, the law holds.

The computation result is given as

$$U^{MU}(t, s, m_i) = 3.53(140 - t)^{0.75} + 2.68(s + 200)^{0.91} + 100(m_i - 25)^{0.40}, \quad (28)$$

and the discrepancy value is $\eta^{MU} = 1.025$. Assuming the incomes are uniformly distributed, we adjusted parameters of the lowest income I_{8957}, the highest income I_1 and the lowest rent p_{18}, and obtained the estimated minimum income as $I_{8957} = 94$ ($\times 1000$yen) and the highest as $I_1 = 1120$.[7]

First, the law of diminishing marginal utility holds for each variable. However, the degree is quite different: the degree of diminishing marginal utility is the largest with consumption, the second with the commuting time-distance, and is the least with the size of an apartment unit.

The fact that it is the least with the size may be caused by our restriction on apartments up to $65\,\mathrm{m}^2$. In Tokyo, we may find a quite small number of apartments larger than $85\,\mathrm{m}^2$, and omitted these "large" apartments, since the number of supply is much smaller than the smaller types. This may be the reason for almost constant marginal utility.

Second, the degree for the commuting time-distance is higher than that for the apartment size. This suggests, perhaps, that the time-distance 70 min to Takao station is already quite large. Our computation result is sensitive with $\beta_1 = 140$, i.e., if we change $\beta_1 = 140$ slightly either up or down, the value of η changed. Thus, this upper limit has a specific meaning; it may be an upper limit for commuting.

Finally, the degree of diminishing marginal utility for consumption is quite large. This means that the choice by a household renting an apartment crucially depends upon its income level. The dependence of willingness-to-pay for an apartment upon income is quite strong: a poor people do not (or cannot) want to pay for a rent for a good apartment, but if they become rich, they would change their attitudes.

Nevertheless, the discrepancy value $\eta^{MU} = 1.025$ for U^{MU} is not very different from $\eta^0 = 1.032$ for U^0; despite of the fact that the latter has 2 parameters controlled and the former has 8. This means that more precision after U^0 does not give much differences. It is more important to see the difference between the discrepancy values for U^0 and U^3 ($\eta^3 = 1.124$) in (21). After all, we conclude that the law of diminishing marginal utility surely holds for consumption, but less for other variables.

---

[7]One possible amendment of our estimation is to change the assumption on the income distribution. We have assumed that the incomes are distributed from the lowest I_{8957} to the highest I_1. The above computation result seems to be quite sensitive by changing these lowest and highest income levels. Hence, it could give a better result if we replace the assumption of a uniform distribution by the data available from the other source. This is an open problem.

This conclusion differs from the estimation result of a utility function in Kanemoto-Nakamura [11] in the hedonic approach (cf., Epple [3]). It is stated in [11, p. 227], that the degree of diminishing marginal utility is very low, for example, consumption term is $x^{0.978}$. The approach itself is totally different from ours. One difference is: all variables take continuous values in the hedonic approach. This approach requires a very large variety of attributes of apartment units. In contrast, the number T of apartment categories should not be so large, because the choice of description criteria is restricted, as discussed in Sect. 1.1.

6   Two Classes of Market Models

First, we argue that Γ_{sep} is too large as a candidate set of models for estimation. Second, we consider the other extreme, i.e., the class of linear utility functions, and show that the Γ-MSE problem is equivalent to linear regression.

6.1 Γ_{sep}-Market Structure Estimation: Ex Post Rationalization

From the viewpoint of mathematical economics, the class Γ_{sep} of market models is quite restrictive. However, the following theorem implies that it is too large to have meaningful estimation. A proof will be given in the end of this section.

Theorem 4 (Ex Post Rationalization)   *Suppose that each D_k^o is nonempty and the average rents $\overline{P}^o(\omega) = (\overline{P}_1^o(\omega), \ldots, \overline{P}_T^o(\omega))$ are positive. Then, there exists a market model $\mathbb{E} = (M, u, I; N, C)$ in the class Γ_{sep} such that for some (x, y), $(\overline{P}^o(\omega), x, y)$ is a maximum competitive equilibrium in $\mathbb{E}$ with $y_k = \left| D_k^o \right| > 0$ for $k = 1, \ldots, T$. This existence assertion holds for any fixed $g : R_+ \to R$ in Condition S2 of Sect. 4.2.*

Within the class Γ_{sep}, we can "fully explain" any data set from housing magazines in the sense that the estimate coincides with the average rents $\overline{P}^o(\omega)$ and the discrepancy measure η takes the exact value 1. The key fact for this is that the number of dependent variables $\overline{P}^o(\omega) = (\overline{P}_1^o(\omega), \ldots, \overline{P}_T^o(\omega))$ is the same as that of independent variables $(h_1, \ldots, h_T)$ in utility function $u(e^k, m_i) = h_k + g(m_i)$. For a different observed $\overline{P}(\omega) = (\overline{P}_1(\omega), \ldots, \overline{P}_T(\omega))$, the theorem gives different $(h_1, \ldots, h_T)$. The $\Gamma(1, 1, \frac{1}{2})$-MSE problem in Sect. 5.1 exhibits a clear-cut contrast: 18 average rents are explained by the choice of parameters by changing essentially 2 parameter values, and $\eta^0 = 1.032$.

Should we be pleased by finding a class to guarantee to "fully explain" each data set? Or should we interpret this theorem as meaning that the true market $\mathbb{E}^0$ is included in the class Γ_{sep}?

Contrary to these interpretations, we regard the above theorem as a negative result. The estimated economic model critically depends upon the observed average rents $\overline{P}^o(\omega) = (\overline{P}_1^o(\omega), \ldots, \overline{P}_T^o(\omega))$. If a different ω' happens and the realized rents

$\overline{P}^o(\omega')$ are different, the estimated model $\mathbb{E}'$ differs, too. This estimation explains the observed rents only after observations; it cannot make any meaningful forecast. In particular, since the assertion is done with an arbitrary given function g, it is totally incapable in talking about the law of diminishing marginal utility.[8]

Perhaps, this is related to the fact that the degree of diminishing marginal utility is very low in the hedonic price approach mentioned in Sect. 5. It allows a great variety of attributes, which is contrary to the above negative interpretation of Theorem 4.

Proof of Theorem 4 We denote $(\overline{P}_1^o(\omega), \dots, \overline{P}_T^o(\omega))$ by $(p_1, \dots, p_T)$, and let $G(k) = \sum_{t=1}^k |D_t^o|$ for $k = 1, \dots, T$. We assume without loss of generality that $p_1 \geq \dots \geq p_T > 0$. First, we let $g : \mathbf{R}_+ \to \mathbf{R}$ be any monotone, strictly concave and continuous function with $\lim_{m_i \to +\infty} g(m_i) = +\infty$.

Let $h_0 = 0$. We choose $I_m, I_{G(T-1)}, \dots, I_{G(1)}$ and define $h_T, h_{T-1}, \dots, h_1$ inductively as follows: the base case is as follows:

$(T - 0)$: choose an income level I_m so that $I_m > p_T > 0$, and then choose $h_T > h_0 + g(I_m) - g(I_m - p_T)$.
The choices of I_m and h_T are possible by the monotonicity of g. Here, $h_T + g(I_m - p_T) > h_0 + g(I_m)$.
Let k be an arbitrary number with $1 \leq k < T$. The inductive hypothesis is that $I_{G(k)}$ and h_k are already defined. First, we choose $I_{G(k-1)}$ so that
$(k - 1)$: $I_{G(k-1)} > p_{k-1}$ and $I_{G(k-1)} > I_{G(k)}$.
This choice is simply possible. Then we define h_{k-1} by
$(k - 2)$: $h_{k-1} = h_k + g(I_{G(k-1)} - p_k) - g(I_{G(k-1)} - p_{k-1})$.
Since $g(I_{G(k-1)} - p_{k-1}) \leq g(I_{G(k-1)} - p_k)$, we have $h_{k-1} \geq h_k$.

By the above inductive definition, we have $I_m, I_{G(T-1)}, \dots, I_{G(1)}$ and $h_T, h_{T-1}, \dots, h_1$. We also choose other I_i's ($i \neq m$ and $i \neq G(k)$ for $k = 1, \dots, T$) so that $I_m \leq I_{m-1} \leq \dots \leq I_1$.

Thus, we have the utility function $u(e^k, m_i) = h_k + g(m_i)$ for $(e^k, m_i) \in X$. By the above inductive definition, $(p_1, \dots, p_T)$ satisfies the recursive Eq. (20).

Let us define the cost function $C_k(\cdot)$ for landlord k. We assume $0 < c_k \leq p_k$ for all $k = 1, \dots, T-1$ and $c_T = p_T$. Then each $C_k(\cdot)$ is defined by (4) for $k = 1, \dots, T$. Then, by Lemma 3, $(p_1, \dots, p_T)$ is the maximum competitive rent vector of $\mathbb{E}$ with $y_k = |D_k^o|$ for $k = 1, \dots, T$. □

---

[8]The reader may recall the Debreu-Mandel-Sonnenshein Theorem in general equilibrium theory (see Mas-Colell et al. [12]) stating that any demand function with a certain required condition is derived from some economic model. It describes the equivalence between the set of demand curves and the set of economic models. In this sense, it gives an important implication to the theory of general equilibrium theory.

6.2   Linear Utility Functions and Linear Regression

Here, we compare our approach with the linear utility assumption to linear regression. We assume that there are L attributes for the base utility function U for each household, and the domain of U is expressed as $Y = \mathbf{R}_+^L \times \mathbf{R}_+$. In the example of Sect. 2.2, there are only two attributes, the commuting time t and the apartment size s. A linear base utility function over Y is expressed as

$$U(a_1, \ldots, a_L, m_i) = \sum_{l=1}^{L} \alpha_l a_l + m_i \text{ for all } (a_1, \ldots, a_L, m_i) \in Y. \tag{29}$$

Here, a_l represents the magnitude of the l-th attribute of an apartment and $\alpha_l \in \mathbf{R}$ is its coefficient for $l = 1, \ldots, L$. We denote the set of all base utility function of the form (29) by $\mathscr{U}_{lin}$. We choose 1 for the coefficient of m_i for a direct comparison to the linear regression analysis, while it was 100 in the previous examples.

An *attribute vector* $\tau^k = (\tau_1^k, \ldots, \tau_L^k) \in \mathbf{R}_+^L$ is given for each $k = 0, 1, \ldots, T$. That is, the choice $\mathbf{e}^k$ gives the attribute vector τ^k, which means that an apartment in category k has the magnitudes $\tau_1^k, \ldots, \tau_L^k$ of attributes $1, \ldots, L$. For $k = 0$, τ^0 is interpreted as the attributes of the outside option. In the example of Sect. 2.2, category $k = 5$ (Nakano, size: 25–45) has the attribute vector $\tau^5 = (18 \text{ min}, 35 \text{ m}^2)$. Then, each U in $\mathscr{U}_{lin}$ determines

$$u(\mathbf{e}^k, m_i) = U(\tau^k, m_i) = \sum_{l=1}^{L} \alpha_l \tau_l^k + m_i \text{ for all } k = 0, 1, \ldots, T. \tag{30}$$

We define the class $\Gamma_{lin} := \{\mathbb{E} \in \Gamma_{sep} : u \text{ is determined by } U \in \mathscr{U}_{lin} \text{ and } \tau^0, \ldots, \tau^T\}$. The boundary condition "$u(0, I_i) > u(\mathbf{e}^k, 0)$ for all $k = 1, \ldots, T$" holds $\mathbb{E} \in \Gamma_{lin}$, because $\mathbb{E} \in \Gamma_{sep}$. Once this class is defined, we have the Γ_{lin}-MSE problem.

The next lemma states that the competitive rents in $\mathbb{E} \in \Gamma_{lin}$ are simply described by the utility from the attributes of an apartment and some constant.

Lemma 4 *Let* $\mathbb{E} \in \Gamma_{lin}$. *If* $p = (p_1, \ldots, p_T)$ *is a maximum competitive rent vector in* $\mathbb{E} \in \Gamma_{lin}$, *then there is some* β *such that*

$$p_k = \sum_l \alpha_l \tau_l^k + \beta > 0 \text{ for } k = 1, \ldots, T \text{ and } \beta < -\sum_{l=1}^{L} \alpha_l \tau_l^0. \tag{31}$$

Proof Let (p, x, y) be any competitive equilibrium in $\mathbb{E} = (M, u, I; N, C)$ in Γ_{lin} with $|D_k^o| = y_k > 0$ for all $k = 1, \ldots, T$. Without loss of generality, we assume that $p_k \geq p_T$ for $k = 1, \ldots, T - 1$. First, we show

$$p_k - p_T = \sum_l \alpha_l \tau_l^k - \sum_l \alpha_l \tau_l^T \text{ for all } k = 1, \ldots, T. \tag{32}$$

Suppose that this is shown. Let $\beta = p_T - \sum_l \alpha_l \tau_l^T$. We have, by (32), $p_k = \sum_l \alpha_l \tau_l^k + \beta$ for $k = 1, \ldots, T$. For each k, since $|D_k^o| = y_k > 0$ and $c_k > 0$, we have $p_k \geq c_k$. Hence $p_k > 0$ for $k = 1, \ldots, T$, which is the first half of (31). Since any household i in D_T^o chooses the T-th apartment rather than $(0, I_i)$, i.e., $U(e^T, I_i - p_T) = \sum_l \alpha_l \tau_l^T + I_i - (\sum_l \alpha_l \tau_l^T + \beta) = I_i - \beta > u(0, I_i) = h_0 + I_i = \sum_l \alpha_l \tau_l^0 + I_i$, which implies $\beta < -\sum_l \alpha_l \tau_l^0$.

Now let us prove (32). Consider any $k = 1, \ldots, T$. Since $|D_k^o| > 1$, we take a household i with $x_i = e^k$. Since he chooses $x_i = e^k$ by utility maximization under $p = (p_1, \ldots, p_T)$, we have $\sum_l \alpha_l \tau_l(k) + I_i - p_k \geq \sum_l \alpha_l \tau_l(T) + I_i - p_T$. By the same argument for a household i' with $x_{i'} = e^T$, we have $\sum_l \alpha_l \tau_l(t) + I_{i'} - p_k \leq \sum_l \alpha_l \tau_l(T) + I_{i'} - p_T$. Equation (32) follows from these two inequalities. $\square$

Now let us turn our consideration to linear regression: the rent of an apartment in category k is assumed to be a linear combination of the magnitudes of attributes and some constant. Mathematically, it is exactly the same as (31) subject to some error, that is,[9]

$$P_k = \sum_{l=1}^{L} \alpha_l \tau_l^k + \beta + \epsilon_k \quad \text{for } k = 1, \ldots, T. \tag{33}$$

The attribute vectors $\tau^0, \ldots, \tau^T$ are fixed. Given the housing magazine $P_D^o(\omega)$ as data, we estimate $\alpha = (\alpha_1, \ldots, \alpha_L)$ and β by minimizing the sum of square residuals, i.e., *the method of least squares*. It is formulated by the following minimization problem:

$$\min_{\alpha, \beta} \sum_{k=1}^{T} \sum_{d \in D_k^o} (P_{kd}^o(\omega) - p_k)^2 = \min_{\alpha, \beta} \sum_{k} \sum_{d} \left(P_{kd}^o(\omega) - (\sum_{l=1}^{L} \alpha_l \tau_l^k + \beta) \right)^2. \tag{34}$$

This is a no-constraint minimization problem and has a solution $(\widehat{\alpha}, \widehat{\beta})$.

The above linear regression problem is very close to the Γ_{lin}-MSE problem. In linear regression, however, neither utility maximization nor profit maximization is included. It would be worth considering the exact relationship.

The minimization (34) is applied to any data set $P_D^o(\omega)$, even if $P_D^o(\omega)$ contains negative elements. On the other hand, the Γ_{lin}-MSE problem may not be if it contains negative elements: if the estimated rent for category k is negative, landlord k provides no apartments, i.e., condition $y_k = |D_k^o|$ is violated. We need a certain condition to avoid such a case. For this, the following condition is enough, though it is not directly on $P_D^o(\omega)$:

---

[9]This is regarded as a linear hedonic price model.

$$\sum_{l=1}^{L} \alpha_l \tau_l^k + \beta > 0 \text{ for all } k = 1, \dots, T \text{ and } \beta < -\sum_{l=1}^{L} \alpha_l \tau_l^0. \qquad (35)$$

Again, this corresponds to (31) in Lemma 4. Using this condition, we can state the equivalence between the Γ_{lin}-MSE problem and the linear regression problem (34).

Theorem 5 (Linear Regression) *Let $(\widehat{\alpha}, \widehat{\beta}) \in \mathbf{R}^L \times \mathbf{R}$. Then, $(\widehat{\alpha}, \widehat{\beta})$ is a solution of the minimization (34) and satisfies (35) if and only if there is a solution model $\widehat{\mathbb{E}}$ in the Γ_{lin}-MSE problem such that $\widehat{u}$ of $\widehat{\mathbb{E}}$ is determined by U of (30) with $\widehat{\alpha}$ and the maximum competitive rent vector $p(\widehat{\mathbb{E}}) = \widehat{p}$ is given as*[10]

$$\widehat{p}_k = \sum_{l=1}^{L} \widehat{\alpha}_l \tau_l^k + \widehat{\beta} \text{ for all } k = 1, \dots, T. \qquad (36)$$

In the example of Chuo line in Sect. 2.2, the base utility function and rents are estimated as follows:

$$U(t, s, m_i) = -0.74t + 1.65s + m_i \text{ and } p_{\lambda_0(t,s)} = -0.74t + 1.65s + 41.3, \qquad (37)$$

where $\lambda_0(t, s)$ is the category function. This U is considered the same as U^3 of (21) in that $U^3 = U/100$. The discrepancy value $\eta = \eta^3 = 1.124$ is larger than the corresponding values given in Sect. 5 except U^1.

The next lemma states that the rent vector given in (35) is sustained as a competitive vector by some $\mathbb{E}$ in Γ_{lin}.

Lemma 5 (Sustainability) *Let (35) hold for $\alpha = (\alpha_1, \dots, \alpha_L)$ and β, and let $p_k = \sum_l \alpha_l \tau_l^k + \beta > 0$ for $k = 1, \dots, T$. Then, there is a model $\mathbb{E}$ in Γ_{lin} such that $p = p(\mathbb{E})$.*

Proof First, we define the base utility function by $U(a_1, \dots, a_L, m_i) = \sum_l \alpha_l a_l + m_i$. Let $I_1, \dots, I_m$ be incomes with $I_1 > \dots > I_m > p_1$. We define cost functions $C_1, \dots, C_{T-1}$ by (4) with $w_k = |D_k^o|$ and $c_k < p_k$ for $k = 1, \dots, T-1$. Define C_T by (4) with $w_T^0 > |D_T^o|$ and $c_T = p_T$. In this case, for each $k = 1, \dots, T$, $y_k = |D_k^o|$ maximizes landlord k's profits.

The rents $p_k = \sum_l \alpha_l \tau_l^k + \beta$ satisfies the rent equation (20). Also, since $\beta < -\sum_l \alpha_l \tau_l^0$, each household i has the utility, $u(e^k, I_i - p_k) = I_i - \beta > I_i + \sum_l \alpha_l \tau_l^0 = u(\mathbf{0}, I_i)$. Hence, his choice of an apartment is better than choosing no apartments. $\square$

Proof of Theorem 5 (Only-If) Let (α, β) be any vector satisfying (35) and let $p_k = \sum_l \alpha_l \tau_l^k + \beta > 0$ for $k = 1, \dots, T$. By Lemma 5, $p = (p_1, \dots, p_T)$ is the maximum

---

[10]In fact, "maximum" can be dropped in here in the sense that each $\mathbb{E}$ has a unique competitive rent vector.

competitive rent vector of some $\mathbb{E} \in \Gamma_{lin}$. Hence, if $(\widehat{\alpha}, \widehat{\beta})$ minimizes the total sum of square errors in (34), then it also minimizes $T_R(P_D^o(\omega), p(\mathbb{E}))$ over Γ_{lin} with $y_k = |D_k^o|$ for $k = 1, \ldots, T$.

(If) Suppose that $\widehat{\mathbb{E}}$ is a solution of the Γ_{lin}-MSE problem, and that its maximum competitive rent vector $\widehat{p} = (\widehat{p}_1, \ldots, \widehat{p}_T)$ is expressed by (36). Let $\widehat{\alpha}$ be the coefficients of the utility function in $\widehat{\mathbb{E}}$ and let $\widehat{\beta}$ be the constant given in (36). For each $k = 1, \ldots, T$, we can assume that $\sum_{l=1}^{L} \widehat{\alpha}_l \tau_l^k + \widehat{\beta} = \widehat{p}_k > 0$, since some unit in category k is supplied in $\widehat{\mathbb{E}}$. Then, $\widehat{\beta} < -\sum_l \widehat{\alpha}_l \tau_l^0$ by the boundary condition in $\widehat{\mathbb{E}}$.

Suppose that $\widehat{p} = (\widehat{p}_1, \ldots, \widehat{p}_T)$ is not a solution of (34). Then, some other $p' = (p'_1, \ldots, p'_T)$ with α' and β' gives a smaller value of the total sum of square errors in (34). Consider the convex combination $\alpha(\pi) = \pi \alpha' + (1 - \pi)\widehat{\alpha}$ and $\beta(\pi) = \pi \beta' + (1 - \pi)\widehat{\beta}$ with $0 \leq \pi \leq 1$. The, $(\alpha(1), \beta(1))$ gives a smaller value of the total sum of square residuals than $(\alpha(0), \beta(0))$. Since the total sum is a convex function of α and β, $(\alpha(\pi), \beta(\pi))$ gives a smaller value than $(\alpha(0), \beta(0))$ for any π $(0 < \pi < 1)$. We can take a small $\pi > 0$ so that $\beta(\pi) < -\sum_l \alpha_l(\pi)\tau_l^0$ and $\sum_l \alpha_l(\pi)\tau_l^k + \beta(\pi) > 0$ for all k. By Lemma 5, there is a model $\mathbb{E}$ in Γ_{lin} such that $p = p(\mathbb{E})$. This is a contradiction to the supposition that $\widehat{\mathbb{E}}$ is a solution of the Γ_{lin}-MSE problem. Hence, $(\widehat{\alpha}, \widehat{\beta})$ is a solution of (34).   □

7 Conclusions

We developed the equilibrium-econometric analysis of rental housing markets. Our analysis provides a bridge between a market equilibrium theory and an econometric analysis. This is built by focusing on housing magazines as serving information about apartment units to economic agents (households, landlords) as well as to the econometric analyzer. We modified the equilibrium theory by incorporating the former aspect, but at the same time, we showed that we can ignore the error terms, which is the convergence theorem (Theorem 3) for equilibrium theory.

Then, we introduced the discrepancy measure as the ratio of the total sum of square residuals from the predicted rents over that from the average rents. In the best estimation we obtained in Sect. 5, the measure takes about the value 1.025. This result has strong implications on the law of diminishing marginal utility. It holds strictly for consumption, less for the commuting time-distance to the office area, and much less for the sizes of apartment units.

We have many untouched problems, which are divided into three classes: we end this paper by mentioning some problems in each class.

(1) *Subjective estimation*: we simply assumed that each economic agent forms an estimate of a rent distribution from housing magazines. Theorem 3 is a study of this subjective estimation. However, a more study is of great interests also from the viewpoint of inductive game theory (Kaneko-Kline [9]): the question

is whether an agent with a limited analytical ability can derive a meaningful estimation. This should be studied not only theoretically but also empirically.

(2) *Applications to housing markets along different railway lines and in different cities*: we discussed only a submarket along the JR Chuo railway line in Tokyo. The authors have been applying the theory to some other railway lines, but those are not more than pilot studies. A more systematic study of rental housing markets in different places and in different time is an important future problem. Then, for example, the law of diminishing marginal utility can be tested in different areas.

   Although there are almost no clear-cut segregations, in the Tokyo area (also in Japan), with different income groups and/or ethnic groups, such segregations are common phenomena in the world. The theory of assignment markets has not been developed to treat such problems. To treat it, we need to develop a more general procedure to calculate a competitive equilibrium than that used in this paper. An application to such cases will make our theory more fruitful.

(3) *Applications to panel data*: this is related to (2). Each housing magazine is issued daily or weekly. Accumulating these housing magazines, we have panel data, and can study the temporal changes of the housing market. One problem is to check the comparative statics results obtained in Kaneko et al. [10] and Ito [6] with those railway lines. In doing so, we may have better understanding of the structure of the housing market.

Appendix: Proof of Theorem 3

Since the condition BDS in $\mathbb{E}$ is preserved to $\mathbb{E}(\epsilon; \epsilon^{M \cup N, \nu})$, we show that the γ-UM and γ-PM hold for $\mathbb{E}(\epsilon; \epsilon^{M \cup N, \nu})$ for all $\nu \geq$ some ν_0, but show it only for a household $i \in M$. It is similar to prove it for $j \in N$; the assumption that the domain of the profit function is finite is used for it.

Now, let γ be an arbitrary positive number, and $P^{i,\nu} = p + \epsilon^{i,\nu}$ for $\nu = 1, \ldots$. Consider any $i \in M$. Let $z^i \in \{\mathbf{0}, \mathbf{e}^1, \ldots, \mathbf{e}^T\}$ with $I_i - pz^i \geq 0$. Then, by UM,

$$u_i(x^i, I_i - px^i) \geq u_i(z^i, I_i - pz^i). \tag{38}$$

We should consider two cases: $x^i = \mathbf{e}^t \ (t \neq 0)$ and $x^i = \mathbf{0}$, but now we consider the case of $x^i = \mathbf{e}^t$.

As $\delta \to 0$, the utility value $u_i(\mathbf{e}^t, I_i - (p_t + \delta)\mathbf{e}^t))$ converges to $u_i(x^i, I_i - px^i) = u_i(\mathbf{e}^t, I_i - p_t)$ by continuity of u_i in Assumption A. Since $\{\epsilon^{i,\nu}\}$ converges to $\mathbf{0}$ in probability, for any $\delta > 0$, there is a $\nu(\delta)$ such that for any $\nu \geq \nu(\delta)$,

$$\mu(\{\omega : \left\| \epsilon^{i,\nu}(\omega) \right\| < \delta\}) < 1 - \frac{\delta}{2}. \tag{39}$$

Since u_i is increasing in consumption by Assumption A, it holds that for all $v \geq v(\delta)$,

$$EU_i(e^t, I_i - P^{i,v} \cdot e^t) \geq (1 - \frac{\delta}{2})u_i(e^t, I_i - (p_t + \delta)e^t)) + \frac{\delta}{2}u_i(e^t, I_i). \qquad (40)$$

Since the right-hand side converges to $u_i(e^t, I_i - p_t e^t)$ as $\delta \to 0$, there is some δ_1 such that for all $\delta \geq \delta_1$,

$$(1 - \frac{\delta}{2})u_i(e^t, I_i - (p_t + \delta)e^t)) + \frac{\delta}{2}u_i(e^t, I_i) \geq u_i(e^t, I_i - p_t e^t) - \frac{\gamma}{2}. \qquad (41)$$

Since δ in (40) is arbitrary, we can take the above δ_1 for δ. From (40) for δ_1 and (41), for any $v \geq v(\delta_1)$, we have

$$EU_i(e^t, I_i - P^{i,v}e^t) \geq u_i(e^t, I_i - p_t e^t) - \frac{\gamma}{2}. \qquad (42)$$

Now, let z^i be in $\{0, e^1, \ldots, e^T\}$. Since u_i is increasing in consumption, we have, using (39) and (38), for all $v \geq v(\delta)$,

$$(1 - \frac{\delta}{2})u_i(e^t, I_i - (p_{t'} - \delta)e^t)) + \frac{\delta}{2}u_i(e^t, I_i) \geq EU_i(e^t, I_i - P^{i,v}e^t) \geq EU_i(z^i, I_i - P^{i,v}z^{it}) \qquad (43)$$

The first term converge to $u_i(e^t, I_i - p_t e^t)$ as $\delta \to 0$. Hence, there is some δ_2 such that for any $\delta \geq \delta_2$,

$$u_i(e^t, I_i - p_t e^t) + \frac{\gamma}{2} \geq (1 - \frac{\delta}{2})u_i(e^t, I_i - (p_t + \delta)e^t)) + \frac{\delta}{2}u_i(e^t, I_i). \qquad (44)$$

Hence, from (43) and (44), it holds that for any $v \geq v(\delta_2)$,

$$u_i(e^t, I_i - p_t e^t) + \frac{\gamma}{2} \geq EU_i(z^i, I_i - P^{i,v}z^i) \qquad (45)$$

Let $\delta_3 = \min(\delta_1, \delta_2)$. Then, it follows from (42) and (45) that for all $v \geq \delta_3$,

$$EU_i(e^t, I_i - P^{i,v}e^t) + \frac{\gamma}{2} \geq u_i(e^t, I_i - p_t e^t) \geq EU_i(z^i, I_i - P^{i,v}z^i) - \frac{\gamma}{2}.$$

Connecting the first term with the last term, we have the final target: $EU_i(e^t, I_i - P^{i,v}e^t) + \gamma \geq EU_i(z^i, I_i - P^{i,v}z^i)$.

In the case $x^i = 0$, the first half of the above proof should be modified.

(2): Suppose the *if clause* of the assertion. Now, let $\{\gamma_\beta\}$ a positive decreasing and converging sequence to 0. For each γ_β, we find a v_β such that for all $v \geq v_\beta$, (p, x, y) is a γ_β-competitive equilibrium in $\mathbb{E}(\epsilon; \epsilon^{MUN,v})$. We show that the utility maximization and profit maximization hold under rent vector p.

Consider utility maximization for x_i. We have, for all β,

$$EU_i(x_i, I_i - P^{i,\nu_\beta} x_i) + \gamma_\beta \geq EU_i(z^i, I_i - P^{i,\nu_\beta} z_i) \text{ for all } z_i \in \{\mathbf{0}, \mathbf{e}^1, \ldots, \mathbf{e}^T\}. \quad (46)$$

Let $z_i \in \{\mathbf{0}, \mathbf{e}^1, \ldots, \mathbf{e}^T\}$ be fixed. Suppose $I_i - pz_i > 0$. Then, both $EU_i(x_i, I_i - P^{i,\nu_\beta} x_i)$ and $EU_i(z^i, I_i - P^{i,\nu_\beta} z_i)$ converge to $u_i(x_i, I_i - px_i)$ and $u_i(z_i, I_i - pz_i)$; by (46), we have $u_i(x_i, I_i - px_i) \geq u_i(z^i, I_i - pz_i)$.

Now, suppose $I_i - pz_i = 0$. Since $u_i(z_i, 0) < u_i(0, I_i)$ by Assumption A, there is a β_0 such that for all $\beta \geq \beta_0$, $EU_i(z^i, I_i - P^{i,\nu_\beta} z_i) > u_i(z_i, 0)$. Hence, by (46), we have $u_i(x_i, I_i - px_i) \geq u_i(z_i, 0) = u_i(z^i, I_i - pz_i)$.

The profit maximization for y_j can be proved even in a simpler manner. $\qquad \square$

Acknowledgements The authors thank Lina Mallozzi for comments on an earlier version of the paper and Ryuichiro Ishikawa for his editorial help. The authors are partially supported by Grant-in-Aids for Scientific Research No.26245026, Ministry of Education, Science and Culture.

References

1. Alonso, W.: Location and Land Use. Harvard University Press, Cambridge (1964)
2. Arnott, R.: Economic theory and housing. In: Mills, E.S. (ed.) Handbook of Regional and Urban Economics, vol. 2. North-Holland, Amsterdam (1991)
3. Epple, D.: Hedonic prices and implicit markets: estimating demand and supply functions for differentiated products. J. Polit. Econ. **95**, 58–80 (1987)
4. Gale, D., Shapley, L.: College admissions and the stability of marriage. Am. Math. Mon. **69**, 9–15 (1962)
5. Hicks, J.: Value and Capital. Carendon Press, Oxford (1939)
6. Ito, T.: Effects of quality changes in rental housing markets with indivisibilities. Reg. Sci. Urban Econ. **37**, 602–617 (2007)
7. Kaneko, M.: The central assignment game and the assignment markets. J. Math. Econ. **11**, 205–232 (1982)
8. Kaneko, M.: Housing market with indivisibilities. J. Urban Econ. **13**, 22–50 (1983)
9. Kaneko, M., Kline, J.J.: Inductive game theory: a basic scenario. J. Math. Econ. **44**, 1332–1363 (2008)
10. Kaneko, M., Ito, T., Osawa, Y.-I.: Duality in comparative statics in rental housing markets with indivisibilities. J. Urban Econ. **59**, 142–170 (2006)
11. Kanemoto, Y., Nakamura, R.: A new approach to the estimation of structural equations in hedonic models. J. Urban Econ. **19**, 218–233 (1986)
12. Mas-Colell, A., Whinston, M.D., Green, J.R.: Microeconomic Theory. Oxford University Press, New York (1995)
13. Miyake, M.: Comparative statics of assignment markets with general utilities. J. Math. Econ. **23**, 519–531 (1994)
14. Ricardo, D.: The Principles of Political Economy and Taxation. J. M. Dent and Sons, London (1965, 1817: original)
15. Sai, S.: The structure of competitive equilibria in an assignment market. J. Math. Econ. **51**, 42–49 (2014)
16. Shapley, L., Shubik, M.: Assignment game I: the core. Int. J. Game Theory **1**, 111–130 (1972)
17. van der Laan, G., Talman, D., Yang, Z.: Existence and welfare properties of equilibrium in an exchange economies with multiple divisible and indivisible commodities and linear production. J. Econ. Theory **103**, 411–428 (2002)

18. van der Vaart, A.W.: Asymptotic Statistics. Cambridge University Press, Cambridge (1998)
19. von Böhm-Bawerk, E.: Positive Theory of Capital (translated by W. Smart, (Original publication in 1891)). Books for Libraries, New York (1921)
20. von Neumann, J., Morgenstern, O.: Theory of Games and Economic Behavior. Princeton University Press, Princeton (1944)
21. Wooldridge, J.M.: Introductory Econometrics: Modern Approach. South-Western College Publishing, Cincinnati (2000)

Large Spatial Competition

Matías Núñez and Marco Scarsini

1 Introduction

Consider a market with consumers and retailers. Suppose that the former ones are distributed on the unit interval and each one of them shops at the closest store whereas the latter ones decide where to locate in order to attract the largest fraction of consumers. This model is called the Pure Location Game and was initially considered by Hotelling [18] for the case of two retailers. This seminal paper has been extended and applied in different fields such as industrial organization or spatial competition (as in [8]), giving rise to an immense literature.

Among the different lessons one can draw from this model, the convergence to the median result is a highly attractive feature. Indeed, with just two players, a unique equilibrium exists. This equilibrium has two main features: (1) it is in pure strategies and (2) both parties locate at the location preferred by the median consumer. Yet, these attractive features are not robust to the introduction of some slight modifications of the model (see the review of the literature for a detailed account). For instance, if one assumes that consumers are distributed on a multidimensional space rather than on the unit interval, a pure equilibrium ceases to exist. Similarly, adding more retailers to the game might imply that a pure strategy equilibrium fails to exist. For instance, a pure equilibrium need not exist with at least four firms [26] when firms can locate over the unit interval. Nuñez and Scarsini [25] prove that, surprisingly a pure equilibrium must exist when the

M. Núñez (✉)
LAMSADE, Université Paris Dauphine, Paris, France
e-mail: matias.nunez@dauphine.fr

M. Scarsini
Dipartimento di Economia e Finanza, LUISS, Rome, Italy
e-mail: marco.scarsini@luiss.it

© Springer International Publishing AG 2017
L. Mallozzi et al. (eds.), *Spatial Interaction Models*, Springer Optimization and Its Applications 118, DOI 10.1007/978-3-319-52654-6_10

number of retailers is large enough as long as firms are restricted to choose from a finite set of locations. More specifically, while the consumers are distributed in a multidimensional space, the retailers can only locate in a finite subset of this space.[1] Moreover, in this pure strategy equilibrium, the distribution of retailers converges towards the distribution of consumers when the number of retailers increases. Note that [25]'s result allows the consumers to be distributed in any multidimensional space and holds independently of the finite set of locations the retailers can choose from.

The current work focuses on a similar framework[2] and attempts to characterize the whole set of symmetric equilibria when the number of retailers becomes large enough. To do so, we first consider a simple version of the model, where all retailers are symmetric. We examine the properties of symmetric mixed strategy equilibria (which must exist since the game is finite and symmetric). We first prove that, as the number of retailers grows large, every symmetric equilibrium must be completely mixed. In other words, in these equilibria, every feasible location is occupied with positive probability. This implies that the expected payoff from choosing each location must be equal for each retailer. A non-trivial consequence of this is that the distribution of retailers induced by the symmetric mixed equilibrium converges to the consumers' distribution.

Once we have considered the simple model with an exogenous number of symmetric retailers, we then examine two extensions. The first extension deals with games with a random number of players and the second one introduces ex-ante asymmetries between the retailers. As far as the first extension is concerned, it is well-known that games with a large number of players can easily produce results that are not robust with respect to the number of players. In order to check this robustness, we consider also a model where the number of players is random, using Poisson games à la [23, 24]. We show that in the unique equilibrium of the Poisson game retailers match consumers when the parameter of the Poisson distribution is large enough, so retailers do not even need to know the exact number of their competitors to play their (mixed) equilibrium strategies.

---

[1]There are several real-life applications where the strategic behavior of the retailers is subject to feasibility constraints as, for instance, when zoning regulations are enforced. Land use regulation has been extensively analyzed in urban economics, mostly from an applied perspective. It is often argued that zoning can have anti-competitive effects and at the same time be beneficial since it might solve problems of externalities [see 31, for a recent work on this area].

[2]Throughout, we assume that competition among retailers is only in terms of location, not price. We do this for several reasons. First, there exist several markets where price is not decided by retailers: think, for instance of newsvendors, shops operating under franchising, pharmacies in many countries, etc. Second, our model without pricing can be used to study other topics, e.g., political competition, when candidates have to take position on several, possibly related, issues. Finally several of the existing models that allow competition on location and pricing are two-stage models, where competition first happens on location and subsequently on price. Our game could be seen as a model of the first stage. It is interesting to notice that the recent paper by Heijnen and Soetevent [16] deals with the second stage in a location model on a graph, assuming that the first has already been solved.

Finally, we consider a richer model where the retailers are of two different types, advantaged and disadvantaged. Consumers prefer advantaged retailers, so they are ready to travel a bit more to shop at one of them rather than at a disadvantaged one. Here we model the comparative advantage of the first type of retailers by an additive constant. This is formally equivalent to the idea of valence in election models [see 2, 3, among others]. We show that, when the number of advantaged players increases, they play as if the disadvantaged retailers did not exist, and these ones get a zero payoff, no matter what they do.

1.1   Review of the Literature

We refer the reader to [13] for a recent survey of the literature on Hotelling games. Here we just mention the articles that are somehow closer to our contribution. Eaton and Lipsey [10] consider a Hotelling-type model with an arbitrary number of players, different possible structures of the space where retailers can locate, and different distributions of the customers. Lederer and Hurter [20] consider a model with two retailers where consumers are non-uniformly distributed on the plane. Aoyagi and Okabe [1] look at a bidimensional market and, through simulation, relate the existence of equilibria and their properties to the shape of the market. Tabuchi [32] considers a two-stage Hotelling duopoly model in a bidimensional market. Hörner and Jamison [17] look at a Hotelling model with a finite number of customers. Note that, with just two retailers, the literature has underlined the existence of a "curse of multidimensionality" (see [5] and [33] for a discussion). This curse implies that there exists no equilibrium in pure strategies for almost all distributions of consumers whenever the competition takes place in a setting with more than one dimension (as first identified by Plott [30]).[3] When the number of retailers becomes large, the location of the retailers at the symmetric mixed equilibrium tends to coincide with the distribution of the consumers on the space. This phenomenon where "retailers match consumers" was first observed by Osborne and Pitchik [26].[4] A similar result is present in [19] and [27] in the context of professional forecasting. The previously mentioned results just focus on the unidimensional space. As far as multidimensional spaces are concerned, [9, 11, 22], and [15] consider a Hotelling model on graphs where retailers can locate only on the vertices of the graph. Pálvölgyi [28], Fournier and Scarsini [13], and Fournier

---

[3]Two main possibilities have been explored to solve for this lack of equilibrium: either alternative candidates' objectives were considered (as in [6]) or the use of mixed strategies (as in [4]).

[4]Formally, [26] prove that the symmetric equilibrium strategies satisfy the claim assuming that the consumers are distributed in the interval [0,1] according to any twice continuously differentiable distribution function.

[12] consider Hotelling games on graphs with an arbitrary number of players. Heijnen and Soetevent [16] extend Hotelling's model of price competition with quadratic transportation costs from a line to graphs. Another model of location-price competition on a graph is studied in [29]. Nuñez and Scarsini [25] prove the existence of pure strategy equilibrium when the number of locations is finite and the number of players is large enough.

Two papers on the optimization literature are related to ours. Crippa et al. [7] focuses on an one-shot optimization problem where several agents, distributed across some space, have access to different services. To use a service, each agent spends some amount of time which is due both to the travel time to the service and to the queue time waiting in the service. This article considers this problem globally and in an equilibrium-like perspective. Mallozzi and Passarelli di Napoli [21] solve a two-stage optimization problem in which a social planner divides the market region into a set of service regions, each served by a single facility, in order to minimize the total cost. More precisely, the social planner decides in the first period the location of the facilities and seeks in the second period an optimal partition of the customers in each of the locations.

The paper is organized as follows. Section 2 introduces the model. Section 3 analyzes its equilibria. Section 4 considers the case of a random number of retailers. Section 5 deals with the case of differentiated retailers. All proofs are in Appendix.

2   The Model

In this section we describe the basic location model, whose different variations will be studied in the rest of the paper. This model falls in the more general framework studied by Nuñez and Scarsini [25].

2.1   *Consumers*

In this model consumers are distributed according to a measure λ on a compact Borel metric space (S, d). For instance S could be a compact subset of $\mathbb{R}^2$ or a compact subset of a 2-sphere, but it could also be a (properly metrized) network.

2.2   *Retailers*

A finite set $N_n := \{1, \ldots, n\}$ of retailers have to decide where to set shop, knowing that consumers choose the closest retailers. Each retailer wants to maximize her market share. The action set of each retailer is a finite subset of S. This means that, unlike what happens in a typical Hotelling-type model, retailers cannot locate anywhere they want, but can choose only one of finitely many possible locations. For instance they can set shop only in one of the existing shopping malls in town.

2.3 Tessellation

More formally, define $K = \{1, \ldots, k\}$ and let $X_K := \{x_1, \ldots, x_k\} \subset S$ be a finite collection of points in S. These are the points where retailers can open a store. For every $J \subset K$ call $X_J := \{x_j : j \in J\}$ and consider the Voronoi tessellation $V(X_J)$ of S induced by X_J. That is, for each $x_j \in X_J$ define the Voronoi cell of x_j as follows:

$$v_J(x_j) := \{y \in S : d(y, x_j) \leq d(y, x_\ell) \text{ for all } x_\ell \in X_J\}.$$

The cell $v_J(x_j)$ contains all points whose distance from x_j is not larger than the distance from the other points in X_J. Call

$$V(X_J) := (v_J(x_j))_{j \in J}$$

the set of all Voronoi cells $v_J(x_j)$. See, for instance, Fig. 1. It is clear that for $J \subset L \subset K$ we have $v_J(x_j) \supset v_L(x_j)$ for every $j \in J$.

Given that λ is the distribution of consumers on the space S, we have that $\lambda(v_J(x_j))$ is the mass of consumers who are weakly closer to x_j than to any other point in X_J. These consumers will weakly prefer to shop at location x_j rather than at other locations in X_J since we assume that all firms offer the same good at the same price.

To simplify the notation and the results, we assume that S is a compact subset of some Euclidean space, that λ is absolutely continuous with respect to the Lebesgue measure on this space and

$$\lambda(v_K(x_j)) > 0 \quad \text{for all } x_j \in X_K. \tag{1}$$

This assumption implies that the set of consumers that belong to r different Voronoi cells $v_J(x_{j_1}), \ldots, v_J(x_{j_r})$ (i.e. are at the same distance of several points in X_K) is of zero measure. This allows us to simplify the payoff functions. More general situations can be considered but they would require more care in handling ties.

2.4 The Game

We will build a game where $N_n := \{1, \ldots, n\}$ is the set of players. For $i \in N_n$ call $a_i \in X_K$ the action of player i. Then $a := (a_i)_{i \in N_n}$ is the profile of actions and $a_{-i} := (a_h)_{h \in N_n \setminus \{i\}}$ is the profile of actions of all the players different from i. Hence $a = (a_i, a_{-i})$.

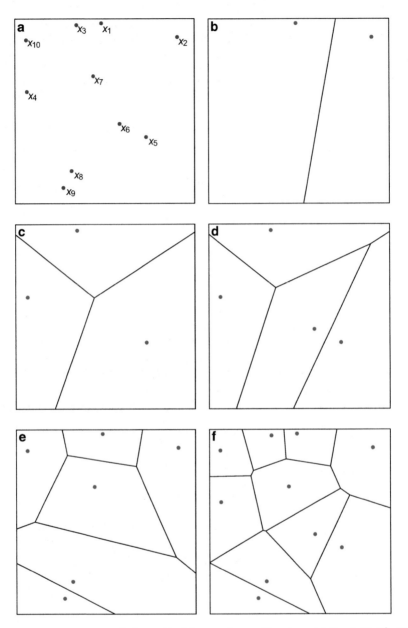

Fig. 1 Various Voronoi tessellations with different subsets of locations. (**a**) $X_K \subset [0, 1]^2$, $K = \{1, \ldots, 10\}$. (**b**) $V(X_J)$, $J = \{1, 2\}$. (**c**) $V(X_J)$, $J = \{3, 4, 5\}$. (**d**) $V(X_J)$, $J = \{3, 4, 5, 6\}$. (**e**) $V(X_J)$, $J = \{1, 2, 7, 8, 9, 10\}$. (**f**) $V(X_J)$, $J = K$

We say that $\boldsymbol{a} := (a_1, \ldots, a_n) \approx X_J$ if for all locations $x_j \in X_J$ there exists a player $i \in N_n$ such that $a_i = x_j$ and for all players $i \in N_n$ there exists a location $x_j \in X_J$ such that $a_i = x_j$. For each $\boldsymbol{a}$, we let $K(\boldsymbol{a})$ denote the subset of K such that $\boldsymbol{a} \approx X_{K(\boldsymbol{a})}$. Therefore, for $i \in N_n$, the payoff of player i is $u_i : X_K^n \to \mathbb{R}$, defined as follows:

$$u_i(\boldsymbol{a}) = \frac{1}{\mathrm{card}\{h : a_h = a_i\}} \lambda(v_{K(\boldsymbol{a})}(a_i)). \tag{2}$$

The idea behind expression (2) is as follows. Player i's payoff is the measure of the consumers that are closer to the location that she chooses than to any other location chosen by any other player, divided by the number of retailers that choose the same action as i. As Fig. 1 shows, some locations may not be chosen by any player, this is why, for every $J \subset K$, we have to consider the Voronoi tessellation $V(X_J)$ with $\boldsymbol{a} \approx X_J$ rather than the finer tessellation $V(X_K)$. We examine a simple example to clarify the idea.

Example 1 Let $S = [0, 1]$, let λ be the Lebesgue measure on $[0, 1]$, and let $X_K = \{0, 1/2, 1\}$. As mentioned before, for any given X_J, the Voronoi cell of location x_j represents the set of points in $[0, 1]$ that are closer to x_j than any other point in X_J.

$$v_J(0) = \begin{cases} [0, 1] & \text{if } X_J = \{0\}, \\ [0, 1/2] & \text{if } X_J = \{0, 1\}, \\ [0, 1/4] & \text{if } X_J = X_K \text{ or } X_J = \{0, 1/2\}. \end{cases}$$

$$v_J(1/2) = \begin{cases} [0, 1] & \text{if } X_J = \{1/2\}, \\ [1/4, 1] & \text{if } X_J = \{0, 1/2\} \\ [0, 3/4] & \text{if } X_J = \{1/2, 1\}, \\ [1/4, 3/4] & \text{if } X_J = X_K. \end{cases}$$

$$v_J(1) = \begin{cases} [0, 1] & \text{if } X_J = \{1\}, \\ [1/2, 1] & \text{if } X_J = \{0, 1\}, \\ [3/4, 1] & \text{if } X_J = X_K \text{ or } X_J = \{1/2, 1\}. \end{cases}$$

See Fig. 2.
Hence

$$\lambda(v_J(0)) = \begin{cases} 1 & \text{if } X_J = \{0\}, \\ 1/2 & \text{if } X_J = \{0, 1\}, \\ 1/4 & \text{if } X_J = X_K \text{ or } X_J = \{0, 1/2\}. \end{cases}$$

$$\lambda(v_J(1/2)) = \begin{cases} 1 & \text{if } X_J = \{1/2\}, \\ 3/4 & \text{if } X_J = \{0, 1/2\} \text{ or } X_J = \{1/2, 1\}, \\ 1/2 & \text{if } X_J = X_K. \end{cases}$$

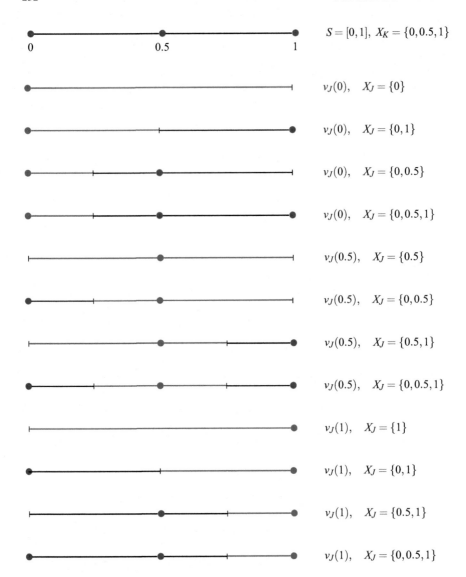

Fig. 2 Voronoi cells with different subsets X_J of locations

$$\lambda(v_J(1)) = \begin{cases} 1 & \text{if } X_J = \{1\}, \\ 1/2 & \text{if } X_J = \{0, 1\}, \\ 1/4 & \text{if } X_J = X_K \text{ or } X_J = \{1/2, 1\}. \end{cases}$$

Therefore the payoff for player i, if she chooses location 0 when the rest of the players' pure actions are $\boldsymbol{a}_{-i}$ is

$$u_i(0, \boldsymbol{a}_{-i}) = \frac{1}{\text{card}\{h : a_h = a_i\}} \phi(\boldsymbol{a}_{-i}),$$

where

$$\phi(\boldsymbol{a}_{-i}) = \begin{cases} 1 & \text{if } \boldsymbol{a} \approx \{0\}, \\ \frac{1}{2} & \text{if } \boldsymbol{a} \approx \{0, 1\}, \\ \frac{1}{4} & \text{if } \boldsymbol{a} \approx X_K \text{ or } \boldsymbol{a} \approx \{0, 1/2\}. \end{cases}$$

The payoffs when she chooses either $1/2$ or 1 can be similarly computed.

Remark 1 As mentioned before, the total demand for a location x_j (i.e. share of consumers that purchase the good from a given location) depends on the location of all the retailers. The minimum value that this demand can assume is equal to $\lambda(v_K(x_j)) > 0$, which happens when there is at least one retailer in each location (i.e. when $\boldsymbol{a} \approx X_K$). This represents one of the main differences with respect to the classical model in which retailers can locate everywhere in the set S. In the classical model the demand for a location could be made arbitrarily small. To see why, consider the classical Downsian model in the interval $[0, 1]$ with three players. Assume, for instance that player 1 locates in x, player 2 locates in $x - \varepsilon$ and player 3 locates in $x + \varepsilon$. Then the total demand for x can be rendered arbitrary small as $\varepsilon \to 0$.

Consider a game where the consumers are distributed on S according to λ, the set of players is N_n, the set of actions for each player is X_K and the payoff of player i is given by (2). Call this game $\mathcal{G}_n = \langle S, \lambda, N_n, X_K, (u_i) \rangle$. Since the set of actions coincides with the set of locations, we will use the two terms interchangeably.

With an abuse of notation, we use the same symbol $\mathcal{G}_n$ for the mixed extension of the game, where, for a mixed strategy profile $\boldsymbol{\sigma} = (\sigma_1, \ldots, \sigma_n)$, the expected payoff of player i is

$$U_i(\boldsymbol{\sigma}) = \sum_{a_1 \in X_K} \cdots \sum_{a_n \in X_K} u_i(\boldsymbol{a}) \sigma_1(a_1) \ldots \sigma_n(a_n).$$

3 Equilibria

In the rest of this section, unless otherwise stated, we consider a sequence $\{\mathcal{G}_n\}$ of games, all of which have the same parameters S, λ, X_K. More precisely, our focus is on the sequence of games when the number of retailers n grows.

We prove that when the number of retailers is large enough the distribution of retailers in equilibrium approaches the distribution of consumers.

3.1 Pure Equilibria

Nuñez and Scarsini [25, Theorem 3.4] prove in a more general setting that, when the number of players is large, the game $\mathscr{G}_n$ admits pure equilibria and the share of players in the different locations in equilibrium is approximately proportional to the measure of the corresponding Voronoi cells. They also show that this is not the case for small n. In our setting their theorem becomes:

Theorem 1 *Consider a sequence of games* $\{\mathscr{G}_n\}_{n\in\mathbb{N}}$, *where* $\mathscr{G}_n = \langle S, \lambda, N_n, X_K, (u_i) \rangle$ *and all the symbols are defined as in Sect. 2. Then there exists* $\bar{n}$ *such that for all* $n \geq \bar{n}$ *the game* $\mathscr{G}_n$ *admits a pure equilibrium* $\boldsymbol{a}^*$. *Moreover, for all* $n \geq \bar{n}$, *any pure equilibrium is such that*

$$\frac{n_j(\boldsymbol{a}^*)}{n_\ell(\boldsymbol{a}^*) + 1} \leq \frac{\lambda(v_K(x_j))}{\lambda(v_K(x_\ell))} \leq \frac{n_j(\boldsymbol{a}^*) + 1}{n_\ell(\boldsymbol{a}^*)}. \tag{3}$$

3.2 Mixed Equilibria

We consider now the mixed equilibria of the game $\mathscr{G}_n$.

Theorem 2 *For every* $n \in \mathbb{N}$ *the game* $\mathscr{G}_n$ *admits a symmetric mixed equilibrium* $\boldsymbol{\gamma}^{(n)} = (\gamma^{(n)}, \ldots, \gamma^{(n)})$ *such that*

$$\lim_{n\to\infty} \gamma^{(n)} = \gamma, \tag{4}$$

with

$$\gamma(x_j) = \frac{\lambda(v_K(x_j))}{\lambda(S)} \quad \text{for all } j \in K. \tag{5}$$

Theorem 2 says that, as the number of players grows, there is a symmetric equilibrium where players mix according to the market share of each location. This result holds only asymptotically. For instance, consider a game $\mathscr{G}_n$ with $n = 2$, $S = [0, 1]$, λ the Lebesgue measure, and $X_K = \{0.45, 0.5, 0.55\}$. Then the only symmetric equilibrium is the pure profile where both players choose the location 0.5.

4 Games with a Random Number of Players

In this section we consider games where the number of players is random and we show how the results of the previous section extend to this case. In particular we focus on Poisson games [see 23, 24, among others]. In these games, the number of players follows a Poisson distribution. We call $\mathscr{P}_n = \langle S, \lambda, N_{\Xi_n}, X_K, (u_i) \rangle$ the game where the cardinality of the players set N_{Ξ_n} is a random variable Ξ_n, with

$$\mathbb{P}(\Xi_n = k) = \frac{e^{-n} n^k}{k!},$$

that is, Ξ_n has a Poisson distribution with parameter n.

Just like in game $\mathscr{G}_n$, in game $\mathscr{P}_n$ all players have the same utility function. So the utility function of player i depends only on i' s action and on the number of players who have chosen x_j for all $j \in K$.

Quoting [23], "population uncertainty forces us to treat players symmetrically in our game-theoretic analysis," so each player choses action x_j with probability $\sigma(x_j)$. As a consequence, all equilibria are symmetric. Properties of the Poisson distribution imply that the number of players choosing action x_j is independent of the number of players choosing action x_ℓ for $j \neq \ell$.

Let $Z(X_K)$ stand for the set of vectors $y = (y(x_i))_{x_i \in X_K}$ such that each component $y(x_i)$ is a nonnegative integer that describes the number of players choosing action x_i. For each mixed strategy σ, the probability that that the actual play equals y for any $y \in Z(X_K)$ equals:

$$\prod_{j \in K} \left(\frac{e^{-n\sigma(x_j)} (n\sigma(x_j))^{y(x_j)}}{y(x_j)} \right),$$

where the product is a consequence of the independence of the different voters choosing a different action. Therefore, the expected utility of each player, when she chooses action x_j and all the other players act according to the mixed strategy σ is

$$U(x_j, \sigma) = \sum_{y \in Z(X_K)} \prod_{j \in K} \left(\frac{e^{-n\sigma(x_j)} (n\sigma(x_j))^{y(x_j)}}{y(x_j)} \right) U(x_j, y).$$

In the rest of this section we consider a sequence $\{\mathscr{P}_n\}$ of games, all of which have the same parameters S, λ, X_K.

Theorem 3 *For every $n \in \mathbb{N}$ the game $\mathscr{P}_n$ admits a symmetric mixed equilibrium $\gamma^{(n)}$ such that*

$$\lim_{n \to \infty} \gamma^{(n)}(x_j) = \frac{\lambda(v_K(x_j))}{\lambda(S)} \quad \text{for all } j \in K. \tag{6}$$

The next example shows that in general the equilibria of $\mathscr{G}_n$ and $\mathscr{P}_n$ do not coincide.

Example 2 Let $S = [0, 1]$ with λ the Lebesgue measure on $[0, 1]$ and $X_K = \{0.1, 0.5, 0.9\}$. We consider the equilibria of the games $\mathscr{G}_3$ (static) and $\mathscr{P}_3$ (Poisson).

In the game $\mathscr{G}_3$, there exists an equilibrium σ^* in which each retailer locates in 0.5. Under σ^* the payoff for each retailer equals $1/3$ since they uniformly split the consumers in S. A deviation towards 0.1 or 0.9 would give a payoff of $0.3 < 1/3$, so σ^* is indeed an equilibrium of $\mathscr{G}_3$.

We now prove that σ^* is not an equilibrium in the game $\mathscr{P}_3$. We have

$$U(\sigma^*) = \frac{1 - e^{-3}}{3} \approx 0.316738,$$

$$U(0.1, \sigma^*) = U(0.9, \sigma^*) = e^{-3} + (0.3)(1 - e^{-3}) \approx 0.334851.$$

This shows that a deviation to either 0.1 or 0.9 is profitable, hence σ^* is not an equilibrium of the game $\mathscr{P}_3$.

5 Competition with Different Classes of Retailers

Up to now, we have considered a model where all retailers are equally able to attract consumers. In other words, a consumer is indifferent between purchasing the good at two different shops if they are equally distant from her location.

In many situations some retailers have a comparative advantage due, for instance, to reputation. Therefore, *ceteris paribus*, a consumer may prefer one retailer over another. Similar models have been studied in the political competition literature with few strategic parties [see 2, among others]. In this literature the term "valence" is used to indicate the competitive advantage of one candidate over another.

In the model that we analyze below, retailers can be of two types: advantaged (A) and disadvantaged (D). We choose this dichotomic model out of simplicity. Results are not qualitatively different when a finite number of types is allowed. More precisely, we have in mind a model with several types of firms ranked by their comparative advantage. If we assume that the number of most advantaged firms goes to infinity (as we do now with just two types), then the most advantaged firms split the consumers among them and the disadvantaged ones get a zero payoff (asymptotically) whatever they do and independently of their comparative advantage.

When choosing between two retailers of the same type, a consumer takes into account only their distance from her and she prefers the closer of the two. When choosing between a retailer of type A located in x^A and a retailer of type D located in x^D, a consumer located in y will prefer the retailer of type A iff

$$d(x^A, y) < d(x^D, y) + \beta, \quad \text{with } \beta > 0.$$

She will be indifferent between the two retailers iff

$$d(x^A, y) = d(x^D, y) + \beta.$$

Obviously the case $\beta = 0$ corresponds to the model examined in Sect. 2.

Different ways to model advantage of one type of players over another have been considered in the literature [see 14, for a discussion].

We now formally define a game $\mathscr{D}_n$ with differentiated retailers. For $j \in \{A, D\}$, call N_n^j the set of retailers of type j and define $n^j = \mathrm{card}(N_n^j)$. Therefore

$$N_n = N_n^A \cup N_n^D,$$
$$n = n^A + n^D.$$

For $j \in \{A, D\}$ and $i \in N_n^j$ call $a_i^j \in X_K$ the action of retailer i. Then the profile of actions is

$$\boldsymbol{a} := (\boldsymbol{a}^A, \boldsymbol{a}^D) := \{(a_i^A)_{i \in N_n^A}, (a_i^D)_{i \in N_n^D}\}.$$

For any profile $\boldsymbol{a} \in X_K^n$ define

$$n_j^A(\boldsymbol{a}) := \mathrm{card}\{i \in N_n^A : a_i^A = x_j\},$$
$$n_j^D(\boldsymbol{a}) := \mathrm{card}\{i \in N_n^D : a_i^D = x_j\}.$$

So n_j^A and n_j^D are the number of A and D players, respectively, who choose action x_j.

We say that $(\boldsymbol{a}^A, \boldsymbol{a}^D) \approx X_{J^A, J^D}$ if for all locations $x_j \in X_{J^A}$ there exists a player $i \in N_n^A$ such that $a_i^A = x_j$ and for all players $i \in N_n^A$ there exists a location $x_j \in X_{J^A}$ such that $a_i^A = x_j$ and for all locations $x_j \in X_{J^D}$ there exists a player $i \in N_n^D$ such that $a_i^D = x_j$ and for all players $i \in N_n^D$ there exists a location $x_j \in X_{J^D}$ such that $a_i^D = x_j$.

Fix $\beta > 0$, and, for $J^A, J^D \subset K$, define

$$v_{J^A, J^D}^A(x_j) := \{y \in S : d(y, x_j) \leq d(y, x_\ell) \text{ for all } x_\ell \in X_{J^A} \text{ and}$$
$$d(y, x_j) \leq d(y, x_\ell) + \beta \text{ for all } x_\ell \in X_{J^D}\}$$
$$v_{J^A, J^D}^D(x_j) := \{y \in S : d(y, x_j) \leq d(y, x_\ell) - \beta \text{ for all } x_\ell \in X_{J^A} \text{ and}$$
$$d(y, x_j) \leq d(y, x_\ell) \text{ for all } x_\ell \in X_{J^D}\}.$$

For $i \in N_n$, the payoff of player i is $u_i : X_K^n \to \mathbb{R}$, defined as follows:

$$u_i(\boldsymbol{a}^A, \boldsymbol{a}^D) =$$

$$\begin{cases} \dfrac{1}{\mathrm{card}\{h : a_h^A = a_i^A\}} \displaystyle\sum_{J^A, J^D \subset K} \lambda(v_{J^A, J^D}^A(a_i^A)) \mathbb{1}((\boldsymbol{a}^A, \boldsymbol{a}^D) \approx X_{J^A, J^D}), & \text{if } i \in N_n^A, \\[2em] \dfrac{1}{\mathrm{card}\{h : a_h^D = a_i^D\}} \displaystyle\sum_{J^A, J^D \subset K} \lambda(v_{J^A, J^D}^D(a_i^D)) \mathbb{1}((\boldsymbol{a}^A, \boldsymbol{a}^D) \approx X_{J^A, J^D}), & \text{if } i \in N_n^D. \end{cases}$$

We call $\mathscr{D}_n := \langle S, \lambda, N_n^A, N_n^D, X_K, \beta, (u_i) \rangle$ a *Hotelling game with differentiated players*.

Note that, in any pure strategy profile of the game $\mathscr{D}_n$, a D-player gets a strictly positive payoff only if she chooses a location that is not chosen by any advantaged players.

The next example shows how substantially different the equilibria of a game $\mathscr{G}_n$ and of a game $\mathscr{D}_n$ can be.

Example 3 Let $S = [0, 1]$ with λ the Lebesgue measure on $[0, 1]$ and $X_K = \{0, 1\}$. The game $\mathscr{G}_2$ admits pure equilibria. Actually any pure or mixed profile is an equilibrium and gives the same payoff $1/2$ to both players.

Consider now the game $\mathscr{D}_2$ with one advantaged and one disadvantaged players. In the unique equilibrium of $\mathscr{D}_2$ both players randomize with probability $1/2$ over the two possible locations.

Indeed, in $\mathscr{D}_2$ there cannot be a pure equilibrium in which both players choose the same location since the disadvantaged player would get 0 and hence would strictly increase her payoff by deviating. Similarly, there cannot be a pure equilibrium in which players choose different locations, since the advantaged player would have an incentive to deviate to the location chosen by the disadvantaged player. Therefore, any equilibrium must be mixed. A simple computation proves that uniform randomization is the unique strategy profile that constitutes an equilibrium.

We now examine the equilibria in this model with differentiated candidates. Given a game $\mathscr{D}_n$, an equilibrium profile $(\boldsymbol{\gamma}^{A,n}, \boldsymbol{\gamma}^{D,n})$ is called (A, D)-symmetric if

$$\boldsymbol{\gamma}^{A,n} = (\gamma^{A,n}, \dots, \gamma^{A,n}), \tag{7}$$

$$\boldsymbol{\gamma}^{D,n} = (\gamma^{D,n}, \dots, \gamma^{D,n}). \tag{8}$$

Theorem 4 *For every $n \in \mathbb{N}$ the game $\mathscr{D}_n$ admits an (A, D)-symmetric equilibrium $(\boldsymbol{\gamma}^{A,n}, \boldsymbol{\gamma}^{D,n})$ such that*

$$\lim_{n^A \to \infty} \gamma^{A,n}(x_j) = \frac{\lambda(v_{K,J^D}^A(x_j))}{\lambda(S)} = \frac{\lambda(v_K(x_j))}{\lambda(S)} \tag{9}$$

for all $x_j \in S$, for all $J^D \subset K$. Moreover, in this equilibrium,

$$\lim_{n^A \to \infty} \sum_{i \in N^D} U_i^D(\boldsymbol{\gamma}^{A,n}, \boldsymbol{\gamma}^{D,n}) = 0. \tag{10}$$

Theorem 4 shows that, as the number n^A of advantaged players grows, they behave as if the disadvantaged players did not exist, so they play the same mixed strategies as in the game $\mathscr{G}_{n^A}$. The disadvantaged players in turn get a zero payoff whatever they do.

Appendix: Proofs

Proofs of Sect. 3

The proof of Theorem 2 requires some preliminary results.

Lemma 1 *Consider a sequence of games* $\{\mathcal{G}_n\}_{n\in\mathbb{N}}$. *There exists* $\bar{n}$ *such that for all* $n \geq \bar{n}$, *if* $\boldsymbol{\gamma}^{(n)}$ *is a symmetric equilibrium of* $\mathcal{G}_n$, *then* $\boldsymbol{\gamma}^{(n)}$ *is completely mixed, i.e.,*

$$\gamma^{(n)}(x_j) > 0 \quad \text{for all } x_j \in X_K.$$

Proof Assume by contradiction that for every $n \in \mathbb{N}$ there exists some $x_j \in X_K$ and a symmetric equilibrium $\boldsymbol{\gamma}^{(n)}$ of $\mathcal{G}_n$ such that $\gamma^{(n)}(x_j) = 0$. Given that $\lambda(S) < \infty$, we have that for all $i \in N_n$,

$$U_i(\boldsymbol{\gamma}^{(n)}) = \frac{\lambda(S)}{n}.$$

If player i deviates and plays the pure action $a_i = x_j$, then she obtains a payoff

$$U_i(a_i, \boldsymbol{\gamma}^{(n)}_{-i}) \geq \lambda(v_K(x_j)) > \frac{\lambda(S)}{n},$$

where the strict inequality holds for n large enough. This contradicts the assumption that $\boldsymbol{\gamma}^{(n)}$ is an equilibrium. $\qquad\square$

Lemma 2 *Let* $(Y_1, \ldots, Y_k)$ *be a random vector distributed according to a multinomial distribution with parameters* $(n - 1; \gamma_1^{(n)}, \ldots, \gamma_k^{(n)})$, *with* $\delta < \gamma_j^{(n)} < 1 - \delta$, *for some* $0 < \delta < 1$ *and for all* $j \in K$. *Then*

$$\lim_{n \to \infty} \frac{\mathbb{E}\left[\dfrac{1}{Y_j + 1} \displaystyle\sum_{J \subset K} \lambda(v_J(x_j))\mathbb{1}(Y_h = 0 \text{ for } h \notin J)\right]}{\mathbb{E}\left[\dfrac{1}{Y_\ell + 1} \displaystyle\sum_{J \subset K} \lambda(v_J(x_\ell))\mathbb{1}(Y_h = 0 \text{ for } h \notin J)\right]} = 1, \quad \text{for all } j, \ell \in K \tag{11}$$

iff

$$\lim_{n \to \infty} \gamma_j^{(n)} = \gamma(x_j) = \frac{\lambda(v_K(x_j))}{\lambda(S)} \quad \text{for all } j \in K. \tag{12}$$

Proof Given $j \in K$, consider all $J \subset K$ such that $j \in J$ and the family $\mathcal{V}_j$ of all corresponding Voronoi tessellations $V(X_J)$. Call $\widetilde{V}_j$ the finest partition of S generated by $\mathcal{V}_j$, that is, the set of all possible intersections of cells $v_J(x_j) \in V(X_J)$ for $V(X_J) \in \mathcal{V}_j$. It is clear that $v_K(x_j) \in \widetilde{V}_j$.

For $A \in \widetilde{V}_j$, call $\widetilde{V}_j(A)$ the class of all cells in $\widetilde{V}_j$ whose intersection with A is nonempty.

Then

$$\mathbb{E}\left[\frac{1}{Y_j + 1} \sum_{J \subset K} \lambda(v_J(x_j)) \mathbb{1}(Y_h = 0 \text{ for } h \notin J)\right]$$

$$= \mathbb{E}\left[\frac{\lambda(v_K(x_j))}{Y_j + 1}\right]$$

$$+ \mathbb{E}\left[\frac{1}{Y_j + 1} \sum_{A \in \widetilde{V}_j} \lambda(A) \mathbb{1}\left(Y_h = 0 \text{ if } v_K(x_j) \cap A \neq \varnothing\right)\right]$$

$$\leq \mathbb{E}\left[\frac{\lambda(v_K(x_j))}{Y_j + 1}\right]$$

$$+ \sum_{A \in \widetilde{V}_j} \lambda(A) \mathbb{P}\left(Y_h = 0 \text{ if } v_K(x_j) \cap A \neq \varnothing\right)$$

$$= \mathbb{E}\left[\frac{\lambda(v_K(x_j))}{Y_j + 1}\right] + o(1/n) \quad \text{for } n \to \infty,$$

since $\mathbb{P}(Y_i = 0) = (1 - \gamma_i^{(n)})^n = o(1/n)$ for $n \to \infty$. Therefore

$$\lim_{n \to \infty} \frac{\mathbb{E}\left[\dfrac{1}{Y_j + 1} \sum_{J \subset K} \lambda(v_J(x_j)) \mathbb{1}(Y_h = 0 \text{ for } h \notin J)\right]}{\mathbb{E}\left[\dfrac{1}{Y_\ell + 1} \sum_{J \subset K} \lambda(v_J(x_\ell)) \mathbb{1}(Y_h = 0 \text{ for } h \notin J)\right]} = \lim_{n \to \infty} \frac{\mathbb{E}\left[\dfrac{\lambda(v_K(x_j))}{Y_j + 1}\right]}{\mathbb{E}\left[\dfrac{\lambda(v_K(x_\ell))}{Y_\ell + 1}\right]}$$

$$= \lim_{n \to \infty} \frac{\lambda(v_K(x_j))}{\lambda(v_K(x_\ell))} \frac{\gamma_\ell^{(n)}}{\gamma_j^{(n)}} \tag{13}$$

$$= \frac{\lambda(v_K(x_j))}{\lambda(v_K(x_\ell))} \frac{\gamma(x_\ell)}{\gamma(x_j)}$$

Given that $\sum_{j=1}^k \gamma(x_j) = 1$, (13) holds if and only if (12) does. $\square$

Proof (Proof of Theorem 2) The game $\mathscr{G}_n$ is finite and symmetric, so it admits a symmetric mixed Nash equilibrium $\boldsymbol{\gamma}^{(n)} = (\gamma^{(n)}, \ldots, \gamma^{(n)})$. Then, given Lemma 1, for all $j, \ell \in K$,

$$U_i(x_j, \boldsymbol{\gamma}_{-i}^{(n)}) = U_i(x_\ell, \boldsymbol{\gamma}_{-i}^{(n)}). \tag{14}$$

Using (2) we obtain

$$
U_i(x_j, \boldsymbol{\gamma}_{-i}^{(n)}) = \sum_{a_1 \in X_K} \cdots \sum_{a_n \in X_K} u_i(a_1, \ldots, a_{i-1}, x_j, a_{i+1}, \ldots, a_n)
$$

$$
\gamma^{(n)}(x_1)^{n_1(a_{-i})} \ldots \gamma^{(n)}(x_j)^{n_j(a_{-i})+1} \ldots \gamma^{(n)}(x_k)^{n_k(a_{-i})}
$$

$$
= \mathbb{E}\left[\frac{1}{Y_j + 1} \sum_{J \subset K} \lambda(v_J(x_j)) \mathbb{1}(Y_h = 0 \text{ for } h \notin J) \right],
$$

where $(Y_1, \ldots, Y_k)$ has a multinomial distribution with parameters $(n - 1; \gamma^{(n)}(x_1), \ldots, \gamma^{(n)}(x_k))$. Notice that $\boldsymbol{a} \approx X_J$ is equivalent to $Y_h = 0$ for all $h \notin J$.

Therefore (14) holds if and only if

$$
\mathbb{E}\left[\frac{1}{Y_j + 1} \sum_{J \subset K} \lambda(v_J(x_j)) \mathbb{1}(Y_h = 0 \text{ for } h \notin J) \right]
$$

$$
= \mathbb{E}\left[\frac{1}{Y_\ell + 1} \sum_{J \subset K} \lambda(v_J(x_\ell)) \mathbb{1}(Y_h = 0 \text{ for } h \notin J) \right],
$$

which implies (11). Lemma 2 provides the result.   □

Proofs of Sect. 4

The next two lemmata are similar to Lemmata 1 and 2, respectively.

Lemma 3 *Consider a sequence of games $\{\mathscr{P}_n\}_{n \in \mathbb{N}}$. There exists $\bar{n}$ such that for all $n \geq \bar{n}$, if $\boldsymbol{\gamma}^{(n)}$ is a symmetric equilibrium of $\mathscr{P}_n$, then $\boldsymbol{\gamma}^{(n)}$ is completely mixed, i.e.,*

$$
\gamma^{(n)}(x_j) > 0 \quad \text{for all } x_j \in X_K.
$$

Proof Assume by contradiction that for every $n \in \mathbb{N}$ there exists some $x_j \in X_K$ and a symmetric equilibrium $\boldsymbol{\gamma}^{(n)}$ of $\mathscr{P}_n$ such that $\gamma^{(n)}(x_j) = 0$. Given that $\lambda(S) < \infty$, we have that for each player i

$$
U_i(\boldsymbol{\gamma}^{(n)}) = \mathbb{E}\left[\frac{\lambda(S)}{\varXi_n} \right],
$$

where $\varXi_n$ has a Poisson distribution with parameter n. If player i deviates and plays the pure action $a_i = x_j$, then she obtains a payoff

$$U_i(a_i, \boldsymbol{\gamma}^{(n)}_{-i}) \geq \lambda(v_K(x_j)) > \mathbb{E}\left[\frac{\lambda(S)}{\Xi_n}\right],$$

where the strict inequality holds for n large enough. This contradicts the assumption that $\boldsymbol{\gamma}^{(n)}$ is an equilibrium.   □

Lemma 4 *Let $(\Xi_1, \ldots, \Xi_k)$ be a random vector of independent random variables where Ξ_j has a Poisson distribution with parameter $n\gamma_j^{(n)}$, with $\delta < \gamma_j^{(n)} < 1 - \delta$, for some $0 < \delta < 1$ and for all $j \in K$. Then*

$$\lim_{n\to\infty} \frac{\mathbb{E}\left[\dfrac{1}{\Xi_j + 1}\displaystyle\sum_{J\subset K}\lambda(v_J(x_j))\mathbb{1}(\Xi_h = 0\,for\,h \notin J)\right]}{\mathbb{E}\left[\dfrac{1}{\Xi_\ell + 1}\displaystyle\sum_{J\subset K}\lambda(v_J(x_\ell))\mathbb{1}(\Xi_h = 0\,for\,h \notin J)\right]} = 1, \quad for\,all\,j, \ell \in K \tag{15}$$

iff

$$\lim_{n\to\infty} \gamma_j^{(n)} = \gamma(x_j) = \frac{\lambda(v_K(x_j))}{\lambda(S)} \quad for\,all\,j \in K. \tag{16}$$

Proof Given $j \in K$, consider all $J \subset K$ such that $j \in J$ and the family $\mathcal{V}_j$ of all corresponding Voronoi tessellations $V(X_J)$. Call $\widetilde{V}_j$ the finest partition of S generated by $\mathcal{V}_j$, that is, the set of all possible intersections of cells $v_J(x_j) \in V(X_J)$ for $V(X_J) \in \mathcal{V}_j$. It is clear that $v_K(x_j) \in \widetilde{V}_j$.

For $A \in \widetilde{V}_j$, call $\widetilde{V}_j(A)$ the class of all cells in $\widetilde{V}_j$ whose intersection with A is nonempty.

Then

$$\mathbb{E}\left[\frac{1}{\Xi_j + 1}\sum_{J\subset K}\lambda(v_J(x_j))\mathbb{1}(\Xi_h = 0\text{ for }h \notin J)\right]$$

$$= \mathbb{E}\left[\frac{\lambda(v_K(x_j))}{\Xi_j + 1}\right]$$

$$+ \mathbb{E}\left[\frac{1}{\Xi_j + 1}\sum_{A\in\widetilde{V}_j}\lambda(A)\mathbb{1}\left(\Xi_h = 0\text{ if }v_K(x_j)\cap A \neq \varnothing\right)\right]$$

$$\leq \mathbb{E}\left[\frac{\lambda(v_K(x_j))}{\Xi_j + 1}\right]$$

$$+ \sum_{A\in\widetilde{V}_j}\lambda(A)\mathbb{P}\left(\Xi_h = 0\text{ if }v_K(x_j)\cap A \neq \varnothing\right)$$

$$= \mathbb{E}\left[\frac{\lambda(v_K(x_j))}{\Xi_j + 1}\right] + o(1/n) \quad \text{for } n \to \infty,$$

since $\mathbb{P}(\Xi_i = 0) = e^{-n} = o(1/n)$ for $n \to \infty$. Therefore

$$
\lim_{n\to\infty} \frac{\mathbb{E}\left[\dfrac{1}{\Xi_j + 1} \sum_{J\subset K} \lambda(v_J(x_j))\mathbb{1}(\Xi_h = 0 \text{ for } h \notin J)\right]}{\mathbb{E}\left[\dfrac{1}{\Xi_\ell + 1} \sum_{J\subset K} \lambda(v_J(x_\ell))\mathbb{1}(\Xi_h = 0 \text{ for } h \notin J)\right]} = \lim_{n\to\infty} \frac{\mathbb{E}\left[\dfrac{\lambda(v_K(x_j))}{\Xi_j + 1}\right]}{\mathbb{E}\left[\dfrac{\lambda(v_K(x_\ell))}{\Xi_\ell + 1}\right]}
$$

$$
= \lim_{n\to\infty} \frac{\lambda(v_K(x_j))}{\lambda(v_K(x_\ell))} \frac{\gamma_\ell^{(n)}}{\gamma_j^{(n)}}
$$

(17)

$$
= \frac{\lambda(v_K(x_j))}{\lambda(v_K(x_\ell))} \frac{\gamma(x_\ell)}{\gamma(x_j)}
$$

Given that $\sum_{j=1}^k \gamma(x_j) = 1$, (17) holds if and only if (16) does. $\square$

Proof (Proof of Theorem 3) Since the number of types and actions is finite, [23, Theorem 3] implies that the Poisson game $\mathscr{P}_n$ admits a symmetric equilibrium $\boldsymbol{\gamma}^{(n)}$. Given Lemma 3, for all $j, \ell \in K$,

$$
U_i(x_j, \boldsymbol{\gamma}_{-i}^{(n)}) = U_i(x_\ell, \boldsymbol{\gamma}_{-i}^{(n)}).
$$

(18)

For $j \in K$ call $n_j(\boldsymbol{a}, \xi)$ the number of players who choose x_j under strategy $\boldsymbol{a}$ when the total number of players in the game is ξ. Using (2) we obtain

$$
U_i(x_j, \boldsymbol{\gamma}_{-i}^{(n)}) = \sum_{\xi=1}^\infty \left[\sum_{a_1\in X_K} \cdots \sum_{a_\xi\in X_K} u_i(a_1, \ldots, a_{i-1}, x_j, a_{i+1}, \ldots, a_\xi) \right.
$$

$$
\left. \gamma^{(n)}(x_1)^{n_1(a_{-i},\xi)} \ldots \gamma^{(n)}(x_j)^{n_j(a_{-i},\xi)+1} \ldots \gamma^{(n)}(x_k)^{n_k(a_{-i},\xi)} \right]
$$

$$
\cdot \frac{e^{-n} n^\xi}{\xi!}
$$

$$
= \mathbb{E}\left[\frac{1}{\Xi_j + 1} \sum_{J\subset K} \lambda(v_J(x_j))\mathbb{1}(\Xi_h = 0 \text{ for } h \notin J)\right],
$$

where $(\Xi_1, \ldots, \Xi_k)$ are independent random variables such that Ξ_j has a Poisson distribution with parameter $n\gamma^{(n)}(x_j)$. Notice that $\boldsymbol{a} \approx X_J$ is equivalent to $\Xi_h = 0$ for all $h \notin J$.

Therefore (18) holds if and only if

$$\mathbb{E}\left[\frac{1}{\varXi_j+1}\sum_{J\subset K}\lambda(v_J(x_j))\mathbb{1}(\varXi_h=0 \text{ for } h\notin J)\right]$$

$$= \mathbb{E}\left[\frac{1}{\varXi_\ell+1}\sum_{J\subset K}\lambda(v_J(x_\ell))\mathbb{1}(\varXi_h=0 \text{ for } h\notin J)\right],$$

which implies (15). Lemma 4 provides the result.   □

Proofs of Sect. 5

Lemma 5 *Consider a sequence of games* $\{\mathscr{D}_n\}_{n\in\mathbb{N}}$. *There exists* $\bar{n}^A$ *such that for all* $n^A \geq \bar{n}^A$, *if* $(\gamma^{A,n}, \gamma^{D,n})$ *is an* (A, D)-*symmetric equilibrium of* $\mathscr{D}_n$, *then* $\gamma^{A,n}$ *is completely mixed, i.e.,*

$$\gamma^{A,n}(x_j) > 0 \quad \text{for all } x_j \in X_K.$$

Proof Assume by contradiction that for every $n \in \mathbb{N}$ there exists some $x_j \in X_K$ and an (A, D)-symmetric equilibrium $(\gamma^{A,n}, \gamma^{D,n})$ of $\mathscr{D}_n$, such that $\gamma^{A,n}(x_j) = 0$. Given that $\lambda(S) < \infty$, we have that for $i \in N_n^A$

$$U_i^A(\gamma^{A,n}, \gamma^{D,n}) \leq \frac{\lambda(S)}{n^A}.$$

If player $i \in N_n^A$ deviates and plays the pure action $a_i = x_j$, then she obtains a payoff

$$U_i^A(a_i, \gamma_{-i}^{A,n}, \gamma^{D,n}) \geq \lambda(v_K(x_j)) \geq \frac{\lambda(S)}{n^A},$$

for n^A large enough. Indeed, note that even if some D-players choose x_j in $\gamma^{D,n}$, the A player attracts all the consumers from x_j. Therefore $(\gamma^{A,n}, \gamma^{D,n})$ is not an equilibrium for n^A large enough.   □

Lemma 6 *Let* $(Y_1, \ldots, Y_k)$ *be a random vector distributed according to a multinomial distribution with parameters* $(n; \gamma_1^{(n)}, \ldots, \gamma_k^{(n)})$, *with* $\delta < \gamma_j^{(n)} < 1-\delta$, *for some* $0 < \delta < 1$ *and for all* $j \in K$. *Then*

$$\lim_{n\to\infty} \mathbb{P}(Y_j = 0) = 0 \quad \text{for all } j \in K.$$

Proof The result is obvious, since

$$\mathbb{P}(Y_j = 0) = (1 - \gamma_j^{(n)})^n \leq (1 - \delta)^n \to 0. \qquad \square$$

Proof (Proof of Theorem 4) Whenever a location x_j is occupied by an advantaged player, any disadvantaged player choosing x_j gets a payoff equal to zero. Therefore (10) is an immediate consequence of Lemmata 5 and 6. Moreover, asymptotically, the actions of disadvantaged players do not affect the payoff of advantaged players. Therefore an application of Lemma 2 with n^A replacing n provides (9). $\qquad \square$

Acknowledgements The authors thank Dimitrios Xefteris for useful discussions and the PHC Galilée G15-30 "Location models and applications in economics and political science" for financial support.

Matías Núñez was supported by the center of excellence MME-DII (ANR-11-LBX-0023-01).

Marco Scarsini was partially supported by PRIN 20103S5RN3 and MOE2013-T2-1-158. This author is a member of GNAMPA-INdAM.

References

1. Aoyagi, M., Okabe, A.: Spatial competition of firms in a two-dimensional bounded market. Reg. Sci. Urban Econ. **23**(2), 259–289 (1993). doi:http://dx.doi.org/10.1016/0166-0462(93)90006-Z. http://dx.doi.org/10.1016/0166-0462(93)90006-Z
2. Aragones, E., Palfrey, T.R.: Mixed equilibrium in a Downsian model with a favored candidate. J. Econ. Theory **103**(1), 131–161 (2002). doi:10.1006/jeth.2001.2821. http://dx.doi.org/10.1006/jeth.2001.2821
3. Aragonès, E., Xefteris, D.: Candidate quality in a Downsian model with a continuous policy space. Games Econ. Behav. **75**(2), 464–480 (2012). doi:10.1016/j.geb.2011.12.008. http://dx.doi.org/10.1016/j.geb.2011.12.008
4. Banks, J.S., Duggan, J., Le Breton, M.: Social choice and electoral competition in the general spatial model. J. Econ. Theory **126**(1), 194–234 (2006). doi:10.1016/j.jet.2004.08.001. http://dx.doi.org/10.1016/j.jet.2004.08.001
5. Bernheim, B.D., Slavov, S.N.: A solution concept for majority rule in dynamic settings. Rev. Econ. Stud. **76**(1), 33–62 (2009). doi:10.1111/j.1467-937X.2008.00520.x. http://dx.doi.org/10.1111/j.1467-937X.2008.00520.x
6. Calvert, R.L.: Robustness of the multidimensional voting model: candidate motivations, uncertainty, and convergence. Am. J. Pol. Sci. **39**(1), 69–95 (1985)
7. Crippa, G., Jimenez, C., Pratelli, A.: Optimum and equilibrium in a transport problem with queue penalization effect. Adv. Calc. Var. **2**(3), 207–246 (2009). doi:10.1515/ACV.2009.009. http://dx.doi.org/10.1515/ACV.2009.009
8. Downs, A.: An Economic Theory of Democracy. Harper and Row, New York (1957)
9. Dürr, C., Thang, N.K.: Nash equilibria in Voronoi games on graphs. In: European Symposium on Algorithms (2007)
10. Eaton, B.C., Lipsey, R.G.: The principle of minimum differentiation reconsidered: some new developments in the theory of spatial competition. Rev. Econ. Stud. **42**(1), 27–49 (1975). http://www.jstor.org/stable/2296817

11. Feldmann, R., Mavronicolas, M., Monien, B.: Nash equilibria for voronoi games on transitive graphs. In: Leonardi, S. (ed.) Internet and Network Economics. Lecture Notes in Computer Science, vol. 5929, pp. 280–291. Springer, Berlin, Heidelberg (2009). doi:10.1007/978-3-642-10841-9_26. http://dx.doi.org/10.1007/978-3-642-10841-9_26

12. Fournier, G.: General distribution of consumers in pure Hotelling games (2016). https://arxiv.org/abs/1602.04851. arXiv 1602.04851

13. Fournier, G., Scarsini, M.: Hotelling games on networks: existence and efficiency of equilibria (2016). http://dx.doi.org/10.2139/ssrn.2423345. sSRN 2423345

14. Gouret, F., Hollard, G., Rossignol, S.: An empirical analysis of valence in electoral competition. Soc. Choice. Welf. **37**(2), 309–340 (2011). doi:10.1007/s00355-010-0495-0. http://dx.doi.org/10.1007/s00355-010-0495-0

15. Gur, Y., Saban, D., Stier-Moses, N.E.: The competitive facility location problem in a duopoly. Mimeo (2014)

16. Heijnen, P., Soetevent, A.R.: Price competition on graphs. Technical Report TI 2014-131/VII. Tinbergen Institute (2014). http://ssrn.com/abstract=2504454

17. Hörner, J., Jamison, J.: Hotelling's spatial model with finitely many consumers. Mimeo (2012)

18. Hotelling, H.: Stability in competition. Econ. J. **39**(153), 41–57 (1929). http://www.jstor.org/stable/2224214

19. Laster, D., Bennet, P., Geoum, I.: Rational bias in macroeconomic forecasts. Q. J. Econ. **45**(2), 145–186 (1999). doi:10.1007/s00355-010-0495-0. http://dx.doi.org/10.1007/s00355-010-0495-0

20. Lederer, P.J., Hurter, A.P., Jr.: Competition of firms: discriminatory pricing and location. Econometrica **54**(3), 623–640 (1986). doi:10.2307/1911311. http://dx.doi.org/10.2307/1911311

21. Mallozzi, L., Passarelli di Napoli, A.: Optimal transport and a bilevel location-allocation problem. J. Glob. Optim. 1–15 (2015) http://dx.doi.org/10.1007/s10898-015-0347-7

22. Mavronicolas, M., Monien, B., Papadopoulou, V.G., Schoppmann, F.: Voronoi games on cycle graphs. In: Mathematical Foundations of Computer Science, vol. 2008. Lecture Notes in Computer Science, vol. 5162, pp. 503–514. Springer, Berlin (2008). doi:10.1007/978-3-540-85238-4_41. http://dx.doi.org/10.1007/978-3-540-85238-4_41

23. Myerson, R.B.: Population uncertainty and Poisson games. Int. J. Game Theory **27**(3), 375–392 (1998). doi:10.1007/s001820050079. http://dx.doi.org/10.1007/s001820050079

24. Myerson, R.B.: Large Poisson games. J. Econ. Theory **94**(1), 7–45 (2000). doi:10.1006/jeth.1998.2453. http://dx.doi.org/10.1006/jeth.1998.2453

25. Nuñez, M., Scarsini, M.: Competing over a finite number of locations. Econ. Theory Bull. **4**(2), 125–136 (2016) http://dx.doi.org/10.1007/s40505-015-0068-6

26. Osborne, M.J., Pitchik, C.: The nature of equilibrium in a location model. Int. Econ. Rev. **27**(1), 223–237 (1986). doi:10.2307/2526617. http://dx.doi.org/10.2307/2526617

27. Ottaviani, M., Sorensen, P.N.: The strategy of professional forecasting. J. Financ. Econ. **81**(2), 441–466 (2006). doi:10.1007/s00355-010-0495-0. http://dx.doi.org/10.1007/s00355-010-0495-0

28. Pálvölgyi, D.: Hotelling on graphs. Mimeo (2011)

29. Pinto, A.A., Almeida, J.P., Parreira, T.: Local market structure in a Hotelling town. J. Dyn. Games **3**(1), 75–100 (2016). doi:10.3934/jdg.2016004. http://dx.doi.org/10.3934/jdg.2016004

30. Plott, C.R.: A notion of equilibrium and its possibility under majority rule. Am. Econ. Rev. **57**(4), 787–806 (1967)

31. Suzuki, J.: Land use regulation as a barrier to entry: evidence from the Texas lodging industry. Int. Econ. Rev. **54**(2), 495–523 (2013). doi:10.1111/iere.12004. http://dx.doi.org/10.1007/s00355-010-0495-0

32. Tabuchi, T.: Two-stage two-dimensional spatial competition between two firms. Reg. Sci. Urban. Econ. **24**(2), 207–227 (1994). doi:http://dx.doi.org/10.1016/0166-0462(93)02031-W. http://dx.doi.org/10.1016/0166-0462(93)02031-W

33. Xefteris, D.: Multidimensional electoral competition between differentiated candidates. Mimeo, University of Cyprus (2015)

Facility Location Situations and Related Games in Cooperation

Osman Palanci and S. Zeynep Alparslan Gök

1 Introduction

Facility location situations are a promising topic in the field of Operations Research (OR), which has many applications to real life. In this type of problems, there exist a given cost for constructing a facility. Further, connecting a player to this facility by minimizing the total cost is necessary.

In cooperative game theory allocating the costs in a fair way is very important, which is known as the cost allocation problem. In facility location situations, two cases can occur. One of them is the case of public facilities (such as libraries, municipal swimming pools, fire stations, etc.) and the other one is the case of private facilities (such as distribution centers, switching stations, etc.).

In a facility location situation, each facility is constructed to please the players. Here, the problem is to minimize the total cost. This cost is composed of both the player distance and the construction of each facility. A facility location game is constructed from a facility location situation [8].

In classical cooperative game theory payoffs to coalitions of players are known with certainty. On the other hand, there are many real-life situations in which people or businesses are uncertain about their coalition payoffs. Situations with uncertain payoffs in which the agents cannot await the realizations of their coalition payoffs cannot be modelled according to classical game theory. Several models that are useful to handle uncertain payoffs exist in the game theory literature [5, 14, 16].

The paper is organized as follows. In Sect. 2, we give some preliminaries about the study. We mention facility location games and PMAS in Sect. 3. Section 4 introduces facility location interval games and their Shapley value.

O. Palanci • S.Z. Alparslan Gök (✉)
Faculty of Arts and Sciences, Department of Mathematics, Suleyman Demirel University, 32260 Isparta, Turkey
e-mail: osmanpalanci@sdu.edu.tr; zeynepalparslan@yahoo.com

© Springer International Publishing AG 2017
L. Mallozzi et al. (eds.), *Spatial Interaction Models*, Springer Optimization and Its Applications 118, DOI 10.1007/978-3-319-52654-6_11

2 Preliminaries

In this section, some terminology on the theory of cooperative games and some useful results from the theory of cooperative interval games are given [3, 4, 8, 15].

A *cooperative (cost) game* in coalitional form is an ordered pair $< N, c >$, where $N = \{1, 2, \ldots, n\}$ is the set of players, and $c : 2^N \to \mathbb{R}$ is a map, assigning to each coalition $S \in 2^N$ a real number $c(S)$, such that $c(\emptyset) = 0$.

We identify a cooperative cost game $< N, c >$ with its characteristic function c. The family of all games with player set N is denoted by G^N. We recall that G^N is a $\left(2^{|N|} - 1\right)$-dimensional linear space for which unanimity games form an interesting basis. The unanimity game based on S, $u_S : 2^N \to \mathbb{R}$ is defined by

$$u_S(T) = \begin{cases} 1 & S \subset T, \\ 0 & \text{otherwise,} \end{cases}$$

where $S \in 2^N \setminus \{\emptyset\}$.

Every coalitional game $< N, c >$ can be written as a linear combination of unanimity games in a unique way such that $c = \sum_{S \in 2^N \setminus \{\emptyset\}} \lambda_S(c) u_S$ [11]. The coefficients $\lambda_S(c)$, $S \in 2^N \setminus \{\emptyset\}$ are called the unanimity coefficients of the game $< N, c >$, where $c \in G^N$ and satisfy

$$\lambda_S(c) = \sum_{T \in 2^S \setminus \{\emptyset\}} (-1)^{|S| - |T|} c(T) \quad \text{for all } S \in 2^N \setminus \{\emptyset\}.$$

Let $c \in G^N$. The potential game $< N, P^{HM}_{(N,c)} >$ associated with $c \in G^N$ is the coalitional game as follows,

$$P^{HM}_{(N,c)}(S) = P\left(S, c_{|S}\right)$$

$\forall S \subseteq N$. Hart and Mas-Colell [7] shows that the characteristic function of the potential game can be expressed in terms of the unanimity coefficients $\lambda_S(c)$ of the game $< N, c >$ which is given by,

$$P^{HM}_{(N,c)} = \sum_{S \in 2^N \setminus \{\emptyset\}} \frac{\lambda_S(c)}{|S|} u_S.$$

Let $\pi(N)$ be the set of all permutations $\sigma : N \to N$ of N and $c \in G^N$. The marginal contribution vector $m^\sigma(c) \in \mathbb{R}^N$ with respect to σ and c has the ith coordinate the value

$$m^\sigma_i(c) := c\left(P^\sigma(i) \cup \{i\}\right) - c\left(P^\sigma(i)\right) \quad \text{for each } S \in 2^N.$$

One of the most important solution concepts in cooperative game theory is the Shapley value [10]. The Shapley value associates to each game $c \in G^N$ one payoff vector in $\mathbb{R}^N$. The Shapley value $\Phi(c)$ of a game $c \in G^N$ is the average of the marginal vectors of the game, i.e.

$$\Phi(c) := \frac{1}{n!} \sum_{\sigma \in \pi(N)} m^\sigma(c).$$

We call a game $< N, c >$ as *concave* iff

$$c(S) + c(T) \geq c(S \cup T) + c(S \cap T) \quad \forall S, T \in 2^N.$$

We denote by CG^N the class of concave games with player set N. It is well known that a concave game has a non-empty core.

In this paper, we consider a (point-valued) solution f on G^N assigns that a payoff vector $f(c) \in \mathbb{R}^N$ to every TU-game $c \in G^N$. Examples of such solutions are the Centre-of-gravity of the Imputation-Set value, shortly denoted by CIS-value, Egalitarian Non-Separable Contribution value, shortly denoted by ENSC-value and the equal division solution (see [6, 17]).

The CIS-value assigns to every player its individual worth, and distributes the remainder of the worth of the grand coalition N equally among all players, i.e.

$$CIS_i(c) = c(\{i\}) + \frac{1}{|N|}(c(N) - \sum_{j \in N} c(\{j\})) \text{ for all } i \in N.$$

The ENSC-value assigns to every player in a game its marginal contribution to the 'grand coalition' and distributes the (positive or negative) remainder equally among the players, i.e.

$$ENSC_i(c) = -c(N \backslash \{i\}) + \frac{1}{|N|}(c(N) + \sum_{j \in N} c(N \backslash \{j\})) \text{ for all } i \in N.$$

The equal division solution (ED-value) just distributes $c(N)$ equally among all players, i.e.

$$ED_i(c) = \frac{1}{|N|}c(N) \text{ for all } i \in N.$$

A cooperative interval (cost) game is an ordered pair $< N, c' >$, where $N = \{1, \ldots, n\}$ is the set of players, and $c' : 2^N \to I(\mathbb{R})$ is the characteristic function such that $c'(\emptyset) = [0, 0]$. Here, $I(\mathbb{R})$ is the set of all nonempty, compact intervals in $\mathbb{R}$. For each $S \in 2^N$, the cost set (or: the cost interval) $c'(S)$ of the coalition S in the interval game $< N, c' >$ is of the form $[\underline{c}'(S), \overline{c}'(S)]$, where $\underline{c}'(S)$ is the minimal cost which coalition S could receive on its own and $\overline{c}'(S)$ is the maximal cost which coalition S could get. The family of all interval games with player set N is denoted by IG^N.

Let $I, J \in I(\mathbb{R})$ with $I = [\underline{I}, \overline{I}]$, $J = [\underline{J}, \overline{J}]$, $|I| = \overline{I} - \underline{I}$ and $\alpha \in \mathbb{R}_+$. Then,

(i) $I + J = [\underline{I}, \overline{I}] + [\underline{J}, \overline{J}] = [\underline{I} + \underline{J}, \overline{I} + \overline{J}]$;

(ii) $\alpha I = \alpha [\underline{I}, \overline{I}] = [\alpha \underline{I}, \alpha \overline{I}]$.

By (i) and (ii) we see that $I(\mathbb{R})$ has a cone structure.

Here, we need a partial substraction operator. We define $I - J$, only if $|I| \geq |J|$, by $I - J := [\underline{I}, \overline{I}] - [\underline{J}, \overline{J}] = [\underline{I} - \underline{J}, \overline{I} - \overline{J}]$. Let us note that $\underline{I} - \underline{J} \leq \overline{I} - \overline{J}$. We recall that I is weakly better than J, which we denote by $I \succcurlyeq J$, if and only if $\underline{I} \geq \underline{J}$ and $\overline{I} \geq \overline{J}$. Furthermore, we use the reverse notation $I \preccurlyeq J$, if and only if $\underline{I} \leq \underline{J}$ and $\overline{I} \leq \overline{J}$. We say that I is better than J, which we denote by $I \succ J$, if and only if $I \succcurlyeq J$ and $I \neq J$.

Finally, let $I, J \in I(\mathbb{R})$ with $I = [\underline{I}, \overline{I}]$, $J = [\underline{J}, \overline{J}]$. We define the minimum of the two intervals, $I \wedge J$, by $I \wedge J = I$ if $I \preccurlyeq J$, and their maximum, $I \vee J$, by $I \vee J = J$ if $I \preccurlyeq J$.

In general, let $I_1, \ldots, I_k \in I(\mathbb{R})$. Suppose that $I_j \succcurlyeq I_r$ for each $r \in \{1, \ldots, k\}$. Then, we say that $I_j := \max \{I_1, \ldots, I_k\}$. If $I_s \preccurlyeq I_r$ for each $r \in \{1, \ldots, k\}$, then $I_s := \min \{I_1, \ldots, I_k\}$. For example, let $I_1 = [0, 1]$, $I_2 = [-1, 2]$ and $I_3 = [3, 5]$. Then, $I_3 = \max \{I_1, I_2, I_3\}$, whereas $\max\{I_1, I_2\}$ does not exist. Similarly, $I_2 = \min \{I_2, I_3\}$, but $\min \{I_1, I_2, I_3\}$ does not exist. For details see [1].

3   Facility Location Games and PMAS

In a facility location game a set $\mathscr{A}$ of agents (also known as cities, clients, or demand points), a set $\mathscr{F}$ of facilities, a facility opening cost f_i for every facility $i \in \mathscr{F}$, and a distance d_{ij} between every pair (i, j) of points in $\mathscr{A} \cup \mathscr{F}$ indicating the cost of connecting j to i are given. We assume that the distances come from a metric space; i.e., they are symmetric and obey the triangle inequality. For a set $S \subseteq \mathscr{A}$ of agents, the cost of this set is defined as the minimum cost of opening a set of facilities and connecting every agent in S to an open facility. More precisely, the cost function c is defined by [8].

$$c(S) = \min_{\mathscr{F}^* \subseteq \mathscr{F}} \{ \sum_{i \in \mathscr{F}^*} f_i + \sum_{j \in S} \min_{i \in \mathscr{F}^*} d_{ij} \} \tag{1}$$

Now, we give an example of facility location game.

Example 3.1 Figure 1 shows a facility location game with three cities {Burdur (Player 1), Antalya (Player 2), Isparta (Player 3)} in Turkey and two hospitals {1, 2}. The cost function is calculated by using (1) the following:

$$c(1) = 5, c(2) = 4, c(3) = 4,$$

$$c(12) = 7, c(23) = 5, c(13) = 9,$$

$$c(123) = 10.$$

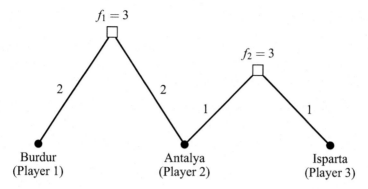

Fig. 1 An example of the facility location game

Table 1 Marginal vectors

σ	$m_1^\sigma(c)$	$m_2^\sigma(c)$	$m_3^\sigma(c)$
$\sigma_1 = (1, 2, 3)$	5	2	3
$\sigma_2 = (1, 3, 2)$	5	1	4
$\sigma_3 = (2, 1, 3)$	3	4	3
$\sigma_4 = (2, 3, 1)$	5	4	1
$\sigma_5 = (3, 1, 2)$	5	1	4
$\sigma_6 = (3, 2, 1)$	5	1	4

Now, we recall the allocation schemes [8, 12]. An allocation scheme is a scheme which provides payoff vectors for a game and all its subgames. Formally, an allocation scheme for a game $< N, c >$ is a vector $(a_{i,S})_{i \in S, S \subseteq N}$. The allocation scheme based on the Shapley value is called the *Shapley allocation scheme*.

Example 3.2 We reconsider the facility location game in **Example 3.1**. The marginal vectors are given in Table 1.

Table 1 illustrates the marginal vectors of the facility location game in **Example 3.2**. The average of the six marginal vectors is the Shapley value of this game, which can be written as:

$$\Phi(c) = (4\tfrac{2}{3}, 2\tfrac{1}{6}, 3\tfrac{1}{6}).$$

The Shapley allocation scheme of $< N, c >$ is represented in Table 2.

PMAS are introduced by Sprumont [13]. Sprumont [13] argues that this requires that the payoff of any player does not decrease as the coalition he belongs to grows larger. An allocation scheme that satisfies this property and that also satisfies efficiency for each subgame is called PMAS [12]. In formula, a vector $a = (a_{i,S})_{i \in S, S \in 2^N \setminus \{\emptyset\}}$ is a population monotonic allocation scheme for a coalitional game $< N, c >$ if it satisfies the following two conditions.

1. $\sum_{i \in S} a_{i,S} = c(S)$ for all $S \in 2^N \setminus \{\emptyset\}$,

2. $a_{i,S} \geq a_{i,T}$ for all $S, T \in 2^N \setminus \{\emptyset\}$ with $S \subset T$ and $i \in S$.

Now, we give a relation between the Shapley allocation scheme being a PMAS and concavity of the associated potential game [8].

Remark 3.1 The Shapley allocation scheme of coalitional game $< N, c >$ is PMAS if and only if the associated potential game $< N, P_{(N,c)}^{HM} >$ is concave.

An illustration of this remark can be found in the following facility location game.

Example 3.3 We continue studying the facility location game in **Example 3.1**. The unanimity game of this facility location game is $< N, c >$ with $N = \{1, 2, 3\}$ and

$$c = 5u_1 + 4u_2 + 4u_3 - 2u_{1,2} - 3u_{2,3} + 2u_{1,2,3}.$$

The Shapley allocation scheme of this game is represented in Table 2. Using the payoffs in this table we find

$$a_{1,\{1,2,3\}} > a_{1,\{1,2\}}$$

Hence, the Shapley allocation scheme is not a population monotonic allocation scheme.

The potential game associated with $< N, P_{(N,c)}^{HM} >$ is described by

$$P_{(N,c)}^{HM} = 5u_1 + 4u_2 + 4u_3 - 1u_{1,2} - \frac{3}{2}u_{2,3} + \frac{2}{3}u_{1,2,3}$$

Consequently, we conclude that $< N, P_{(N,c)}^{HM} >$ is not concave because of the following result.

$$c(12) + c(23) = 8 + \frac{13}{2} < 4 + \frac{67}{6} = c(2) + c(123).$$

Table 2 The Shapley allocation scheme

Coalition	Player 1	Player 2	Player 3
$\{1\}$	5	*	*
$\{2\}$	*	4	*
$\{3\}$	*	*	4
$\{1, 2\}$	4	3	*
$\{1, 3\}$	5	*	4
$\{2, 3\}$	*	$2\frac{1}{2}$	$2\frac{1}{2}$
$\{1, 2, 3\}$	$4\frac{2}{3}$	$2\frac{1}{6}$	$3\frac{1}{6}$

Remark 3.2 The facility location game is not concave. Then, the Shapley allocation scheme of the facility location game is not PMAS.

Now, we study the other allocation schemes on the facility location game. The allocation scheme based on CIS-value, ENSC-value and ED-value is called the *CIS, ENSC and ED allocation scheme* respectively.

Example 3.4 We use the facility location game in **Example 3.1** again. The CIS-value is defined by

$$CIS_i(c) = c(\{i\}) + \frac{1}{|N|}(c(N) - \sum_{j \in N} c(\{j\})) \quad \text{for all } i \in N.$$

Then,

$$CIS_1(c) = c(\{1\}) + \frac{1}{3}(c(\{123\}) - (c(\{1\}) + c(\{2\}) + c(\{3\})))$$

$$= 4,$$

$$CIS_2(c) = c(\{2\}) + \frac{1}{3}(c(\{123\}) - (c(\{1\}) + c(\{2\}) + c(\{3\})))$$

$$= 3,$$

$$CIS_3(c) = c(\{3\}) + \frac{1}{3}(c(\{123\}) - (c(\{1\}) + c(\{2\}) + c(\{3\})))$$

$$= 3$$

So, the CIS-value of this game is obtained by

$$CIS(c) = (4, 3, 3).$$

The CIS-values of the subgames of $< N, c >$ are easily computed. The CIS allocation scheme of $< N, c >$ is represented in Table 3.

Using the payoffs in this table we find

$$a_{2,\{1,2,3\}} > a_{2,\{2,3\}}$$

Table 3 The CIS allocation scheme

Coalition	Player 1	Player 2	Player 3
$\{1\}$	5	*	*
$\{2\}$	*	4	*
$\{3\}$	*	*	4
$\{1, 2\}$	4	3	*
$\{1, 3\}$	5	*	4
$\{2, 3\}$	*	$2\frac{1}{2}$	$2\frac{1}{2}$
$\{1, 2, 3\}$	4	3	3

Hence, the CIS allocation scheme is not a population monotonic allocation scheme. Let us compute the ENSC-value of this game. The ENSC-value is defined by

$$ENSC_i(c) = -c(N\setminus\{i\}) + \frac{1}{|N|}(c(N) + \sum_{j\in N} c(N\setminus\{j\})) \quad \text{for all } i \in N.$$

Then,

$$ENSC_1(c) = -c(\{2,3\}) + \frac{1}{3}(c(\{123\}) + c(\{1,2\}) + c(\{1,3\}) + c(\{2,3\}))$$

$$= \frac{16}{3},$$

$$ENSC_2(c) = -c(\{1,3\}) + \frac{1}{3}(c(\{123\}) + c(\{1,2\}) + c(\{1,3\}) + c(\{2,3\}))$$

$$= \frac{4}{3},$$

$$ENSC_3(c) = -c(\{1,2\}) + \frac{1}{3}(c(\{123\}) + c(\{1,2\}) + c(\{1,3\}) + c(\{2,3\}))$$

$$= \frac{10}{3}$$

So, the ENSC-value of this game is obtained by

$$ENSC(c) = (5\frac{1}{3}, 1\frac{1}{3}, 3\frac{1}{3}).$$

The ENSC-values of the subgames of $< N, c >$ are easily computed. The ENSC allocation scheme of $< N, c >$ is represented in Table 4.

Hence, the ENSC allocation scheme is not a population monotonic allocation scheme. Finally we compute the ED-value of this game. The ED-value is defined by

$$ED_i(c) = \frac{1}{|N|}c(N) \quad \text{for all } i \in N.$$

Table 4 The ENSC allocation scheme

Coalition	Player 1	Player 2	Player 3
{1}	5	*	*
{2}	*	4	*
{3}	*	*	4
{1, 2}	4	3	*
{1, 3}	5	*	4
{2, 3}	*	$2\frac{1}{2}$	$2\frac{1}{2}$
{1, 2, 3}	$5\frac{1}{3}$	$1\frac{1}{3}$	$3\frac{1}{3}$

Then,

$$ED_1(c) = \frac{c(\{1,2,3\})}{3}$$

$$= \frac{10}{3}$$

$$ED_2(c) = \frac{c(\{1,2,3\})}{3}$$

$$= \frac{10}{3}$$

$$ED_3(c) = \frac{c(\{1,2,3\})}{3}$$

$$= \frac{10}{3}$$

So, the ED-value of this game is obtained by

$$ED(c) = (3\tfrac{1}{3}, 3\tfrac{1}{3}, 3\tfrac{1}{3}).$$

The ED-values of the subgames of $< N, c >$ are easily computed. The ED allocation scheme of $< N, c >$ is represented in Table 5.

Using the payoffs in this table we find

$$a_{2,\{1,2,3\}} > a_{2,\{2,3\}}$$

Hence, the ED allocation scheme is not a population monotonic allocation scheme.

As you can see that in facility location games the *CIS, ENSC and ED allocation scheme* does not form population monotonic allocation scheme. Now, we give main result of this paper.

Remark 3.3 The three allocation schemes CIS, ENSC and ED used in Example 3.4 do not generate PMAS.

Table 5 The ED allocation scheme

Coalition	Player 1	Player 2	Player 3
{1}	5	*	*
{2}	*	4	*
{3}	*	*	4
{1, 2}	$3\frac{1}{2}$	$3\frac{1}{2}$	*
{1, 3}	$4\frac{1}{2}$	*	$4\frac{1}{2}$
{2, 3}	*	$2\frac{1}{2}$	$2\frac{1}{2}$
{1, 2, 3}	$3\frac{1}{3}$	$3\frac{1}{3}$	$3\frac{1}{3}$

Remark 3.4 When we check the results in Table 2 (Shapley allocation scheme), only the player 2 is satisfied from cooperation. On the other hand we consider the results in Table 3 (CIS allocation scheme), the player 2 and player 3 are satisfied from cooperation. Additionally, we can see that the results in Table 4 (ENSC allocation scheme), only the player 3 is satisfied from cooperation. Finally, we take the results in Table 5 (ED allocation scheme), only the player 1 is satisfied from cooperation.

4 Facility Location Interval Games and Their Interval Shapley Value

In this section, we introduce the facility location interval games inspired by Nisan [8]. In a facility location interval game, a set $\mathscr{A}$ of agents (also known as cities, clients, or demand points), a set $\mathscr{F}$ of facilities, a facility opening interval cost f_i' for every facility $i \in \mathscr{F}$, and a distance d_{ij}' between every pair (i, j) of points in $\mathscr{A} \cup \mathscr{F}$ indicating the interval cost of connecting j to i are given. Here, $f_i' := \left[\underline{f_i}, \overline{f_i}\right]$, $d_{ij}' := \left[\underline{d_{ij}}, \overline{d_{ij}}\right] \in I(\mathbb{R})$. The distances are supposed to come from a metric space. So, these distances are symmetric and satisfy the triangle inequality. For a set $S \subseteq \mathscr{A}$ of agents, the interval cost of this set is defined as the minimum interval cost of opening a set of facilities and connecting every agent in S to an open facility. More precisely, the interval cost function c' is defined by

$$c'(S) = \left[\min_{\mathscr{F}^* \subseteq \mathscr{F}} \{ \sum_{i \in \mathscr{F}^*} \underline{f_i} + \sum_{j \in S} \min_{i \in \mathscr{F}^*} \underline{d_{ij}} \}, \min_{\mathscr{F}^* \subseteq \mathscr{F}} \{ \sum_{i \in \mathscr{F}^*} \overline{f_i} + \sum_{j \in S} \min_{i \in \mathscr{F}^*} \overline{d_{ij}} \} \right] \in I(\mathbb{R})$$

(2)

Now, we give the example of facility location interval game.

Example 4.1 Figure 2 shows a facility location interval game with three cities {Burdur (Player 1), Antalya (Player 2), Isparta (Player 3)} in Turkey and two hospitals {1, 2}. The interval cost function is calculated by using (2) the following:

$$c'(1) = [5, 5.5], c'(2) = [4, 4.4], c'(3) = [4, 4.4],$$
$$c'(12) = [7, 7.7], c'(23) = [5, 5.5], c'(13) = [9, 9.9],$$
$$c'(123) = [10, 11].$$

Now, we calculate the interval Shapley value of the facility location interval game. Firstly, we recall the definition of the interval Shapley value. For this, we need to recall some notions from the theory of cooperative interval games [2].

Interval solutions are useful to solve reward/cost sharing problems with interval data using cooperative interval games as a tool. The interval payoff vectors, which are the building blocks for interval solutions, are the vectors whose components belong to $I(\mathbb{R})$. We denote by $I(\mathbb{R})^N$ the set of all such interval payoff vectors.

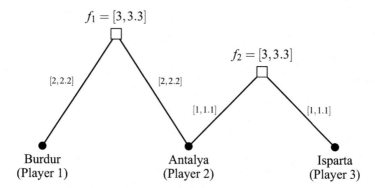

Fig. 2 An example of the facility location interval game

We call a game $< N, c' >$ *size monotonic* if $< N, |c'| >$ is monotonic, i.e., $|c'| (S) \leq |c'| (T)$ for all $S, T \in 2^N$ with $S \subset T$. For further use we denote by $SMIG^N$ the class of size monotonic interval games with player set N.

The following theorem shows that the facility location interval games are size monotonic.

Theorem 4.1 *The facility location interval game* $< N, c' >$ *belongs the class of* $SMIG^N$.

Proof We show that the facility location interval game c' belongs to the class of $SMIG^N$. For this,

$$\left| c' \right| (S) \leq \left| c' \right| (T) \text{ for all } S, T \in 2^N \text{ with } S \subset T.$$

It can be seen that $< N, c' >$ belongs to the class of $SMIG^N$. For details see [9].

We know that if an interval game is belonging to $SMIG^N$, then the interval Shapley value is always given [2].

Remark 4.1 The interval Shapley value of the facility location interval games always exists.

The interval marginal operators and the interval Shapley value were defined on $SMIG^N$ in [2] as follows.

Denote by $\Pi(N)$ the set of permutations $\sigma : N \to N$ of $N = \{1, 2, \ldots, n\}$. The interval marginal operator $m^\sigma : SMIG^N \to I(\mathbb{R})^N$ corresponding to σ, associates with each $c' \in SMIG^N$ the interval marginal vector $m^\sigma (c')$ of c' with respect to σ, defined by $m_i^\sigma (c') = c'(P^\sigma (i) \cup \{i\}) - c'(P^\sigma (i))$ for each $i \in N$, where $P^\sigma (i) := \{r \in N | \sigma^{-1}(r) < \sigma^{-1}(i)\}$. Here, $\sigma^{-1}(i)$ denotes the entrance number of player i.

For size monotonic games $< N, c' >$, $c'(T) - c'(S)$ is defined for all $S, T \in 2^N$ with $S \subset T$, since $|c'(T)| = |c'| (T) \geq |c'| (S) = |c'(S)|$. Now, we notice that for each $c' \in SMIG^N$ the interval marginal vectors $m^\sigma (c')$ are defined for each $\sigma \in \Pi(N)$, because the monotonicity of $|c'|$ implies $\overline{c'}(S \cup \{i\}) - \underline{c'}(S \cup \{i\}) \geq \overline{c'}(S) - \underline{c'}(S)$, which can be rewritten as $\overline{c}'(S\cup\{i\})-\overline{c}'(S) \geq \underline{c}'(S\cup\{i\})-\underline{c}'(S)$. So, $c'(S\cup\{i\})-c'(S)$ is defined for each $S \subset N$ and $i \notin S$.

The interval Shapley value assigns to each cooperative interval game a payoff vector whose components are compact intervals of real numbers. Cooperative games in the additive cone on which we use the interval Shapley value arise from several OR and economic situations with interval data.

The *interval Shapley value* $\Phi : SMIG^N \to I(\mathbb{R})^N$ is defined by

$$\Phi(c') := \frac{1}{n!} \sum_{\sigma \in \Pi(N)} m^\sigma(c'), \text{ for each } c' \in SMIG^N.$$

The following example shows the calculation of the interval Shapley value in the facility location interval game.

Example 4.2 Consider $< N, c' >$ as the facility location interval game in **Example 4.2**. Here, $N = \{1, 2, 3\}$ and the characteristic function c' is given as

$$c'(1) = [5, 5.5], c'(2) = [4, 4.4], c'(3) = [4, 4.4],$$

$$c'(12) = [7, 7.7], c'(23) = [5, 5.5], c'(13) = [9, 9.9],$$

$$c'(123) = [10, 11].$$

Then, the interval marginal vectors are given in the Table 6. The set of permutations of N is

$$\pi(N) = \left\{ \begin{array}{l} \sigma_1 = (1, 2, 3), \sigma_2 = (1, 3, 2), \sigma_3 = (2, 1, 3), \\ \sigma_4 = (2, 3, 1), \sigma_5 = (3, 1, 2), \sigma_6 = (3, 2, 1) \end{array} \right\}.$$

Firstly, for $\sigma_2 = (1, 3, 2)$, we calculate the interval marginal vectors. Then,

$$m_1^{\sigma_2}(c') = c'(1) = [5, 5.5],$$

$$m_2^{\sigma_2}(c') = c'(123) - c'(13) = [10, 11] - [9, 9.9] = [1, 1.1],$$

$$m_3^{\sigma_2}(c') = c'(13) - c'(1) = [9, 9.9] - [5, 5.5].$$

The others can be calculated similarly, which is shown in Table 1.

Table 6 Interval marginal vectors

σ	$m_1^\sigma(c')$	$m_2^\sigma(c')$	$m_3^\sigma(c')$
$\sigma_1 = (1, 2, 3)$	$[5, 5\frac{1}{2}]$	$[2, 2\frac{1}{5}]$	$[3, 3\frac{3}{10}]$
$\sigma_2 = (1, 3, 2)$	$[5, 5\frac{1}{2}]$	$[1, 1\frac{1}{10}]$	$[4, 4.4]$
$\sigma_3 = (2, 1, 3)$	$[3, 3\frac{3}{10}]$	$[4, 4\frac{2}{5}]$	$[3, 3\frac{3}{10}]$
$\sigma_4 = (2, 3, 1)$	$[5, 5\frac{1}{2}]$	$[4, 4\frac{2}{5}]$	$[1, 1\frac{1}{10}]$
$\sigma_5 = (3, 1, 2)$	$[5, 5\frac{1}{2}]$	$[1, 1\frac{1}{10}]$	$[4, 4\frac{2}{5}]$
$\sigma_6 = (3, 2, 1)$	$[5, 5\frac{1}{2}]$	$[1, 1\frac{1}{10}]$	$[4, 4\frac{2}{5}]$

Table 6 illustrates the interval marginal vectors of the facility location interval game in **Example 4.2**. The average of the six interval marginal vectors is the interval Shapley value of this game, which can be written as:

$$\Phi(c') = ([4\tfrac{2}{3}, 5\tfrac{2}{15}], [2\tfrac{1}{6}, 2\tfrac{23}{60}], [3\tfrac{1}{6}, 3\tfrac{29}{60}]).$$

5   Conclusion and Outlook

The objective of cooperative game theory is to study ways to enforce and sustain cooperation among agents willing to cooperate. A central question in this field is how the benefits (or costs) of a joint effort can be divided among participants, taking into account individual and group incentives, as well as various fairness properties.

In this paper, we study some results related with facility location situations and games. After that, we introduce facility location interval games. Further, we show that some allocation schemes of facility games do not have PMAS.

For future studies, allocation schemes of other solutions can be studied and interpreted. In this study, we introduce facility location interval games. Similarly, potential interval games can be studied.

References

1. Alparslan Gök, S.Z.: Cooperative interval games. Ph.D. dissertation, Institute of Applied Mathematics, Middle East Technical University (2009)
2. Alparslan Gök, S.Z., Branzei, R., Tijs, S.: Convex interval games. J. Appl. Math. Decis. Sci. (2009). Article ID 342089
3. Alparslan Gök, S.Z., Miquel, S., Tijs, S.: Cooperation under interval uncertainty. Math. Method Oper. Res. **69**, 99–109 (2009)
4. Branzei, R., Branzei, O., Alparslan Gök, S.Z., Tijs, S.: Cooperative interval games: a survey. Cent. Eur. J. Oper. Res. **18**(3), 397–411 (2010)
5. Charnes, A., Granot, D.: Prior solutions: extensions of convex nucleus solutions to chance constrained games. In: Proceedings of the Computer Science and Statistics Seventh Symposium at Iowa State University, pp. 323–332. Ames, Iowa (1973)
6. Driessen, T.S.H., Funaki, Y.: Coincidence of and collinearity between game theoretic solutions. Oper. Res. Spektr. **13**(1), 15–30 (1991)
7. Hart, S., Mas-Colell, A.: Potential, value, and consistency. Econometrica **57**(3), 589–614 (1989)
8. Nisan, N., Roughgarden, T., Tardos, É., Vazirani, V.V.: Algorithmic Game Theory. Cambridge University Press, Cambridge (2007). ISBN 0-521-87282-0
9. Palanci, O., Alparslan Gök, S.Z., Olgun, M.O., Weber, G.-W.: Transportation interval situations and related games. OR Spectr. **38**(1), 119–136 (2016)
10. Shapley, L.S.: A value for n-person games. Ann. Math. Stud. **28**, 307–317 (1953)
11. Shapley, L.S.: Additive and non-additive set functions. Ph.D. thesis, Princeton University (1953)
12. Slikker, M., van den Nouweland, A.: Social and economic networks in cooperative game theory. Theor. Decis. Library **27**, 294 (2001)

13. Sprumont, Y.: Population monotonic allocation schemes for cooperative games with transfer-
    able utility. Games Econ. Behav. **2**(4), 378–394 (1990)
14. Suijs, J., Borm, P., DeWaegenaere, A., Tijs, S.: Cooperative games with stochastic payoffs. Eur.
    J. Oper. Res. **113**(1), 193–205 (1999)
15. Tijs, S.: Introduction to Game Theory. Hindustan Book Agency, Haryana (2003)
16. Timmer, J., Borm, P., Tijs, S.: Convexity in stochastic cooperative situations. Int. Game Theory
    Rev. **7**(1), 25–42 (2005)
17. van den Brink, R., Funaki, Y.: Axiomatizations of a class of equal surplus sharing solutions for
    TU-games. Theor. Decis. **67**, 303–340 (2009)

Sequential Entry in Hotelling Model with Location Costs: A Three-Firm Case

Stefano Patrí and Armando Sacco

1 Introduction

In Industrial Organization Theory, the Horizontal Product Differentiation emerges in the market when consumers do not agree on the preference ordering. Price and location can be viewed as a metaphor of products characteristics. In this perspective, Hotelling [12] provide a two-stage model where two firms first choice their location and then fix the price of an homogenous good. Consumers are uniformly distributed along a segment of unitary length and their utility is a function of prices and transportation costs. In this context Hotelling derive the principle of minimum differentiation, for which the equilibrium solution is that both firms choose their location in the middle of the road.

Since this seminal work, several types of extensions have been provided in the literature, regarding the number of firms, the structure of the city and the functional forms of the model. The first attempts are Chamberlin [3] and Lerner and Singer [16], in which models with more than two firms are considered. Eaton and Lipsey [5] study a model with an arbitrary number of players, several possible structures of the city and different distributions of the consumers, and they conclude that the principal of minimum differentiation holds just under strong hypothesis. Moreover, d'Aspremont et al. [4] show that neither this strategy neither any other possible location are subgame perfect, because they fail to imply an equilibrium in price in every subgame. If the utility function is altered, assuming a quadratic transportation cost, then the principle of maximum differentiation is established. That means that the two firms maximize the distance between them. Economides [7] consider a linear utility for consumers and more than two firms, supporting a non-cooperative

S. Patrí (✉) • A. Sacco
Department of Methods and Models for Economics, Territory and Finance, Sapienza University, Rome, Italy
e-mail: stefano.patri@uniroma1.it; armando.sacco@uniroma1.it

© Springer International Publishing AG 2017
L. Mallozzi et al. (eds.), *Spatial Interaction Models*, Springer Optimization and Its Applications 118, DOI 10.1007/978-3-319-52654-6_12

equilibrium in prices for each subgame, but the model fails to imply an equilibrium in locations. More recently, Stuart [27] uses the core of cooperative games in the price stage, while Brenner [1] analyze a multi-firm unit interval Hotelling model under quadratic transportation costs. Peters et al. [25] add the expected waiting time in consumers' utility function, which depends on the number of consumer that choose the same firm.

Salop [28] and Economides [6] analyze models with linear utility and multiple firms located on a circumference. In particular, Salop shows that when firms are equidistant then exists an equilibrium in prices. In the model of Lederer and Hurter [15] firms are different and consumers are distributed non uniformly on the plane, while Eiselt and Laporte [9] consider a model where three firms are on a tree. Papers where consumers are distributed on a graph start to appear in literature just in the last years (see e.g., Mavronicolas et al. [19], Nuñez and Scarsini [22, 23]). Simultaneously entry is a common assumption in this kind of literature.

For that reason Prescott and Visscher [26] introduce the concept of sequential location games, where firms enter the market sequentially, paying a fixed setup cost and have no possibilities to relocate as response to new firm entry. Neven [21] proposes a model of sequential entry in a standard Hotelling framework with quadratic transportation costs. Extensions of this model are in Economides, Howell and Meza [8], in which they also analyze consumer welfare and calculate various measures of degree of asymmetry among firms, while Götz [10] reexamines the results of both papers. Moreover, Palfrey [24], Weber [29], Callander [2] and Loertscher and Muehlheusser [17] combine the sequential entry with a not uniform distribution of consumers. Jost et al. [13] consider a model of sequential location in which the two incumbent firms can react to the entry of the third, choosing a new location.

An implicit assumption in the Hotelling model is that the cost of location is independent of location and normalized to zero. Also, the literature followed this idea that location itself is free good. That means that the effects of location costs on equilibrium are still poorly explored. An attempt in this sense is Mayer [20], in which the production cost is conditioned on firm's location, while Hinloopen and Martin [11] consider the geographic interpretation of Hotelling model adding in firms profit function a location cost that is independent on production. Mallozzi [18] uses the cooperative game tools to address the problem of a single facility location, when an installation cost that depends on the region occurs.

The aim of this paper is to make another step in exploring the effects of costs of location in the Hotelling model. We are in the classical linear city, where consumers have to face a quadratic transportation cost. We solve the game backward, assuming a general form of costs of locations in the price stage. Also, we will show that this terms can lead to multiple equilibrium in the location stage, when a third firm enters in the market. To solve the location problem we consider two different functional forms, to take into account that the costs of location are function of location themselves. For example, in Europe centre-city locations are more expensive than the periphery, while in the USA is just the opposite (see Karmon [14]). Moreover, we want to analyze the impact of cost of locations in a dynamic framework, then we divide the paper in two step. In the first one two firms play the classical location-cum-price Hotelling game. In the second step a third firm enters the market choosing price and location, and we allow the two incumbent players to react.

The rest of the paper is organized as follows: in Sect. 2 we present the model; in Sect. 3 the case of two firms is considered, while in Sect. 4 we analyze the impact of the entry of a third firm. Section 5 concludes.

2   The Model

As in d'Aspremont et al. [4] and Brenner [1], a classical Hotelling game, where consumer are uniformly distributed on the unitary interval, is considered. Assuming that each consumer i, located in a point s, has a quadratic transportation cost, the utility function can be written as follows

$$u_i(x_j, p_j) = k - p_j - (x_j - s)^2,$$

where x_j and p_j are, respectively, the position and the price charged by the firm j. The parameter $k > 0$ represents the reservation price and it is assumed to be high enough to guarantee that every consumer buys a unit of product.

Following Jost et al. [13], the model considers a dynamic version of the Hotelling problem, in which the game is played into two steps. In both steps the game is a two stage, in which the firms choose location in the first stage and fix the prices in the second stage. Also, in both steps the two stage game is solved backward, addressing the price game first and using the optimal solutions to solve the location game.

- **Step One.** In the first step the set of players I is composed of two firms, denoted by A and B, that play simultaneously a location-cum-price game. The set of strategies in the first stage are the locations (x_A, x_B), where $x_j \in [0, 1]$, while strategies are given by prices (p_A, p_B) in the second stage. This step is a classical Hotelling situation.
- **Step Two.** In the second step a third firm, denoted by C, enters the market. The difference in this step is that the location stage is not played simultaneously by every players. The stage is played as follows

    1. Firm C maximizes its own profit function to find the optimal location x_C.
    2. The incumbent firms A and B internalize the optimal location x_C and evaluate their profit function taking it into account. Then, the incumbents are allowed to react to the entry of firm C changing the locations (x_A, x_B) chosen in the first step.
    3. The price stage is played simultaneously by the three firms.

Respect Jost et al. [13] in this model the location is not a free good. A cost of location function affects the payoff of each player. This function is assumed to be continuous and depending only on the locations. Then, it has no effects in the price stage, but it can change significantly the choice of the optimal location.

3 Two-Firm Case

In this section the first step with two firms is considered. The two players, denoted by A and B, address a two stage game in which they have to choice first their locations (x_A, x_B), where the admissible positions are on a road of unitary length. In the second stage the firms fix the prices (p_A, p_B). In order to find the Nash equilibrium the game is solved backward, then the price stage is solved first, taking as given the locations.

The payoff of the firm j, denoted by $\Pi_j(\cdot, \cdot)$, depends on the vector of locations $\bar{x} = (x_A, x_B)$, on the vector of prices $\bar{p} = (p_A, p_B)$ and on the costs of location, as follows

$$\Pi_j(\bar{x}, \bar{p}) = p_j D_j(\bar{x}, \bar{p}) - l(x_j), \tag{1}$$

where $D_j(\bar{x}, \bar{p})$ is the demand function and $l(x_j)$ is the location cost of the firm j.

Given the uniformly distribution of consumers on the unitary segment, the two demand functions are determined by the indifferent consumer, that is the individual for which is indifferent to buy from firm A or firm B. The location s_{AB} of the indifferent consumer is given by solving the equation $u(x_A, p_A) = u(x_B, p_B)$, as follows

$$s_{AB} = \frac{p_B - p_A}{2(x_B - x_A)} + \frac{x_A + x_B}{2}.$$

Then, the demand functions are $D_A(\bar{x}, \bar{p}) = s_{AB}$ and $D_B(\bar{x}, \bar{p}) = 1 - s_{AB}$.

The functional form of $l(x_j)$ is not established in the price stage. The backward approach requires to find first the equilibrium in prices, that is given by solving the system of maximization problems

$$\max_{p_A} \Pi_A(\bar{x}, \bar{p}) = \max_{p_A} p_A \left[\frac{p_B - p_A}{2(x_B - x_A)} + \frac{x_A + x_B}{2} \right] - l(x_A),$$

$$\max_{p_B} \Pi_B(\bar{x}, \bar{p}) = \max_{p_B} p_B \left[1 - \frac{p_B - p_A}{2(x_B - x_A)} - \frac{x_A + x_B}{2} \right] - l(x_B).$$

Computing the first order conditions the optimal prices are given by

$$p_A^\star = \frac{1}{3}(2 + x_A + x_B)(x_B - x_A),$$

$$p_B^\star = \frac{1}{3}(4 - x_A - x_B)(x_B - x_A). \tag{2}$$

In order to solve the first stage of the game and find the optimal locations, it is necessary to substitute the solutions (2) in the payoff function (1) and derive respect the variables x_A and x_B as follows

$$\frac{\partial}{\partial x_A} \left\{ \frac{1}{3}(2 + x_A + x_B)(x_B - x_A) \left[\frac{1}{3}(1 - x_A - x_B) + \frac{x_A + x_B}{2} \right] - l(x_A) \right\} = 0,$$

(3)

$$\frac{\partial}{\partial x_B} \left\{ \frac{1}{3}(4 - x_A - x_B)(x_B - x_A) \left[1 - \frac{1}{3}(1 - x_A - x_B) - \frac{x_A + x_B}{2} \right] - l(x_B) \right\} = 0.$$

The system (3) can be reduce to

$$\frac{\partial \Pi_A}{\partial x_A} = -\frac{1}{18}(3x_A^2 + 8x_A + 2x_A x_B - x_B^2 + 4) - l'(x_A) = 0,$$

(4)

$$\frac{\partial \Pi_B}{\partial x_B} = \frac{1}{18}(3x_B^2 - 16x_B + 2x_A x_B - x_A^2 + 16) - l'(x_B) = 0.$$

(5)

The system of Eqs. (4)–(5) is non linear in the variables x_A and x_B and require a specific functional form of the cost of location to be solved. The next proposition characterizes the solution for some classes of cost of location.

Proposition 1 *If the cost of location function is constant or increasing going toward the centre, then the maximum differentiation principle holds.*

Proof Consider a function $l(x_j) = q$, where q is a constant, so that $l'(x_A) = l'(x_B) = 0$. In this case the system (4)–(5) become

$$\frac{\partial \Pi_A}{\partial x_A} = -\frac{1}{18}(3x_A^2 + 8x_A + 2x_A x_B - x_B^2 + 4) = 0,$$

$$\frac{\partial \Pi_B}{\partial x_B} = \frac{1}{18}(3x_B^2 - 16x_B + 2x_A x_B - x_A^2 + 16) = 0.$$

After some algebras it is possible to show that $\frac{\partial \Pi_A}{\partial x_A} < 0$ and $\frac{\partial \Pi_B}{\partial x_B} > 0$, for every allowed values of x_A and x_B. Consequently, the firm A chooses the minimum value possible, that is $x_A = 0$, while the firm B chooses the maximum value possible, that is $x_B = 1$. For the second part of the proof consider a location cost function that is not constant, but that increase going toward the center.

This function $l(x_j)$ is assumed to be always positive in the interval $[0, 1]$, and that it has a maximum in $x_j = 1/2$. In order to proof the proposition the only condition required is that $l(0) = l(1) = \min_x l(x)$.

Consider first the firm A. Its payoff is given by the function

$$\Pi_A = R(x_A) - C(x_A) = p_A D_A(x_A) - l(x_A),$$

where $R(\cdot)$ represents the revenues and $C(\cdot)$ represents the costs. Assuming that payoff is always positive, the aim is to prove that

$$R(0) - C(0) \geq R(x_A) - C(x_A), \qquad \forall x_A \in [0, 1]. \tag{6}$$

From Eq. (4) it is known that $R(0) \geq R(x_A)$, while for hypothesis holds $C(0) \leq C(x_A)$, $\forall x_A \in [0, 1]$. As consequence, the inequality $R(0) - R(x_A) \geq C(0) - C(x_A)$ always holds and the proposition is proved for firm A.

Heuristically, it is possible to say that the location $x_A = 0$ maximize the revenues and minimize the location cost, and for that is the optimal location for the firm A.

A specular reasoning leads to conclude that the location $x_B = 1$ is the optimal one for firm B.

Then, the two firms has no incentives to move and the equilibrium of the game is given by

$$(p_A^\star, x_A^\star) = (1, 0), \qquad (p_B^\star, x_B^\star) = (1, 1).$$

Remark 1 (Costly Periphery) The locations are not more expensive in the centre-city everywhere. For that reason it is interesting to analyze also the case of costs of location that increase in the periphery. Then the function $l(x_j)$ is chosen as follows

$$l(x_j) = \left(x_j - \frac{1}{2}\right)^2 + r,$$

where r is a constant. The derivative of this function is

$$l'(x_j) = 2x_j - 1. \tag{7}$$

Substituting (7) in the system (4)–(5), the optimal locations solve the following

$$\frac{\partial \Pi_A}{\partial x_A} = -\frac{1}{18}[3x_A^2 - 2x_A(14 - 2x_B) - x_B^2 + 22] = 0, \tag{8}$$

$$\frac{\partial \Pi_B}{\partial x_B} = \frac{1}{18}[3x_B^2 - 2(26 - x_A)x_B - x_A^2 + 34] = 0. \tag{9}$$

This system has a unique solution in the set of admissible values that is given by the couple $(x_A^\star, x_B^\star) = (0.31, 0.69)$. Then the game has a unique Nash equilibrium, in which prices are still symmetric, given by

$$(p_A^\star, x_A^\star) = (0.38, 0.31), \qquad (p_B^\star, x_B^\star) = (0.38, 0.69).$$

4   Three-Firm Case

In this section the second step of the game is presented. Consider a third firm, denoted by C, that enters in the market in position x_C, such that $x_A \leq x_C \leq x_B$, and that offers the product at the price p_C.

Then, there are two indifferent consumers that are located in s_{AC} and s_{CB}, given by the points

$$s_{AC} = \frac{p_C - p_A}{2(x_C - x_A)} + \frac{x_A + x_C}{2},$$

$$s_{CB} = \frac{p_B - p_C}{2(x_B - x_C)} + \frac{x_C + x_B}{2}.$$

The indifferent consumer positions imply the demand functions of the three firms, as follows

$$D_A(x_A, x_C, p_A, p_C) = s_{AC},$$

$$D_C(x_A, x_B, x_C, p_A, p_B, p_C) = s_{CB} - s_{AB},$$

$$D_B(x_C, x_B, p_C, p_B) = 1 - s_{CB}.$$

As in the first step, the three firms have to maximize their own payoff, respect price and location, that are given by

$$\Pi_A = p_A D_A(x_A, x_C, p_A, p_C) - l(x_A),$$

$$\Pi_C = p_C D_C(x_A, x_B, x_C, p_A, p_B, p_C) - l(x_C),$$

$$\Pi_B = p_B D_B(x_A, x_C, p_A, p_C) - l(x_B).$$

4.1   Price Stage

Also in the second step the game is solved backward looking first to the equilibrium in prices. Then, the three firms maximize the profit function respect the vector of prices. The optimal prices are the triple $(p_A^\star, p_C^\star, p_B^\star)$ that solves the system

$$\frac{\partial}{\partial p_A}[p_A D_A(x_A, x_C, p_A, p_C) - l(x_A)] = 0,$$

$$\frac{\partial}{\partial p_C}[p_C D_C(x_A, x_B, x_C, p_A, p_B, p_C) - l(x_C)] = 0, \qquad (10)$$

$$\frac{\partial}{\partial p_B}[p_B D_B(x_A, x_C, p_A, p_C) - l(x_B)] = 0.$$

In this stage players are supposed to play simultaneously, then substituting the demand functions and solving the system (10), the prices of equilibrium are given by

$$p_A^\star = \frac{x_C - x_A}{6(x_B - x_A)}[3(x_C + x_A)(x_B - x_A) + (x_B - x_C)(2 + x_B - x_A)],$$

$$p_C^\star = \frac{(x_B - x_C)(x_C - x_A)(x_B - x_A + 2)}{3(x_B - x_A)},$$ (11)

$$p_B^\star = \frac{x_B - x_C}{6(x_B - x_A)}[3(x_B - x_A)(2 - x_B - x_C) + (x_C - x_A)(2 + x_B - x_A)].$$

4.2 Location Stage

The location game is solved first by the entrant firm C. The payoff of this player is given at this stage by the function

$$\Pi_C(x_A, x_B, x_C) = \frac{(x_B - x_A + 2)^2}{18(x_B - x_A)}(x_C - x_A)(x_B - x_C) - l(x_C).$$

Computing the first order conditions the solution $x_C^\star$ solves the equation

$$\frac{(x_B - x_A + 2)^2}{18(x_B - x_A)}(x_A + x_B - 2x_C) - l'(x_C) = 0.$$ (12)

Again, the effective positioning of the firm depends on the functional form of the location cost function. If it is constant, so that $l'(x_C) = 0$, then the firm C chooses a location exactly halfway between firms A and B:

$$x_C^\star = \frac{x_A + x_B}{2}.$$

4.2.1 Costly Centre-City

In the first step, in case of locations that are more expensive in the centre of the city the maximum differentiation principle holds and the equilibrium locations are $(x_A^\star, x_B^\star) = (0, 1)$. What happens if a third firm is present in the market?

A quadratic function is assumed to represent the location costs, as follows

$$l(x_j) = x_j - x_j^2.$$

Substituting the first derivatives of $l(x_C)$ in Eq. (12) and solving respect x_C, the optimal location is given by

$$x_C^\star = \frac{(x_B - x_A + 2)^2 (x_A + x_B) - 18(x_B - x_A)}{2[(x_B - x_A + 2)^2 - 18(x_B - x_A)]}. \tag{13}$$

The incumbent firms react to the entrance of the firm C internalizing its optimal position (13) and maximizing their payoff. The payoff functions of firms A and B are as follows

$$\Pi_A = p_A^\star D_A(x_A, x_B, x_C^\star) - x_A(1 - x_A),$$

$$\Pi_B = p_B^\star D_B(x_A, x_B, x_C^\star) - x_B(1 - x_B).$$

The first order conditions, $\frac{\partial \Pi_A}{\partial x_A} = 0$ and $\frac{\partial \Pi_B}{\partial x_B} = 0$, constitute the system that brings to the optimal response. Solving this system leads to multiple Nash equilibria in the location game, more specifically there are two solutions $(x_A^\star, x_B^\star)$ allowed by the model:

$$(x_A^\star, x_B^\star) = (0.586184, 0.600714),$$

$$(x_A^\star, x_B^\star) = (0.391533, 0.584469).$$

Substituting these solutions in Eq. (13), the complete Nash equilibria for the location stage are given by

$$(x_A^\star, x_C^\star, x_B^\star) = (0.586184, 0.599886, 0.600714),$$

$$(x_A^\star, x_C^\star, x_B^\star) = (0.391533, 0.456813, 0.584469). \tag{14}$$

Both solutions highlight the tendency of firms A and B to deal with higher location costs to react to the entry of firm C. From these results seems that the incumbent firms prefer to pay more for location and stay close to the entrant instead of minimize the cost of location giving more space to the firm C. That is clearly a result that overturn the solution of the first stage in which the two firms prefer to maximize the distance between them.

4.2.2 Costly Periphery

The entry of the firm C in case of more expensive centre locations leads the incumbents to pay more for location in order to relocate close to the entrant. In the first step higher costs of location in the periphery lead the firm A and B to move toward the centre. Then, it is interesting to analyze what equilibrium is achieved in the second step, also in this last case. Consider the following location cost function

$$l(x_j) = \left(x_j - \frac{1}{2}\right)^2 + r,$$

where r is a constant. The computational process is the same of Sect. 4.2.1.

The first move is made by the firm C that addresses the payoff maximization and looks for the optimal location that is given by the solution of the equation

$$\frac{(x_B - x_A + 2)^2}{18(x_B - x_A)}(x_A + x_B) - 2x_C + 1 = 0, \tag{15}$$

respect the variable x_C. The two incumbent firms internalize $x_C^\star$ solution of (15) and maximize their own welfare, in order to choose the best response to the entry of firm C. In this case there is a unique Nash equilibrium allowed by the model, that is given by the triple

$$(x_A^\star, x_C^\star, x_B^\star) = (0.437419, 0.558699, 0.823202). \tag{16}$$

Figure 1 shows how the incumbent firms reply to the entry of firm C changing their position and prices. On the horizontal axis there are locations, while on the vertical one there are prices. In the second step (the blue lines), the two incumbent firms react moving toward the end of the street respect the first step (the green lines). Moreover, the entry on the firm C, cause a sensible decreasing in prices, that are given by the triple

$$(p_A^\star, p_C^\star, p_B^\star) = (0.0935, 0.0661, 0.1148). \tag{17}$$

Independently on the location cost function, the entrant firm C chooses a location that is around the middle of the road. Instead, the reaction of the incumbents changes as the location costs change. When the structure of this costs causes higher costs in the centre of the city the firms A and B prefer to pay these higher costs and relocate closer to the firm C. When the location costs are higher in the periphery, the incumbents move toward the end of the line, that means that the firm A pays lower location costs and relocate itself near the entrant, while the firm B pays higher location costs and choose to move away from the firm C.

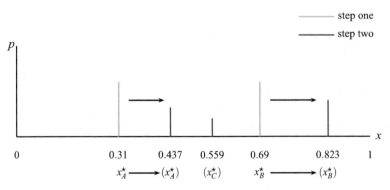

Fig. 1 Reaction of the incumbent firms when $l(x_j) = (x_j - \frac{1}{2})^2 + r$

5 Conclusions

The effects of costly locations were largely not considered in the literature. The aim of this paper was to address the location-cum-price problem when the location is not a free good. Moreover, we wanted to consider also how it can affects the choice of the firms when a dynamic interaction is allowed. Then the game was played in two different step, a first one in which a classical Hotelling duopoly is played and a second one in which a third firm enters the market producing the reaction of the incumbents. In the first step the classical result of maximum differentiation is confirmed, if the location costs are constants or more expensive in the centre-city. Nevertheless, we showed that when the periphery is more expensive the firms tend to move toward the centre.

In the second step a third firm enters in the market and chooses the location. As consequence, the incumbents reconsider their optimization problem internalizing the presence of the entrant and react to the new situation. The results are different and depend on the functional form of the location cost function. In both cases considered in the paper, the entrant choose to locate itself around the middle of the road, but the reaction of the incumbents change as the structure of location cost change. Moreover, we showed that in some cases multiple equilibria may be present in the location stage.

References

1. Brenner, S.: Hotelling games with three, four, and more players. J. Reg. Sci. **45**(4), 851–864 (2005)
2. Callander, S.: Electoral competition in heterogeneous districts. J. Polit. Econ. **113**, 1116–1144 (2005)
3. Chamberlin, E.H.: The Theory of Monopolistic Competition. Harvard University Press, Cambridge (1933)
4. d'Aspremont, C., Gabszewicz, J.J., Thisse, J.-F.: On Hotelling's "stability in Competition". Econometrica **47**, 1145–1150 (1979)
5. Eaton, B.C., Lipsey, R.G.: The principle of minimum differentiation reconsidered: some new developments in the theory of spatial Competition. Rev. Econ. Stud. **42**, 27–49 (1975)
6. Economides, N.: Symmetric equilibrium existence and optimality in differentiated product markets. J. Econ. Theory **47**, 178–194 (1989)
7. Economides, N.: Hotelling's "main street" with more than two competitors. J. Reg. Sci. **33**, 303–319 (1993)
8. Economides, N., Howell, J., Meza, S.: Does it Pay to be the First? Sequential Location Choice and Foreclosure, EC-02-19. Stern School of Business, New York University, New York (2002)
9. Eiselt, H.A., Laporte, G.: The existence of equilibria in the 3-facility Hotelling model in a tree. Transp. Sci. **27**, 39–43 (1993)
10. Götz, G.: Endogenous sequential entry in a spatial model revisited. Int. J. Ind. Organ. **23**, 249–261 (2005)
11. Hinloopen, J., Martin, S.: Costly location in Hotelling duopoly. Tinbergen Institute Discussion Paper. http://works.bepress.com/hinloopen/8/ (2013)
12. Hotelling, H.: Stability in competition. Econ. J. **3**, 41–57 (1929)

13. Jost, P.-J., Schubert, S., Zschoche, M.: Incumbent positioning as a determinant of strategic response to entry. Small Bus. Econ. **44**, 577–596 (2015)
14. Karmon, J.: Rental costs, city vs. suburbs: a handy infographic. https://www.yahoo.com/news/blogs/spaces/rental-costs-city-vs-suburbs-handy-infographic-225331978.html (2012)
15. Lederer, P.J., Hurter, A.P., Jr.: Competition of firms: discriminatory pricing and location. Econometrica **54**, 623–640 (1986)
16. Lerner, A.P., Singer, H.W.: Some notes on duopoly and spatial competition. J. Polit. Econ. **45**, 145–186 (1937)
17. Loertscher, S., Muehlheusser, G.: Sequential location games. RAND J. Econ. **42**(4), 639–663 (2011)
18. Mallozzi, L.: Cooperative games in facility location situations with regional fixed costs. Optim. Lett. **5**, 173–181 (2011)
19. Mavronicolas, M., Monien, B., Papadopoulou, V.G., Schoppmann, F.: Voronoi games on cycle graphs. In: Mathematical Foundations of Computer Sciences 2008. Lecture Notes in Computer Science, vol. 5162, pp. 503–514. Springer, New York (2008)
20. Mayer, T.: Spatial Cournot competition and heterogeneous production costs across locations. Reg. Sci. Urban Econ. **30**, 325–352 (2000)
21. Neven, D.J.: Endogenous sequential entry in a spatial model. Int. J. Ind. Organ. **5**, 419–434 (1987)
22. Nuñez, M., Scarsini, M.: Competing over a finite number of locations. Econ. Theory Bull. (2015, forthcoming)
23. Nuñez, M., Scarsini, M.: Large location models. Technical report, SSRN 2624304. http://ssrn.com/abstract=2624304 (2015)
24. Palfrey, T.: Spatial equilibrium with entry. Rev. Econ. Stud. **51**, 139–156 (1984)
25. Peters, H., Schröder, M., Vermeulen, D.: Waiting in the queque on Hotelling's main street. Technical report RM/15/040, Maastricht University (2015)
26. Prescott, E.C., Visscher, M.: Sequential location among firms with foresight. Bell J. Econ. **8**, 705–729 (1977)
27. Stuart, H.W., Jr.: Efficient spatial competition. Games Econ. Behav. **49**, 345–362 (2004)
28. Salop, S.C.: Monopolistic competition with outside goods. Bell J. Econ. **10**, 141–156 (1979)
29. Weber, S.: On Hierarchical spatial competition. Rev. Econ. Stud. **59**, 407–425 (1992)

Nash Equilibria in Network Facility Location Under Delivered Prices

Blas Pelegrín, Pascual Fernández, and Maria D. García

1 Introduction

Some competitive location models attempt to find the locations of facilities at which profit is maximized. Profit is strongly affected by both the locations of facilities of the competing firms and the price set by firms in each customer area. If the firms enter simultaneously in the market, the maximization of their profit can be seen as a two-stage game. In the first stage, the firms simultaneously choose their facility locations. In the second stage, the firms will compete on price. The division into two stages is motivated by the fact that the choice of location is usually prior to the decision on price. Observe that location decision is relatively permanent whereas price decision can be easily changed. The two stage game can be reduced to a location game if there exists a price equilibrium in the second stage which is determined by the locations chosen by the firms in the first stage. Once the facility locations are chosen, the firms would set the equilibrium prices, and then their profit would be determined. Thus, the location-price problem could be considered as a game in which firms decide only on facility location. Other similar location games where the payoffs are given by market share or profit can be seen in [1, 7, 17, 20, 27].

The existence of a price equilibrium in the second stage of the game depends on the price policy to be considered, among other factors. Most of the papers dealing

B. Pelegrín (✉)
University of Murcia, Murcia, Spain
e-mail: pelegrin@un.es; pfdez@um.es

P. Fernández
Department of Statistics and Operations Research, University of Murcia, Murcia, Spain
e-mail: mdolores@pdi.ucam.edu

M.D. García
San Antonio Catholic University of Murcia, Murcia, Spain

© Springer International Publishing AG 2017
L. Mallozzi et al. (eds.), *Spatial Interaction Models*, Springer Optimization
and Its Applications 118, DOI 10.1007/978-3-319-52654-6_13

with the location-price problem consider two competing firms under any of the two following policies: mill pricing and delivered pricing. With *mill pricing*, a price equilibrium rarely exists (see for instance [4, 13, 14, 24]). Then the location-price problem has been studied as a location game by taking prices as parameters. For two competing firms on a tree network, it has been proved that a Nash equilibrium (NE) exists with locations at the median nodes if both firms set equal prices (see [9, 10]). For more than two firms, a location NE on a tree may not exist for equal prices as it has been proved in [11]. The profit maximization problem for an entering firm has been studied on a general network (see [26, 28]), but existence of a Nash equilibrium on a general network has not been proved when firms compete simultaneously on location. With *delivered pricing*, a price equilibrium always exists under quite general conditions. The existence of a price equilibrium was shown for the first time by Hoover [19], who analyzed spatial discriminatory pricing for firms with fixed locations and concluded that the local price set by a firm serving a particular market will be constrained by the delivery cost of the other firms serving that market. In situations where demand elasticity is 'not too high', the equilibrium price at a given market is equal to the delivery cost of the firm with the second lowest delivery cost. This result was extended later to spatial duopoly (see [21, 22]) and to spatial oligopoly for different types of location spaces (see [9, 14]).

Under delivered pricing, the equilibrium prices are usually determined by the locations of the facilities, then the location-price problem can be reduced to a location game. This location game has been scarcely studied in the location literature. For completely inelastic demand, the existence of a location Nash equilibrium has been proved. In a duopoly with constant marginal production costs, Lederer and Thisse [22] showed that a location Nash equilibrium exists which is a global minimizer of the social cost. The social cost is defined as the total delivered cost if each customer were served with the lowest marginal delivered cost. In oligopoly, the same result is obtained in [6], where the authors present a model in which firms compete with delivery pricing and locate single facilities on a network of connected but spatially separated markets. If demand is price sensitive or marginal production costs are not constant, the minimizers of social cost may not be a location NE (see [16, 18]). The profit maximization problem for an entering firm has been studied with price sensitive demand (see [15]), but existence of a location Nash equilibrium has not been proved when firms compete simultaneously on location.

The problem of minimizing the social cost on a network has been studied for two competing firms when marginal delivered costs are concave. This problem is equivalent to the r-median problem if the marginal delivered cost from each site location to each demand point is the same for all competing firms (see [25]). There is an extensive literature on algorithms to solve the r-median problem on networks which can be used to find a location NE (see for instance [2]). If marginal delivery cost from each site location to each demand point is different for each competitor the problem has been solved by using a Mixed Integer Linear Programming (MILP) formulation (see [25]). The problem of minimizing the social cost on a plane has been solved for two competing firms which locate single facilities (see [5, 12]).

The aim of this chapter is to extend the main results on the above mentioned location-price problem with delivered pricing to a general framework where there are more than two competing firms, each of them locating multiple facilities. The problem is studied for constant and variable demand which is located at the nodes of a transportation network. The chapter is organized as follows: In Sect. 2, it is shown how the location-price problem is reduced to a game with decisions on location. In Sect. 3, the existence and determination of location NE is studied for constant demand. In Sect. 4, the existence and determination of location NE is studied for variable demand. Finally, the selection of a location NE when there are multiple location NE is discussed in Sect. 5.

2 The Location-Price Problem

Let us consider N firms that sell an homogeneous product and compete for demand in a certain region. The firms manufacture and deliver the product to the customers, which buy from the firm that offers the lowest price. The firms have to choose their facility locations in some predetermined location space. Once their facility locations are fixed, the firms will set delivered prices at each customer area. Thus, each firm has to make decisions on location and price in order to maximize its profit.

As location space we will take a transportation network $G = (V, E, l)$, where V is the set of nodes, E is the set of edges, and $l : E \to \mathbb{R}$ with $l(e)$ being the length of edge e. Distance between two points a and b in the network is measured as the length of the shortest path linking the two points and it is denoted by $d(a, b)$. It is assumed that customers are grouped at the nodes, then the set of customer areas is given by $V = \{1, 2, \ldots, m\}$. The firms are supposed to locate their facilities at points on the network, then the set of location candidates for each firm is $L = V \cup E$.

The following notation will be used:

Indices

n index of the firms, $n = 1, \ldots, N$
k index of the nodes, $k = 1, \ldots, m$.

Data

$q_k(p)$	demand function at node k
c_x^n	marginal production cost of firm n at location x
t_{xk}^n	marginal transportation cost of firm n from location x to node k
$C_{xk}^n = c_x^n + t_{xk}^n$	marginal delivered cost (or minimum delivered price) of firm n from location x to node k

Decision variables

X^n	set of facility locations for firm n
p_k^n	price the firm n sets at node k

Miscellaneous

$$C_k^n(X^n) = min \{C_{xk}^n : x \in X^n\}$$ minimum price the firm n can set at node k
$$C_k(X) = min\{C_{xk} : x \in X\}$$ minimum price the facilities in the set X
                                                            can set at node k

Let $q_k(p)$ be continuous and strictly decreasing at all p in $[0, p_k^{max}]$, where p_k^{max} is the maximum price that customers in market k are willing to pay for the product. We consider that the demand function $q_k(p)$ in market k may be different from the demand function in other markets. In order to make competition effective in each market k, we assume that the competing firms are able to price below the maximum price , i.e. $C_k^n(X^n) < p_k^{max}$ for all X^n, $n = 1, 2, \ldots, N$.

Marginal delivered costs are supposed to be independent of the amounts delivered and firms use linear prices. Thus, the profit any firm gets from market k, serving the full market at price p, is $\Pi_k(p) = q_k(p)(p - c)$, where c is the marginal delivered cost of the firm. Then the *monopoly price* in market k is the optimal solution to the problem:

$$\max \{\Pi_k(p) : c \leq p \leq p_k^{max}\}$$

and it will be denoted by $p_k^{mon}(c)$.

The following assumptions concerning the previous maximization problem are considered:

Assumption 1 $\Pi_k(p)$ *is a unimodal quasi-concave function in* $[0, p_k^{max}]$.

Assumption 2 $c < p_k^{mon}(c)$ *, for each* $c \geq 0$.

Assumption 3 $p_k^{mon}(c)$ *is a continuous increasing function at all c in* $[0, p_k^{max}]$.

The first assumption guarantees the existence of a unique maximizer of the profit function, and therefore a unique monopoly price for each c value. The second assumption avoids trivial cases in which the optimal price is the marginal delivered cost, and consequently the profit is zero. The third assumption will be used to prove a convexity property of the maximum profit. There exists a variety of demand functions for which the previous assumptions are verified. Some examples are shown in Table 1.

The previous location-price problem can be seen as a two-stage game. In the first stage the firms compete on location. In the second stage, once the facility locations are fixed, the firms will compete on price.

2.1 The Second Stage of the Game

First, we will show the existence of a unique price equilibrium for any set of facility locations. Let us consider that customers do not have any preference concerning the supplier and they buy from the firm that offers the lowest price. It is assumed that

Table 1 Some demand functions and their monopoly prices

Demand	$q_k(p)$	$p_k^{mon}(c)$
Linear	$\alpha_k - \beta_k p$	$\frac{1}{2}(c + \frac{\alpha_k}{\beta_k})$
	$0 \le p \le \frac{\alpha_k}{\beta_k}$	
Quadratic	$\alpha_k - \beta_k p^2$	$\frac{1}{3}(c + \sqrt{c^2 + 3\frac{\alpha_k}{\beta_k}})$
	$0 \le p \le \sqrt{\frac{\alpha_k}{\beta_k}}$	
Exponential	$\alpha_k e^{-\beta_k p}$	$c + \frac{1}{\beta_k}$
	$0 \le p < \infty$	
Hyperbolic	$\alpha_k p^{-\beta_k}$	$\frac{c\beta_k}{\beta_k - 1}$
	$0 \le p < \infty, \beta_k > 1$	

each firm cannot offer a price below its marginal delivered cost and each facility can supply all demand placed on it. Thus, each firm n will set a price at node k which is greater than, or equal to, $C_k^n(X^n)$ for any set of facility locations X^n, $n = 1, 2 \ldots, N$.

If two firms offer a minimum price at node k, the one with the minimum marginal delivered cost can lower its price and it obtains all the demand in node k. Then we consider that ties in price are broken in favour of the firm with the lowest marginal delivered cost. If the tied firms have the same marginal delivered cost in node k, no tie breaking rule is needed to share demand at node k because they will obtain zero profit from node k as a result of price competition.

In the long-term competition, customers at node k will not buy from firm n if $C_k^n(X^n) > min\{C_k^u(X^u) : u = 1, 2, \ldots, N\}$. Therefore, each node will be served by the firm with the minimum marginal delivered cost and such a firm will set a price which maximizes its profit. Let $X = (X^1, X^2, \ldots, X^N)$ denote the set of fixed facility locations. For $n = 1, 2, \ldots, N$, let $C_k^{com}(X^n) = min\{C_{xk}^u : x \in X^u, u = 1, \ldots, N, u \ne n\}$ denote the minimum delivered cost of the competitors of firm n.

The price competition is as follows:

1. If $C_k^n(X^n) < C_k^{com}(X^n)$, then firm n obtains a maximum profit from node k by offering a price equal to the optimal solution of the following problem:

$$Max \{\Pi_k^n(p) = q_k(p)(p - C_k^n(X^n)) : C_k^n(X^n) \le p \le C_k^{com}(X^n)\}$$

The optimal solution to this problem is unique and it depends on the set of facility locations X. The solution is given by:

$$\hat{p}_k^n(X) = \begin{cases} p_k^{mon}(C_k^n(X^n)) & \text{if } p_k^{mon}(C_k^n(X^n)) < C_k^{com}(X^n) \\ \\ C_k^{com}(X^n) & \text{if } p_k^{mon}(C_k^n(X^n)) \ge C_k^{com}(X^n) \end{cases}$$

2. If $C_k^n(X^n) \geq C_k^{com}(X^n)$, then firm n obtains zero profit from node k. In this case, firm n sets a price $\hat{p}_k^n(X) = C_k^n(X^n)$ to make its competitors obtain a minimum profit from node k.

It is clear that no firm n can get a greater profit from node k by changing the price $\hat{p}_k^n(X)$ while the other firms keep such prices. Then $\hat{p}_k^n(X)$, $n = 1, 2, \ldots, N$, are the unique *equilibrium prices* in market k.

2.2 The First Stage of the Game

Let us assume that for any fixed set $X = (X^1, X^2, \ldots, X^N)$, the firms will set the equilibrium prices $\hat{p}_k^n(X)$, $n = 1, 2, \ldots, N$. Observe that price competition lead to each firm n will monopolize a group of nodes from which the firm gets a positive profit. This group of nodes not only depends on the locations of the facilities of firm n, but it also depends on the locations of the facilities of its competitors. Such group of nodes is denoted by $M^n(X)$ and it is given by:

$$M^n(X) = \{k : C_k^n(X^n) < C_k^{com}(X^n)\}$$

Then, the profit obtained by firm n is:

$$\Pi^n(X) = \sum_{k=1}^m q_k(\hat{p}_k^n(X))(\hat{p}_k^n(X) - C_k^n(X)) =$$

$$\sum_{k \in M^n(X)} q_k(\hat{p}_k^n(X))(\hat{p}_k^n(X) - C_k^n(X))$$

If the competing firms set the equilibrium prices, the location-price problem can be seen as a location game $LG = \{N, X^n, \Pi^n : n = 1, \ldots, N\}$, where N is the number of firms (players), X^n represents the set of facility locations chosen by firm n, and Π^n is the payoff firm n obtains. This game captures the idea that, when firms select their locations, they all anticipate the consequences of their choice on price.

For simplicity, given $X = (X^1, X^2, \ldots, X^N)$, we will use the notation $X = (X^n, X^{-n})$, where X^{-n} is the set of locations of the competing firms but n. Then a location Nash equilibrium (NE) is defined as a set of locations $\hat{X} = (\hat{X}^1, \hat{X}^2, \ldots, \hat{X}^N)$ such that for any n it is verified that:

$$\Pi^n(\hat{X}^n, \hat{X}^{-n}) \geq \Pi^n(X^n, \hat{X}^{-n}), \quad \forall X^n$$

In the following we will study the problem of existence of location NE, and the problem of finding such equilibria if they exist. We will distinguish between essential and non essential products.

3 Location Nash Equilibria with Essential Products

Let us assume that firms sell essential products. This means that demand does not change when price changes. Then the amount of demand at each node k is given by a constant function, $q_k(p) = Q_k, k = 1, \ldots, m$.

3.1 Existence of NE

For constant demand functions, the existence of a location NE can be proved by using the concept of social cost. The social cost is defined as the total cost incurred to supply demand to customers if each customer would pay for the product the minimum delivered cost. Then, for any fixed set of locations $X = (X^1, X^2, \ldots, X^n)$, the social cost is given by:

$$S(X) = \sum_{k=1}^{m} Q_k \min \left\{ C_k^1(X^1), C_k^2(X^2), \ldots, C_k^N(X^N) \right\}$$

Firstly, it is shown that the profit obtained by any firm is the total cost that would be experienced by its competitors serving the entire market with the minimum delivered cost minus the social cost. Secondly, a characterization of location NE is obtained. Finally, the existence of a location NE is proved.

Property 1 *If the firms set the equilibrium prices in each market, then for* $n = 1, \ldots, N$, *it is verified that:*

$$\Pi^n(X) = \sum_{k=1}^{m} C_k^{com}(X^n) Q_k - S(X)$$

Proof Since $q_k(p) = Q_k$, the equilibrium prices are given by:

$$\hat{p}_k^n(X) = \begin{cases} C_k^{com}(X^n) & \text{if } C_k^n(X^n) < C_k^{com}(X^n) \\ C_k^n(X^n) & \text{otherwise.} \end{cases}$$

Then the profit obtained by firm n can be expressed as follows:

$$\Pi^n(X) = \sum_{k \in M^n(X)} Q_k \left(C_k^{com}(X^n) - C_k^n(X^n) \right)$$

$$= \sum_{k \in M^n(X)} Q_k \left(C_k^{com}(X^n) - C_k^n(X^n) \right)$$

$$+ \sum_{k \notin M^n(X)} Q_k C_k^{com}(X^n) - \sum_{k \notin M^n(X)} Q_k C_k^{com}(X^n)$$

$$= \sum_{k=1}^{m} Q_k C_k^{com}(X^n) - \sum_{k=1}^{m} Q_k \min \left\{ C_k^n(X^n), C_k^{com}(X^n) \right\}$$

$$= \sum_{k=1}^{m} Q_k C_k^{com}(X^n) - S(X).$$

□

Property 2 $\hat{X}$ *is a location NE if and only if for* $n = 1, \ldots, N$, *it is verified that:*

$$S(\hat{X}^n, \hat{X}^{-n}) \leq S(X^n, \hat{X}^{-n}) \quad \forall X^n.$$

Proof Note that $\hat{X}$ is a location NE if and only if for $n = 1, 2, \ldots, N$, it is verified that:

$$\Pi^n(\hat{X}^n, \hat{X}^{-n}) \geq \Pi^n(X^n, \hat{X}^{-n}) \quad \forall X^n.$$

From Property 1, these inequalities are equivalent to the following ones:

$$S(\hat{X}^n, \hat{X}^{-n}) \leq S(X^n, \hat{X}^{-n}) \quad \forall X^n.$$

□

Property 3 *Any global minimizer of* $S(X)$ *is a location NE.*

Proof It follows from Property 2.   □

The existence of a global minimizer of social cost is proved by considering the following assumptions:

Assumption 4 *For* $n = 1, 2, \ldots, N$ *the marginal production cost,* c_x^n, *is a positive concave function when x varies along any edge in the network, and it is independent of the quantity produced.*

Assumption 5 *For* $n = 1, 2, \ldots, N$ *the marginal transportation cost,* t_{xk}^n , *is a positive, concave and increasing function with respect to the distance from x to each node k.*

Concavity of marginal production cost and marginal transportation cost is realistic in certain situations as it has been remarkable by many authors (see for instance [20, 22, 27]). Under such assumptions, as d_{xk} is a concave function at x, for any node k and x varying along any fixed edge, it is verified that the marginal delivered cost, $C_{xk}^n = C_x^n + t_{xk}^n$, is also a concave function for any node k and x varying along any fixed edge.

Property 4 *Under Assumptions 4 and 5, there exists a set of nodes which is a global minimizer of the social cost.*

Proof Let $X = (X^1, X^2, \ldots, X^N)$ be an arbitrary set of facility locations on the network. If $x \in X^n$ is not a node, then x is in the interior of some edge $e = (a, b) \in E$. Assume that all points in X are fixed, but the point x, which varies on the edge e. Under Assumptions 1 and 2, it results that the minimum price to serve market k, $\min \{C_k^1(X^1), C_k^2(X^2), \ldots, C_k^N(X^N)\}$, is a concave function when x varies on the edge e and the other locations are fixed. Since the sum of weighted concave functions, with non-negative weights, is also concave, it follows that the social cost, $S(X)$, is concave when x varies on the edge e and the other locations are fixed. Therefore, the social cost reaches its minimum value on edge e for $x = a$ or $x = b$.

Therefore, if we replace each non-node point in X by the corresponding minimizer node of the social cost when the other locations of the facilities are fixed, we will obtain sets of nodes $V^1, V^2, \ldots, V^N$ for which $S(V^1, V^2, \ldots, V^N) \leq S(X^1, X^2, \ldots, X^N)$. Consequently, there exists a set of nodes $\hat{V} = (V^1, V^2, \ldots, V^N)$ which minimizes the social cost.   □

3.2   Finding a Location NE

From Property 4, it follows that a location NE can be found by minimizing the social cost on the set of nodes. For any set of nodes $X = (X^1, X^2, \ldots, X^N)$ every set X^n can be represented by a vector $x^n = (x_1^n, x_2^n, \ldots, x_m^n)$ with components:

$$x_i^n = \begin{cases} 1 & \text{if node } i \in X^n \\ 0 & \text{otherwise.} \end{cases}$$

Let $x = (x^1, x^2, \ldots, x^N)$. With this representation of X, the social cost $S(x)$ is given by the optimal value of the following optimization problem:

$$SC(x) = \text{Min} \sum_{k=1}^m Q_k \left(\sum_{i=1}^m C_{ik}^1 z_{ik}^1 + \sum_{i=1}^m C_{ik}^2 z_{ik}^2 + \ldots + \sum_{i=1}^m C_{ik}^N z_{ik}^N \right)$$

$$\text{s.t.} \sum_{i=1}^m z_{ik}^1 + \sum_{i=1}^m z_{ik}^2 + \ldots + \sum_{i=1}^m z_{ik}^N = 1; \ k = 1, \ldots, m \qquad (1)$$

$$z_{ik}^n \leq x_i^n; n = 1, 2, \ldots, N; \ i, k = 1, 2, \ldots, m \qquad (2)$$

$$z_{ik}^n \in \{0, 1\}; \ n = 1, 2, \ldots, N; \ i, k = 1, 2, \ldots, m$$

Constraints (1) mean that for any k, only one variable z_{ik}^n will be equals to 1, the one corresponding to the minimum delivered cost from the locations of the facilities to node k. Constraints (2) mean that variable z_{ik}^n may take the value 1 if $x_i^n = 1$.

For each k, note that the optimal solution to the previous problem is obtained by assigning the value 1 to one variable z_{ik}^n for which the following two conditions are verified: $C_{ik}^n = min\{C_{jk}^n : j = 1, 2, \ldots, m$ and $x_j^n = 1\}$ and $C_{ik}^n \leq C_k^{com}(X^n)$. The value 0 is assigned to the other variables.

Therefore, the problem of minimizing the social cost when locations are nodes becomes into the problem:

$$(SCM) : \text{Min} \sum_{k=1}^m Q_k [\sum_{n=1}^N \sum_{i=1}^m C_{ik}^n z_{ik}^n]$$

$$\text{s.t.} \sum_{n=1}^N \sum_{i=1}^m z_{ik}^n = 1; \ k = 1, \ldots, m \tag{3}$$

$$z_{ik}^n \leq x_i^n; \ n = 1, \ldots, N; \ i, k = 1, \ldots, m \tag{4}$$

$$\sum_{i=1}^m x_i^n = r_n; \ n = 1, \ldots, N \tag{5}$$

$$z_{ik}^n, x_i^n \in \{0, 1\}; \ n = 1, \ldots, N; \ i, k = 1, \ldots, m$$

Constraints (5) show that each firm n selects r_n facility locations, where r_n is the number of facilities to be located by firm n. Let $\hat{x}_i^n$ be the optimal values of variables x_i^n, then a location NE is given by $\hat{X} = (\hat{X}^1, \hat{X}^2, \ldots, \hat{X}^N)$ where $\hat{X}^n = \{i : \hat{x}_i^n = 1\}$, $n = 1, 2, \ldots, N$.

Problem (SCM) can be solved by any standard ILP-optimizer (Xpress, Cplex, ...). However, computational difficulties may occur when the number of binary variables, which is $Nm(m + 1)$, is large. To solve more efficiently problem (SCM), the constraints $z_{ik}^n \in \{0, 1\}$ can be replaced by $z_{ik}^n \geq 0$. Note that for both sets of constraints the same value of $SC(x)$ is obtained. Therefore, an optimal solution of (SCM) can be obtained by talking either the sets of constraints $z_{ik}^n \in \{0, 1\}$, or the set of constraints $z_{ik}^n \geq 0$.

3.3   Firms with Equal Marginal Delivered Costs

If the marginal delivered costs are equal for all the firms, $C_{ik}^n = C_{ik}$ for $n = 1, 2, \ldots, N$, the previous formulation of the social cost minimization problem can be simplified. In fact, once the facility locations are fixed, note that any node k is served from the facility with the minimum delivered cost. Since the marginal delivered cost from each node is the same for all the firms, the minimum delivered cost to node k only depends on the nodes where the facilities are located, no matter which of the firms is the owner of the facility. Thus, if we consider the following variables:

$$x_i = \begin{cases} 1 & \text{if a facility is located at node } i \\ 0 & \text{otherwise} \end{cases}$$

$$z_{ik} = \begin{cases} 1 & \text{if node } k \text{ is served from node } i \\ 0 & \text{otherwise} \end{cases}$$

then the social cost minimization problem becomes into the following problem:

$$(SCM1) : \text{Min} \sum_{k=1}^{m} Q_k [\sum_{i=1}^{m} C_{ik} z_{ik}]$$

$$\text{s.t.} \sum_{i=1}^{m} z_{ik} = 1; \; k = 1, \ldots, m \tag{6}$$

$$z_{ik} \le x_i; \; i, k = 1, \ldots, m \tag{7}$$

$$\sum_{i=1}^{m} x_i = r \tag{8}$$

$$z_{ik} \ge 0, x_i \in \{0, 1\}; \; i, k = 1, \ldots, m$$

Constraints (6) mean that each node will be served by one facility, the one with the minimum delivered cost. Constraints (7) show that node k can be served from node i if $x_i = 1$. Constraint (8), where $r = r_1 + \ldots + r_N$, represents the total number of facilities to be located. The constraints $z_{ik} \in \{0, 1\}$ have been replaced by the constraints $z_{ik} \ge 0$. The number of variables is now equal to $m(m + 1)$, where m of them are binary and the other are non negative. Then large size problems can be solved by using standard optimizers. Observe that $(SCM1)$ is a formulation of the well known r−median problem (see [2, 23]).

If $\hat{X}$ is the set of nodes corresponding to the optimal solution $\hat{x} = (\hat{x}_1, \ldots, \hat{x}_m)$ of problem (SCM1), i.e., $\hat{X} = \{i : \hat{x}_i = 1\}$, then any partition $(\hat{X}^1, \hat{X}^2, \ldots, \hat{X}^N)$ of $\hat{X}$ verifying $|\hat{X}^n| = r_n, n = 1, 2, \ldots, N$, is a location NE. This is true due to $\hat{X}$ is a global minimizer of social cost. Consequently, there exist a large number of location NE. The problem of selecting one of such equilibria is considered in Sect. 5.

4   Location Nash Equilibria with Non Essential Products

Let us now consider that demand is sensible to price. This happens for products considered as not necessary to the customer. The demand at each node k is given by a function $q_k(p)$. We first prove a convexity property of the maximum profit that is obtained by any firm at each node. This property will be used to show that for any firm n, there is a set of optimal locations at the nodes for any fixed locations of its competitors.

4.1 Convexity of the Maximum Profit at a Node

Let us consider that the locations of the facilities of firm n, X^n, may change, but the locations of the facilities of its competitors are fixed. Given a node k, for simplicity let $c = C_k^n(X^n)$ and $c_k^{com} = C_k^{com}(X^n)$. The maximum profit of firm n at node k, as a function of the marginal delivered cost, is given by:

$$\Pi_k^n(c) = \begin{cases} \max\{q_k(p)(p-c) : c \le p \le c_k^{com}\} & \text{if } c < c_k^{com} \\ 0 & \text{if } c \ge c_k^{com} \end{cases}$$

Since $p_k^{mon}(c)$ is a continuous increasing function (Assumption 3), it follows that $p_k^{mon}(c) < c_k^{com}$ if and only if $c < c_k$ for someone threshold value c_k. Then the maximum profit in market k is given by the following function (see Fig. 1):

$$\Pi_k^n(c) = \begin{cases} q_k(p_k^{mon}(c))(p_k^{mon}(c) - c) & \text{if } c < c_k \\ q_k(c_k^{com})(c_k^{com} - c) & \text{if } c_k \le c < c_k^{com} \\ 0 & \text{if } c_k^{com} \le c \end{cases}$$

Observe that $\Pi_k^n(c)$ depends on the demand function $q_k(p)$ and it can be nonlinear in the interval $[0, c_k]$, but it is always linear in $[c_k, c_k^{com}]$.

Property 5 $\Pi_k^n(c)$ is a decreasing convex function in $[0, p_k^{max}]$.

Proof It is clear that function $\Pi_k^n(c)$ is decreasing at c, so we will show that it is convex. Let c_1 and c_2 be in $[0, p_k^{max}]$, and $c_\lambda = \lambda c_1 + (1 - \lambda)c_2$, $0 < \lambda < 1$. For simplicity, let $p_1 = p_k^{mon}(c_1)$, $p_2 = p_k^{mon}(c_2)$, $p_\lambda = p_k^{mon}(c_\lambda)$. From Assumption 3, as $c_1 < c_\lambda < c_2$, it follows that $p_1 < p_\lambda < p_2$.

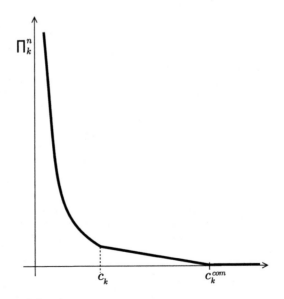

Fig. 1 Maximum profit function

We have to prove that $\Pi_k^n(c_\lambda) \leq \lambda\Pi_k^n(c_1)+(1-\lambda)\Pi_k^n(c_2)$, for which we consider the three following possible cases:

i) If $c_\lambda < c_k$, then:

$$\Pi_k^n(c_\lambda) = q_k(p_\lambda)(p_\lambda - c_\lambda) = q_k(p_\lambda)(p_\lambda - \lambda c_1 - (1 - \lambda)c_2)$$
$$= \lambda q_k(p_\lambda)(p_\lambda - c_1) + (1 - \lambda)q_k(p_\lambda)(p_\lambda - c_2)$$

Since $c_\lambda < c_k$ and $c_1 < p_k^{mon}(c_1) = p_1 < p_\lambda$, it follows that $c_1 < p_\lambda < c_k^{com}$, and therefore $q_k(p_\lambda)(p_\lambda - c_1) \leq \Pi_k^n(c_1)$. Since $p_\lambda < c_k^{com}$, it is verified that $q_k(p_\lambda)(p_\lambda - c_2) \leq \Pi_k^n(c_2)$. Then we obtain:

$$\Pi_k^n(c_\lambda) \leq \lambda\Pi_k^n(c_1) + (1 - \lambda)\Pi_k^n(c_2)$$

ii) If $c_k \leq c_\lambda < c_k^{com}$, then:

$$\Pi_k^n(c_\lambda) = q_k(c_k^{com})(c_k^{com} - c_\lambda) = q_k(c_k^{com})(c_k^{com} - \lambda c_1 - (1 - \lambda)c_2)$$
$$= \lambda q_k(c_k^{com})(c_k^{com} - c_1) + (1 - \lambda)q_k(c_k^{com})(c_k^{com} - c_2).$$

Since $c_1 < c_\lambda$ and $c_\lambda < c_k^{com}$, then $c_1 < c_k^{com}$. Therefore, $q_k(c_k^{com})(c_k^{com} - c_1) \leq \Pi_k^n(c_1)$. On the other hand, we have that $q_k(c_k^{com})(c_k^{com} - c_2) \leq \Pi_k^n(c_2)$ if $c_2 < c_k^{com}$ and $q_k(c_k^{com})(c_k^{com} - c_2) \leq 0 \leq \Pi_k^n(c_2)$ if $c_2 \geq c_k^{com}$. Then we obtain:

$$\Pi_k^n(c_\lambda) \leq \lambda\Pi_k^n(c_1) + (1 - \lambda)\Pi_k^n(c_2)$$

iii) If $c_k^{com} \leq c_\lambda$, then: $\Pi_k^n(c_\lambda) = 0 \leq \lambda\Pi_k^n(c_1) + (1 - \lambda)\Pi_k^n(c_2)$. □

4.2   *Existence of Location NE*

For variable demand functions $q_k(p)$, $k = 1, \ldots, m$, the social cost is given by:

$$S(X) = \sum_{k=1}^{m} q_k(C_k(X))\, C_k(X)$$

where $C_k(X) = min\{C_k^1(X^1), C_k^2(X^2), \ldots, C_k^N(X^N)\}$. Contrary to what happens for constant demand functions, a minimizer of the social cost may not be a location NE, as it is shown by the following example.

Consider two competing firms on the network shown in Fig. 2, each firm locating one facility. The number in each edge (i, k) is the marginal delivered cost, C_{ik}^n, between i and k, being $C_{ik}^1 = C_{ik}^2$. Demand in each node k is linear and given by $q_k(p) = 4 - p, 0 \leq p \leq 4$.

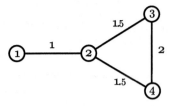

Fig. 2 Transportation network

Table 2 Social cost and profits

X	S(X)	$\Pi^1(X)$	$\Pi^2(X)$
1,2	7.5	7.125	4
1,3	6.75	7.25	4
1,4	6.75	7.25	4
2,3	7	5	4.75
2,4	7	5	4.75
3,4	5.5	3	3

Let $X = (i, j)$ be facility locations, where node i is the facility location for firm 1, and node j is the facility location for firm 2. Since $C_{ik}^1 = C_{ik}^2$, it is verified that $S(i, j) = S(j, i)$ and $\Pi^1(i, j) = \Pi^2(j, i)$ for all (i, j). In Table 2, the values $S(X), \Pi^1(X)$ and $\Pi^2(X)$ are shown for the different combinations (i, j), $i < j$. Note that pairs $(2, 3)$ and $(2, 4)$ are location NE while the minimizer of social cost, the pair $(3, 4)$, it is not a location NE.

The previous example shows that social cost cannot be used to obtain a location NE if demand is sensible to price. In this case, to our knowledge, no proof has been given to guarantee the existence of a location NE.

4.3 Finding a Location NE

We propose to use the best response procedure to find a location NE. This procedure has extensively been used in Game Theory to find NE when they exist (see [8]).

In our location game, the best response function is obtained as follows:

- Given a set $X = (X^1, X^2, \ldots, X^N)$ of facility locations, for each firm n the following optimization problem is solved:

$$P^n(X^{-n}) : Max\{\Pi^n(Y^n, X^{-n}) : |Y^n| = r_n, Y^n \subset L\}$$

- Let $\hat{X}^n$ be an optimal solution of problem $P^n(X^{-n})$, then the best response of firm n to the locations of the facilities of its competitors X^{-n} is defined as follows:

$$R^n(X) = \begin{cases} \hat{Y}^n & \text{if } \Pi^n(\hat{Y}^n, X^{-n}) > \Pi^n(X^n, X^{-n}) \\ X^n & \text{otherwise} \end{cases}$$

- The best response function is $R(X) = (R^1(X), R^2(X), \ldots, R^N(X))$.

It is clear that X is a Nash equilibrium if and only if $R(X) = X$. Therefore, the following algorithm can be used to obtain a location NE.

Algorithm BR

Step 1: Start with any feasible set of facility locations,
$X = (X^1, X^2, \ldots, X^N)$.

Step 2: For each n:
i) Find an optimal solution $\hat{Y}^n$ to problem $P^n(X^{-n})$.
ii) Determine $R^n(X)$.
  Set $R(X) = (R^1(X), R^2(X), \ldots, R^N(X))$.

Step 3: If $X = R(X)$, X is a location NE, STOP.
Otherwise, set $X = R(X)$ and go to Step 2.

Algorithm BR requires to solve problem $P^n(X^{-n})$ for each firm n. The following property will be used to solve such a problem.

Property 6 *Under Assumptions 4 and 5, there exists a set of nodes which is an optimal solution to problem $P^n(X^{-n})$ for each n.*

Proof Let $\hat{X}^n$ be an optimal solution to $P^n(X^{-n})$. Assume that there is a location $x_i \in \hat{X}^n$ which is an interior point on some edge (a, b). Let us consider all locations in $\hat{X}^n$ are fixed but x_i which is assumed to vary in (a, b).

Since the minimum of concave functions is also concave and C_{x_ik} is a concave function at x_i in (a, b), then $C_k^n(\hat{X}^n) = \min\{C_{x_ik}, C_k^n(\hat{X}^n \setminus x_i)\}$ is also concave at x_i in (a, b). From Property 5 we have that $\Pi_k^n(C_k^n(\hat{X}^n))$, as function of $C_k^n(\hat{X}^n)$, is decreasing and convex. From the theorem of composition of convex functions (see [3]) we obtain that $\Pi_k^n(C_k^n(\hat{X}^n))$ is convex at x_i in (a, b) if $\hat{X}^n \setminus x_i$ is fixed.

The sum of convex functions is also convex, therefore the profit function defined as $\Pi(\hat{X}^n) = \sum_{k=1}^m \Pi_k^n(C_k^n(\hat{X}^n))$ is convex at x_i in (a, b) if $\hat{X}^n \setminus x_i$ is fixed. This function reaches a maximum value in an extreme point of the edge (a, b) when x_i varies in (a, b). Then the location set $\hat{X}^n$ can be improved by replacing point x_i by one of the nodes a or b (the one for which a maximum profit is obtained).

Therefore, if the set of optimal locations $\hat{X}^n$ contains non nodes points, each non node point can be replaced by one node so that a new set of locations V^n is obtained whose points are nodes and $\Pi(\hat{X}^n) = \Pi(V^n)$. Consequently, there exists a set of nodes which is an optimal solution to $P^n(X^{-n})$. $\qquad\square$

If Assumptions 4 and 5 hold, from Property 6 an optimal solution to problem $P^n(X^{-n})$ can be found as follows:

Let us define the following sets and variables:

$$L_k^n = \{i : C_{ik}^n < C_k(X^{-n})\}$$

$$M^n = \{k : L_k^n \neq \emptyset\}$$

$$x_i^n = \begin{cases} 1 \text{ if a facility is located at node } i \\ 0 \text{ otherwise} \end{cases}$$

$$z_{ik}^n = \begin{cases} 1 \text{ if node } k \text{ is served by firm } n \text{ from node } i \\ 0 \text{ otherwise} \end{cases}$$

Note that L_k^n is the set of locations at which firm n can price below its competitors at node k. M^n is the set of nodes where firm n can get a positive profit. x_i^n and z_{ik}^n are location and allocation variables, respectively.

If node k is served from node $i \in L_k^n$, the equilibrium price is:

$$\hat{p}_k^n(i) = \begin{cases} p_k^{mon}(C_{ik}^n) \text{ if } p_k^{mon}(C_{ik}^n) < C_k(X^{-n}) \\ \\ C_k(X^{-n}) \text{ if } p_k^{mon}(C_{ik}^n) \geq C_k(X^{-n}) \end{cases}$$

Then the problem $P^n(X^{-n})$ can be formulated as follows:

$$P^n(X^{-n}) : \quad \max \sum_{k \in M^n} \sum_{i \in L_k^n} q_k(\hat{p}_k^n(i))(\hat{p}_k^n(i) - C_{ik}^n) z_{ik}^n$$

$$\text{s.t.} \quad \sum_{i \in L_k^n} z_{ik}^n \leq 1, \, k \in M^n \tag{9}$$

$$z_{ik}^n \leq x_i^n, \, k \in M^n, i \in L_k^n \tag{10}$$

$$\sum_{i \in L_k^n} x_i^n = r_n, \, k \in M^n \tag{11}$$

$$x_i^n, z_{ik}^n \in \{0, 1\}, \, k \in M^n, i \in L_k^n$$

The objective function of problem $P^n(X^{-n})$ represents the profit of firm n. Observe that the prices $\hat{p}_k^n(i)$ depend on the set X^{-n}. Constraints (9) mean that each node $k \in M^n$ can be served from at most one of the facilities of firm n (the facility with the minimum marginal delivered cost in the optimal solution). Constraints (10) imply that variable z_{ik}^n may be positive only if firm n locates a facility at i. Constraint (11) represents the number of facilities to be located by firm n.

The above problem is a Binary Integer Linear Programming (BILP) problem which contains a lot of binary variables. However, the number of binary variables can be notably reduced as follows. Let $\hat{x}_i^n$ denote an optimal solution for variables x_i^n, then an optimal solution for variables z_{ik}^n is given by:

$$\hat{z}_{ik}^n = \begin{cases} 1 \text{ if } c_{ik} = \min\{c_{hk} : h \in L_k^n, x_h^n = 1\} \\ 0 \text{ otherwise} \end{cases}$$

As the allocation variables are determined by the decision variables in the optimal solution, z_{ik}^n can be taken as non negative variable instead of a binary variable. Then, replacing constraints $z_{ik}^n \in \{0, 1\}$ by $z_{ik}^n \geq 0$ in the above formulation, we obtain an equivalent problem which is a Mixed Integer Linear Programming (*MILP*) problem.

It may occur that Algorithm BR does not stop, and therefore it does not find a NE. It may also occur that a location NE does not exist but if it exists Algorithm BR could find it.

5 Existence of Multiple Location Nash Equilibria

In the case of an essential product, social cost minimization at the nodes of the network is a combinatorial optimization problem that may have multiple global optima. Then more than one location NE could exist. Furthermore, if X is a global minimizer of social cost and $C_{ik}^n = C_{ik}$ for all n, then any partition of the set of optimal locations X into sets $X^1, \ldots, X^N$ such that $|X^n| = r_n$, $n = 1, \ldots, N$, is a location NE. Thus, the number of location NE corresponding to such partitions is $\frac{r!}{r_1! r_2! \cdots r_N!}$.

In the case of a non essential product, minimizers of social cost may not be location NE and the previous result does not hold, but it is possible the existence of more than one location NE as it was shown in the example of Sect. 4.

When more than one location NE are found, the competing firms could agree to select a Pareto optimum equilibrium. Thus, if $\hat{X}$ and $\tilde{X}$ are location NE and it is verified that $\Pi^n(\hat{X}) \geq \Pi^n(\tilde{X})$, $n = 1, \ldots, N$, with at least one strict inequality, then the firms could agree to select $\hat{X}$ better than $\tilde{X}$.

5.1 Aggregated Profit Maximization

Let X be a set of nodes corresponding to an optimal solution to problem (SCM1). We want to determine a partition of the set X into N subsets $X^1, \ldots, X^N$ such that the aggregated profit obtained by the firms is maximized.

Then we have to solve the following problem:

$$P(X) : Maximize \sum_{n=1}^{N} \Pi^n(X^n)$$

$$s.t. \qquad X^n \subset X \text{ and } |X^n| = r_n, n = 1, \ldots, N.$$

The aggregated profit associated to X is given by:

$$\sum_{n=1}^{N} \Pi^n(X) = \sum_{n=1}^{N} (\sum_{k=1}^{m} Q_k C_k^{com}(X^n) - S(X)) =$$

$$= \sum_{k=1}^{m} Q_k(C_k^{com}(X^1) + C_k^{com}(X^2) + \ldots + C_k^{com}(X^N)) - NS(X)$$

We consider the following variables:

$$y_{jn} = \begin{cases} 1 \text{ if } j \in X^n \\ 0 \text{ otherwise} \end{cases}$$

$$C_k^n = min\{C_{jk} : j \in X, y_{jn} = 0\}$$

Note that variables y_{jn} define a partition $(X^1, \ldots, X^N)$ of set X, and each variable C_k^n takes the value $C_k^{com}(X^n)$ associated to the partition $(X^1, \ldots, X^N)$.

Then the problem of determining the partition of X with the maximum aggregated profit can be formulated as follows:

$$P(X) : Maximize \sum_{k=1}^{m} Q_k(C_k^1 + C_k^2 + \ldots + C_k^N) - NS(X)$$

$$s.t. \quad \sum_{j \in X} y_{jn} = r_n; \; n = 1, \ldots, N \tag{12}$$

$$C_k^n \le C_{jk}(1 - y_{jn}) + Dy_{jn}; \; n = 1, \ldots, N, k = 1, \ldots, m, j \in X \tag{13}$$

$$y_{jn} \in \{0, 1\}, C_k^n \ge 0; \; n = 1, \ldots, N, k = 1, \ldots, m, j \in X$$

Constraints (12) mean that each firm n locates r_n facilities which are selected from the set X. Constraints (13) guarantee that each variable C_k^n will take the value $C_k^{com}(X^n)$ corresponding to the optimal partition $(X^1, \ldots, X^N)$ of X. D is a fixed positive value greater than any cost C_{jk}.

Let $\hat{y}_{jn}$ be the values of variables y_{jn} corresponding to an optimal solution to problem $P(X)$. Then the location NE which maximize the aggregated profit is $\hat{X}^n = \{j \in X : \hat{y}_{jn} = 1\}, n = 1, \ldots, N$.

5.2 Equity Constraints

Any firm n could disagree with the partition $(\hat{X}^1, \ldots, \hat{X}^N)$ of X which maximizes the aggregated profit if $\Pi^n(\hat{X}^n)$ is not high enough. An alternative way of selecting a partition of X is by including *equity constraints*. The aim of such constraints is to determine a location equilibrium, so that the firms get similar profits per facility.

Let $\hat{\Pi}$ denote the maximum aggregated profit, which can be obtained by solving problem $P(X)$. For a fixed value λ, $0 \le \lambda \le 1$, any firm n could agree on selecting a partition of X (location NE) if *the average profit per facility the firm obtains is greater than, or equal to, $\lambda\hat{\Pi}/r$*, where r is the total number of facilities. A location equilibrium verifying the equity constraints, for which the aggregated profit is maximum, could be obtained by solving the following *MILP* problem:

$$P_\lambda(X) : Maximize \sum_{k=1}^{m} Q_k(C_k^1 + C_k^2 + \ldots + C_k^N)$$

$$s.t. \quad C_k^n \le C_{jk}(1 - y_{jn}) + Dy_{jn}; \ n = 1, \ldots, N, k = 1, \ldots, m, j \in X$$

$$\sum_{j \in X} y_{jn} = r_n; \ n = 1, \ldots, N$$

$$\frac{1}{r_n}\left(\sum_{k=1}^{m} Q_k C_k^n - S(X)\right) \ge \lambda\frac{\hat{\Pi}}{r}; \ n = 1, \ldots, N$$

$$y_{jn} \in \{0, 1\}, C_k^n \ge 0; \ n = 1, \ldots, N, k = 1, \ldots, m, j \in X$$

Observe that $P_\lambda(X)$ reduces to $P(X)$ for $\lambda = 0$. For small values of λ problem $P_\lambda(X)$ is feasible, but it could be unfeasible for values of λ close to 1. In order to select a location equilibrium, a sequence of problems $P_\lambda(X)$ can be solved for fixed increasing λ values until one not feasible problem is found. Let $\overline{\lambda}$ be the greater value of λ for which $P_\lambda(X)$ is feasible, then firms could select the location equilibrium given by the following partition:

$$\hat{X}^n = \{j \in X : \hat{y}_{jn} = 1\}$$

where $\hat{y}_{jn}$ are the optimal values for variables y_{jn} in problem $P_{\overline{\lambda}}(X)$.

Acknowledgements This research has been supported by the Ministry of Economy and Competitiveness of Spain under the research project MTM2015-70260-P, and the Fundación Séneca (The Agency of Science and Technology of the Region of Murcia) under the research project 19241/PI/14.

References

1. Abellanas, M., Lillo, I., López, M.D., Rodrigo, J.: Electoral strategies in a dynamical democratic system. Geometric models. Eur. J. Oper. Res. **175**(2), 870–878 (2006)
2. Avella, P., Sassano, A., Vasilev, I.: Computational study of large-scale p-Median problems. Math. Program. Ser. A **109**, 89–114 (2007)
3. Bazaraa, M.S., Sherali, H.D., Shetty, C.M.: Nonlinear Programming: Theory and Algorithms. Wiley, New York (2006)

4. De Palma, A., Ginsburgh, V., Thisse, J.F.: On existence of location equilibria in the 3-firm Hotelling problem. J. Ind. Econ. **36**, 245–252 (1987)
5. Díaz-Bañez, J.M., Heredia, M., Pelegrín, B., Pérez-Lantero, P., Ventura, I.: Finding all pure strategy Nash Equilibria in a planar location game. Eur. J. Oper. Res. **214**, 91–98 (2011)
6. Dorta-González, P., Santos-Peñate, D.R., Suárez-Vega, R.: Spatial competition in networks under delivered pricing. Pap. Reg. Sci. **84**, 271–280 (2005)
7. Dürr, C., Thang, N.K.: Nash equilibria in Voronoi games on graphs. Lect. Notes Comput. Sci **4698**, 17–28 (2007)
8. Eichberger, J.: Game Theory for Economics. Academic, New York (1993)
9. Eiselt, H.A.: Hotelling's duopoly on a tree. Ann. Oper. Res. **40**, 195–207 (1992)
10. Eiselt, H.A., Laporte, G.: Locational equilibrium of two facilities on a tree. Recherche Operationnelle **25**, 5–18 (1991)
11. Eiselt, H.A., Laporte, G.: The existence of equilibria in the 3-facility Hotelling model in a tree. Transp. Sci. **27**, 39–43 (1993)
12. Fernández, J., Salhi, S., Toth, B.G.: Location equilibria for a continuous competitive facility location problem under delivered pricing. Comput. Oper. Res. **4**, 185–195 (2014)
13. Gabszewicz, J.J., Thisse, J.F.: Location. In: Aumann, R., Hart, S. (eds.) Handbook of Game Theory with Economic Applications, pp. 281–304. Elsevier Science, New York (1992)
14. García, M.D., Fernández, P., Pelegrín, B.: On price competition in location-price models with spatially separated markets. TOP **12**, 351–374 (2004)
15. García, M.D., Pelegrín, B., Fernández, P.: Location strategy for a firm under competitive delivered prices. Ann. Reg. Sci. **47**, 1–23 (2011)
16. Gupta, P.: Competitive spatial price discrimination with strictly convex production costs. Reg. Sci. Urban Econ. **24**, 265–272 (1994)
17. Hakimi, S.L.: Locations with spatial interactions: competitive locations and games. In: Francis, R.L., Mirchandani, P.B. (eds.) Discrete Location Theory. Wiley, New York (1990)
18. Hamilton, J.H., Thisse, J.F., Weskamp, A.: Spatial discrimination, Bertrand vs. Cournot in a model of location choice. Reg. Sci. Urban Econ. **19**, 87–102 (1989)
19. Hoover, E.M.: Spatial price discrimination. Rev. Econ. Stud. **4**, 182–191 (1936)
20. Labbe, M., Hakimi, S.L.: Market and location equilibrium for two competitors. Oper. Res. **39**, 749–756 (1991)
21. Lederer, P.J., Hurter, A.P.: Competition of chains, discriminatory pricing and location. Econometrica **54**, 623–640 (1986)
22. Lederer, P.J., Thisse, J.F.: Competitive location on networks under delivered pricing. Oper. Res. Lett. **9**, 147–153 (1990)
23. Marianov, V., Serra, D.: Median problems in networks. In: Eiselt, H.A., Marianov, V. (eds.) Foundations of Location Analysis, pp. 39–59. Springer, Berlin (2011)
24. Osborne, M.J., Pitchik, C.: Equilibrium in Hotelling's model of spatial competition. Econometrica **55**, 911–922 (1987)
25. Pelegrín, B., Dorta, P., Fernández, P.: Finding location equilibria for competing firms under delivered pricing. J. Oper. Res. Soc. **62**, 729–741 (2011)
26. Plastria, F., Vanhaverbeke, L.: Maximal covering location problem with price decision for revenue maximization in a competitive environment. OR Spectr. **31**, 555–571 (2009)
27. Sarkar, J., Gupta, B., Pal, D.: Location equilibrium for Cournot oligopoly in spatially separated markets. J. Reg. Sci. **2**, 195–212 (1997)
28. Serra, D., ReVelle, C.S.: Competitive location and pricing on networks. Geogr. Anal. **31**, 109–129 (1999)

Sharing Costs in Some Distinguished Location Problems

Justo Puerto

1 Introduction

Location Analysis is an important area within Operations Research that has attracted a lot of attention in the last decades by its theoretical implications and its many real world applications [28]. A lot of effort has been devoted to its modeling aspects and algorithmic developments with special emphasis on complexity issues and links with other areas such as logistics, network design or transportation (see e.g. [21]).

Apart from the above mentioned subjects there is another one, namely *cost sharing*, that is becoming more and more appealing in Location Analysis. In a standard location problem one has to determine the placement of some facilities to deliver a service to a set of users. Provided that this placement is done to minimize the overall cost and assuming that the cost must be assumed by the users, there is an additional question that should be addressed: How to share this cost in such a way that users do not have any incentives to break apart and leave the group. This problem is not new and, in general, it is a central question in cooperative game theory (see [30]). To each O.R. model that leads to an optimization problem one can associate a cooperative game where the different entities in the model are the players and the characteristic function of each group of players (coalition) is given as the optimal value of the solution of the model applied to that group [3]. This way one defines a Transferable Utility (TU) game and then the remaining question is how to find fair cost shares among the users. This issue can be addressed by different approaches: using some distinguished allocations (as for instance the Shapley value

J. Puerto (✉)
Instituto de Investigación Matemática de la Universidad de Sevilla (IMUS), Edificio Celestino Mutis, Universidad de Sevilla, Sevilla, Spain
e-mail: puerto@us.es

© Springer International Publishing AG 2017
L. Mallozzi et al. (eds.), *Spatial Interaction Models*, Springer Optimization and Its Applications 118, DOI 10.1007/978-3-319-52654-6_14

or the nucleolus...) or by means of sets of stable allocations [31]. In this chapter we will restrict ourselves to this latter approach and among the different choices of solution sets we shall focus on the concept of core.

The core of a TU game is the set of cost shares that are efficient and coalitionally stable. Efficiency ensures that the overall cost incurred by the complete set of users is fully distributed. This is always implicitly assumed. Therefore, an allocation x is in the core if, for any group of users S, the cost supported by S, namely $x(S) = \sum_{i \in S} x_i$, is not greater than the characteristic value of that group, $c(S)$. This property ensures that no group of users will have any incentive to break apart while sharing the cost by any allocation scheme in the core. The reason is clear: any group is entitled to pay less than what it would have paid by the characteristic value. This condition is rather desirable because it enforces a stability property to cost shares in the core. However, the core is a polyhedral set that may be empty depending on the characteristic function of the game.

The goal of this chapter is to analyze some location models under the perspective of their cost sharing aspects. We will consider some standard location problems and will associate cost TU games to each of them. The final goal is to provide non emptiness conditions for the core and to give its geometrical or analytical description. In our way we will recall some already known and also introduce new classes of location games focussing essentially on continuous location problems. Unlike the class of discrete location problems whose cost sharing analysis has been well-analyzed in [4, 14], cost sharing aspects of the class of continuous location problems is not well-understood and core existence results are only known for some single facility situations [24, 38, 39] which are easier than the related multifacility problems.

The contribution of this work is to unify the analysis of cost sharing in continuous location models. In doing that, we revisit some results from the literature that cover continuous single facility location games as in [24, 38] and the diameter and radius games [39, 40]. In addition, we present a new framework to handle continuous multifacility location games based on the analysis of some mixed-integer programs resulting from their respective optimization problems.

The rest of the chapter is organized as follows. In Sect. 2 we recall the main elements describing a location problem and introduce some basic location models that will be analyzed in the following sections. Section 3 is devoted to revise the literature of location games and to build the links between the optimization and the cost sharing models. The next three sections analyze cost sharing problems on continuous location problems. Specifically, Sect. 4 considers single-and-multifacility location problems and Sect. 5 a continuous set covering location model. Section 6 deals with diameter Steiner location and radius location problems on general metric spaces. The chapter ends with the acknowledgements and the list of references.

2   A Primer on Location Theory

A location problem consists of determining the position of one or more facilities in order to optimize a measure of effectiveness with respect to a set of known demand locations. Location theory, as any other discipline in Operations Research, develops mathematical models to represent in the best possible way the real situation and to give adequate solutions to the problem under study. This area of research has already a long history and it is now in full expansion since a lot of methods and procedures can successfully be adapted in order to solve complex problems belonging to other knowledge areas [21, 28, 37].

Roughly speaking, location problems can be classified into three categories: discrete, network and continuous location. Discrete location imposes that the set of candidate locations for placing the new facility(ies) is finite. Network location problems consider that the demand points are in a graph and facilities have to be located at the nodes or at the interior points of edges of the graph. Finally, continuous location considers problems where the demand points and the service-facility locations belong to a continuous space, typically the Euclidean space. Excellent references that cover these fields in Location Theory are [6, 22, 23, 27]. Other references covering all fields are [7, 8, 21, 35].

In order to get a better understanding of the location problems structure, we briefly describe, next, the common elements to all these problems.

2.1   The Solution Space

The solution space is the framework where the problem is defined. It contains as elements the existing facilities and the new facility(ies). The choice of an appropriate solution space is crucial, because it determines important aspects such as the accuracy and efficiency of the model. Some usual solution spaces are:

- *Discrete spaces*: When there exists a finite number of potential locations for the new facilities.
- *Networks:* The solution candidates lie within a graph, usually representing a communication network. Nodes represent important elements, such as cities or crossroads. Arcs represent connections between nodes, like roads, streets, cables, etc. A kind of network that has received considerable attention is the "tree network". This is due mainly to the uniqueness of a path between pairs of points.
- *Euclidean space* $\mathbb{R}^q$: It is used when the problem presents regional aspects that cannot be discretized. In addition, it can be used to approximate networks when the number of nodes and arcs is large.

    The cases $q = 2$ and $q = 3$ have a clear physical meaning. Cases where $q \geq 4$ have been used to model and solve estimation problems in Statistics.
- *Sphere*: It is useful for those real situations that cope with large scale distances.

– *Embedded network in a continuous space*: This is the solution space where a network, that represents high speed connections, overlaps on a general continuous framework, the most common are the Euclidean space or the sphere.

2.2 Existing Locations

In terms of Location Theory, *existing facilities* are the users that require to be served. Therefore, they are called demand points. Usually, they are modeled by means of a set N and an *intensity* function to weight the elements of N.

There exist two main ways of representing demand in the solution space: by a finite set of points and by regions (see e.g. [29, 36]). In the first case, a set of points $N = \{a_1, \ldots, a_n\}$ is considered as well as a set of *weights* $\{\omega_1, \ldots, \omega_n\}$ that represent the importance (or intensity) of the demand generated at each point. In the regional model, demand is represented by means of a region $\mathscr{R}$ (not necessarily connected) included in the solution space and it is a probability measure which gives importance to each measurable subset of $\mathscr{R}$.

2.3 The New Facility(ies)

The location of the new facility is the decision variable of the general location problem. This variable is characterized by

(a) Number and quality of the service provided. If more than one facility is to be located, it will be necessary to specify the characteristics of each one of them. When they are identical, as for instance mail boxes, we face with a multifacility problem; otherwise as in the case of health services, we may face hierarchical location problems.
(b) Nature of the service. Not all the services are attractive for the community where they will be located. For instance, nuclear plants, solid waste disposals or garbage plants are usually refused by population. Therefore, in modeling a problem it is very important to determine the attractiveness of the service.

2.4 The Objective Function

Location problems mentioned in this chapter have the following objective function in common:

$$\text{opt}_{X=\{x_1,\ldots,x_m\}\subset\mathbb{X}} F\left(d(X,a)_{a\in N}\right),$$

where

F is a globalizing function,
"opt" means optimize, either minimize or maximize,
$\mathbb{X}$ is the solution space,
$X = \{x_1, \ldots, x_m\} \subset S$ is the new facility(ies), either single $m = 1$ or multiple $m > 1$.
N is the set of existing facilities (demand points),
a is a general existing facility,
$d(\cdot, \cdot)$ is a measure of distances. In general, $d(X, a)$ stands for the distance between demand point a and the set of facilities $(x_1, \ldots, x_m)$, i.e. $d(X, a) = \min_{b \in X} d(b, a)$.

Determining which objective function has to be used is sometimes a hard task. It should be noted that the final solution strongly depends on that choice. Therefore, it is important to devote some effort to this part of the modelling process. Some of the most common objective functions in the literature of location analysis are described below [21].

1. *The p-median problem or "minisum"*. The p-median problem [16, 17] searches for the location of m facilities with the goal of minimizing the weighted sum of distances between the demand points and the facilities to which they are allocated. A general p-median formulation is the following:

$$\min_{X \subset \mathbb{X}} \sum_{a \in D} \omega_a d(X, a).$$

2. *The p-center problem or minmax.* The p-center problem [16] assumes that all the demand is covered with p facilities and minimizes the coverage distance for doing so: the maximum weighted distance between a demand and its nearest facility is minimized

$$\min_{X \subset \mathbb{X}} \max_{a \in N} \omega_a d(X, a).$$

The minmax model can be interpreted as an equity based criterion.

3. *Cent-dian problem.* Given a positive scalar $\lambda \in (0, 1)$, this objective function corresponds to a convex combination of the minisum and minmax criteria. That is, the problem is:

$$\min_{X \subset \mathbb{X}} (\lambda \sum_{a \in N} d(X, a) + (1 - \lambda) \max_{a \in D} d(X, a)).$$

The cent-dian model corresponds to a compromise between the center and median criteria, that are conflicting criteria in most of the cases [44].

4. *Ordered median problem.*
   Given a finite number of existing facilities $N = \{a_1, \ldots, a_n\}$ and nonnegative weights $\mu_1, \ldots, \mu_n$, the goal is to find the location of X minimizing an ordered weighted average of distances, i.e.,

$$\min_{X \subset \mathbb{X}} \sum_{i=1}^{m} \mu_i d_{(i)}(X).$$

   Here, $d_{(i)}(X) = d(X, a_{\sigma_i})$ is the ith element in the list of sorted distances

$$d(X, a_{\sigma_1}) \leq \ldots \leq d(X, a_{\sigma_n}),$$

   where σ is a permutation of $\{1, \ldots, n\}$. Note that this objective function is pointwise defined, because its expression changes when the order between distances is modified. This function is somehow similar to the p-median, but is more general because depending on the choice of the parameter μ, it includes as particular instances the minsum, minmax and centdian, among many others [28, 37].

5. *Set covering problem.* In this problem, the number of facilities to be located is not fixed *a priori*, that is, the cardinality of X (denoted by $|X|$) has to be minimized and determined together with its elements. The requirement is that each existing facility a must have a server within a specified distance, r_a. The goal is to find the lowest number of facilities and their location satisfying the above constraint [21, 43]. Thus, the problem can be written as:

$$\min_{X \subset \mathbb{X}: d(X,a) \leq r_a, a \in N} |X|.$$

3   From Location Problems to Location Games

As we already mentioned in the introduction, the last decades have been a florist period in the development of Location Analysis. Among the many fields that have attracted the interest of location analysts one of them is Game Theory. This interest has covered noncooperative and cooperative game theory and specifically, cost sharing in locational decisions. The first historical reference to location games goes back to the duopoly model by Hotelling in [20]. Since then several extensions and further results have appeared in the specialized literature. The interested readers are referred to [9, 12] and to the good surveys on this topic by Owen and Daskin [32] and Fragnelli and Gagliardo [11].

One of the main goals of an optimization problem defined on an applied mathematical model is cost reduction. One way to achieve this goal, in location problems, is to incentive collaboration of groups of users (demand points) that might form coalitions to diminish costs. These coalitions should induce individual and collective cost reductions; thus, stability must be achieved in the process of promoting cooperation. In a location problem the main costs are set up and transportation. This implies that to reduce expenses, users may share the service facilities and their operation and construction costs.

In our framework a coalition allows each of its users (demand points) to have access to the same service facilities. Obviously, this planning process makes sense only throughout a long term time horizon since building a service location is a strategic decision that is made to last for a long period of time. The model that represents the above described situation can happen in most location problems considered in the literature. The aim of any group of demand points (users) is to satisfy their service needs at a minimum cost. Depending on the framework the former question gives rise to well-known optimization problems that can appear in the discrete, network and continuous fields. The optimal solutions of these problems lead to the best locational decisions for the group of demand points and these policies generate an optimal operation cost for the group as a whole. The question is what portion of this cost is to be supported by each demand point. Cooperative game theory provides the natural tools for answering this question.

The analysis of cost sharing in location problems is not new. Some of the first papers considering cost allocation games coming from location problems are [5, 15, 41–43]. The first paper, namely [15], analyzes single facility location problems on a tree. The paper by Tamir [43] considers cost sharing problems on coverage location models. In [5] several discrete multifacility location problems from a cooperative point of view are described, whereas [41] considers cooperative games based on Hub-location problems. Many more references can be found in the literature and the reader is again referred to the recent survey [11] for a comprehensive list of references.

Our goal in this chapter is to analyze some continuous location problems under the cost sharing perspective. We have chosen, to be included in our presentation, three single facility games, namely the single facility game proposed in [38], the Steiner diameter [40] and the radius games [39]. Moreover, motivated by the lack of references, we have also included two new classes of continuous multifacility games: the multifacility single allocation ordered median game and the multifacility coverage game. The reason for these choices is that all of them provide interesting results that shed light on the structure of their core sets. Moreover, the last two classes of multifacility games are new and extend some results known for their discrete version to the continuous framework. The main difference between discrete and continuous problems rests on their sets of feasible solutions. In the former case, feasible solutions are finite and known a priori whereas in the latter feasible solutions belong to a set with a continuum of points, one has to choose the candidates and only a description of the set is provided. Usually, this fact introduces a first degree of difficulty on the problem. In addition, there exists another difference between single facility and multifacility problems that is inherited from their respective optimization models. Multifacility problems are often more complex than their single facility counterpart since they incorporate a combinatorial aspect due to the *patronizing rule* used to assign the users to their service facility (in our problems we shall assign demand points to *the cheapest* or *the closest* service facility). This feature also introduces a second degree of difficulty on the problems.

For the sake of completeness the last part of this section is devoted to recall the definition of some concepts that will be referred to extensively in the rest of the chapter.

First, we recall that a generic finite cooperative game is a pair (N, c), where N is a finite set of n players and c is the characteristic function defined from 2^N to $\mathbb{R}$, which satisfies $c(\emptyset) = 0$, and assigns to each coalition $S \subseteq N$ a real value (it can be a benefit or a cost). The game (N, c) is called *monotone* if for any pair of subsets $S_1 \subseteq S_2 \subseteq N$, $c(S_1) \leq c(S_2)$. It is called *subadditive* if for any pair of subsets $S_1, S_2 \subseteq N$, $S_1 \cap S_2 = \emptyset$, $c(S_1 \cup S_2) \leq c(S_1) + c(S_2)$, and it is called *submodular* if for any pair of subsets $S_1, S_2 \subseteq N$, $c(S_1 \cup S_2) + c(S_1 \cap S_2) \leq c(S_1) + c(S_2)$.

The core of (N, c) (in the case of a cost game) is the set

$$C(N, c) = \{x \in \mathbb{R}^n : x(N) = c(N), x(S) \leq c(S), \forall\, S \subseteq N\}, \tag{1}$$

where $x(S) = \sum_{j:a_j \in S} x_j$, for any $S \subseteq N$.

4 Cost Sharing of Some Location Situations on a Continuous Framework

This section considers cost sharing situations that are based on two basic facility location problems, namely the continuous single facility location and the continuous single allocation multifacility location models. The first one was already considered in [24, 38] whereas the analysis of the second one is new. For the presentation of the results on the single facility case we follow the notation and material in [38].

4.1 The Single Facility Case

To start with, we consider the continuous single facility location problem. Informally, in such a problem we have a set of n users of a certain facility, placed in n different points in the space $\mathbb{R}^q$ with $q \geq 1$. The problem consists of finding a location for the facility which minimizes the transportation cost (which depends on the distances from the users to the facility).

Formally, a continuous single facility location problem is a triplet (N, Φ, d) where $N = \{a_1, \ldots, a_n\}$ is a set of n different points in $\mathbb{R}^q$ (with $n \geq 2$), $\Phi : \mathbb{R}^n \to \mathbb{R}$ is a lower semicontinuous globalizing function that satisfies (1) $\Phi(x) = 0$ if and only if $x = 0$; and (2) $\Phi(x) \leq \Phi(y)$ whenever $x \leq y$, and $d : \mathbb{R}^q \times \mathbb{R}^q \to \mathbb{R}$ is a measure of distance, satisfying that, for every $r, s \in \mathbb{R}^q$, $d(r, s) = f(\| r - s \|)$, where f is a lower semicontinuous, non decreasing and non negative map from $\mathbb{R}$ to $\mathbb{R}$ with $f(0) = 0$, and $\| \|$ is a norm on $\mathbb{R}^q$.

Solving the continuous single facility location problem (N, Φ, d) for $S \subset N$ means to find a $\bar{x} \in \mathbb{R}^q$ minimizing $\Phi(d^S(x))$, where $d^S(x)$ is the vector in $\mathbb{R}^n$ whose ith component is equal to $d(x, a_i)$ if $a_i \in S$, and equal to zero otherwise. We denote

$L(S) = \min_{x \in \mathbb{R}^q} \Phi(d^S(x))$. It is worth noting that this problem always has a solution for every $S \subset N$ (see, for instance, [34]).

This is the classical version of the continuous single facility location problem. Here we consider a natural variant of this problem in which the users in N are interested not only in finding an optimal location of the facility, but also in sharing the corresponding total costs. By total costs we mean the sum of the variable costs (depending on the users and on the location of the facility; they are mostly transportation costs), plus the fixed costs (independent of the number of users and of the location of the facility; they are mostly installation costs). Formally, a continuous single facility location situation is a 4-tuple (N, Φ, d, K) where (N, Φ, d) is a continuous single facility location problem and $K \in \mathbb{R}$, $K \geq 0$, is the fixed installation cost of the facility. Note that we can associate with (N, Φ, d, K) a cost TU-game (N, c_{SF}) whose characteristic function c_{SF} is defined, for every $S \subset N = \{a_1, \ldots, a_n\}$, by:

$$c_{SF}(S) = \begin{cases} K + L(S) & \text{if } S \neq \emptyset \\ 0 & \text{if } S = \emptyset. \end{cases}$$

Every cost TU-game defined in this way is what we call a continuous single facility location game.

In a location situation, the goal of the users is to find a location for the facility which minimizes the total cost, and to allocate the corresponding minimal total cost.

An interesting problem which arises now is to study under what conditions there exists a stable allocation of the minimal total costs in a location situation, i.e., under what conditions the core of the corresponding location game is non empty.

First, we include some preliminary properties of the single facility location games.

Proposition 1 *The game (N, c_{SF}) corresponding to (N, Φ, d, K) is monotonic (i.e., $c_{SF}(S) \leq c_{SF}(T)$ for all $S, T \subset N$ with $S \subset T$).*

Proof Let $S \subset T$ be two coalitions. By definition $d_i^S(x) \leq d_i^T(x)$ for all i and x. Then, since Φ is monotone, $\Phi(d^S(x)) \leq \Phi(d^T(x))$. Hence, the result follows. □

Proposition 2 *If $L(N) \leq K$ then the game (N, c_{SF}) corresponding to (N, Φ, d, K) satisfies that $c_{SF}(S \cup T) \leq c_{SF}(S) + c_{SF}(T)$ for all $S, T \subset N$.*

Proof Let S, T be two coalitions. Then, by the monotonicity of L (see the proof of Proposition 1) and the properties of Φ,

$$L(S \cup T) - (L(S) + L(T)) \leq L(S \cup T) \leq L(N).$$

Now, since $L(N) \leq K$, then $L(S \cup T) \leq K + L(S) + L(T)$ and

$$c_{SF}(S \cup T) = K + L(S \cup T) \leq K + L(S) + K + L(T) = c_{SF}(S) + c_{SF}(T).$$

□

Note that the result above implies the subadditivity of c_{SF} because in particular, if $L(N) \leq K$, then $c_{SF}(S \cup T) \leq c_{SF}(S) + c_{SF}(T)$ for any pair of coalitions S and T disjoint.

The next example, taken from [38] shows a subadditive location game with an empty core. It motivates the development of sufficient conditions that ensure the non emptiness of the core of this location game.

Example 1 Let $N = \{a_1, a_2, a_3\}$ be the set of players, located on the vertices of an equilateral triangle of side l. Consider that the globalizing function is the sum and d is the Euclidean distance to the power of b ($b \geq 2$). Then

$$\Phi(d^S(x)) = \sum_{a_i \in S} \| x - a_i \|_2^b$$

for every $S \subset N$ and every $x \in \mathbb{R}^m$. It is easy to check that the location game associated with (N, Φ, d, K) is given by:

$$c_{SF}(a_1) = c_{SF}(a_2) = c_{SF}(a_3) = K,$$

$$c_{SF}(a_1a_2) = c_{SF}(a_1a_3) = c_{SF}(a_2a_3) = K + 2(l/2)^b,$$

$$c_{SF}(a_1a_2a_3) = K + 3(\frac{\sqrt{3}}{3}l)^b.$$

After some algebra, it can be checked that this game is subadditive if and only if $K \geq (l^b/\sqrt{3}^{b-2}) - (l^b/2^{b-1})$. However, taking for instance $K = (l^b/\sqrt{3}^{b-2}) - (l^b/2^{b-1})$, it can be seen that the resulting location game has an empty core. Namely, since all its players are symmetric, a necessary and sufficient condition for the non emptiness of its core is that the egalitarian allocation $(c_{SF}(N)/3, c_{SF}(N)/3, c_{SF}(N)/3)$ belongs to it. After some algebra it can be checked that this is not the case when $b > 2$.

The rest of this section presents a sufficient condition for the non emptiness of the core of the single facility location game. First, we include a technical lemma concerning the sum of the balancing coefficients of a balanced family of coalitions. Recall that a collection of coalitions $\mathscr{B} \subset 2^N$ is balanced if and only if there exists a set of positive real coefficients $\{\gamma_S/ S \in \mathscr{B}\}$ (balancing coefficients) satisfying that $\sum_{S:a_i \in S} \gamma_S = 1$ for every $a_i \in N$. The set of balancing coefficients associated with a balanced collection needs not to be unique. However, every minimal balanced collection of coalitions (in the sense that it does not properly contain another balanced collection) has a unique set of balancing coefficients (see [31]). It is a well-known result that a cost game (N, c) has a non empty core if and only if, for every minimal balanced collection $\mathscr{B}$ with balancing coefficients $\{\gamma_S/S \in \mathscr{B}\}$, it satisfies that $\sum_{S \in \mathscr{B}} \gamma_S c(S) \geq c(N)$ (again, see [31]).

Note that the only balanced collection with balancing coefficients summing up to one is $\mathscr{B} = \{N\}$. We say that $\mathscr{B} = \{N\}$ is the trivial collection. The next result establishes bounds on the sum of the balancing coefficients for any non trivial balanced collection.

Lemma 1 *Let $\mathscr{B}$ be a non trivial balanced collection with balancing coefficients $\{\gamma_S \, / \, S \in \mathscr{B}\}$. Then,*

$$\frac{n}{n-1} \le \sum_{S \in \mathscr{B}} \gamma_S \le n.$$

Proof Let us consider the following linear programming problem (2):

$$
\begin{aligned}
&\min_{\substack{S \in 2^N \setminus \{N\}}} \sum \gamma_S \\
&s.t. : \sum_{\{S \in 2^N \setminus \{N\}: a_i \in S\}} \gamma_S = 1 \quad \forall \, a_i \in N \qquad (2) \\
&\qquad \gamma_S \ge 0 \quad \forall S \in 2^N \setminus \{N\}.
\end{aligned}
$$

A solution to this problem is a set of balancing coefficients of a non trivial balancing collection with a minimal sum ($\mathscr{B} = \{S \in 2^N / \gamma_S > 0\}$). Let us denote the coalition $N \setminus \{a_j\} = \{a_1, a_2, ..., a_{j-1}, a_{j+1}, ..., a_n\}$ by $-j$. Consider the basis B of Problem (2) of the columns which correspond to $\gamma_{-1}, \gamma_{-2}, ..., \gamma_{-n}$. In this problem the matrix of B, its inverse B^{-1} and the transformed right-hand side $B^{-1}b$ are:

$$
B = \begin{bmatrix} 0 & 1 & \dots & 1 \\ 1 & 0 & \dots & 1 \\ \vdots & \vdots & \ddots & \vdots \\ 1 & 1 & \dots & 0 \end{bmatrix}, B^{-1} = \frac{1}{n-1} \begin{bmatrix} -(n-2) & 1 & \dots & 1 \\ 1 & -(n-2) & \dots & 1 \\ \vdots & \vdots & \ddots & \vdots \\ 1 & 1 & \dots & -(n-2) \end{bmatrix},
$$

and $B^{-1}b = \left[\frac{1}{n-1}, \cdots, \frac{1}{n-1}\right]^t$. The reduced costs for any coalition S with $1 \le k \le n-1$ players are:

$$
\begin{aligned}
c_B B^{-1} a_S - c_S &= \frac{k}{n-1} - 1 < 0 \text{ iff } k < n-1, \\
c_B B^{-1} a_S - c_S &= \frac{n-1}{n-1} - 1 = 0 \text{ iff } k = n-1.
\end{aligned}
$$

Then B is a basis associated with an optimal solution of Problem (2), which proves the lower bound. The proof for the upper bound is straightforward and it follows taking the collection whose elements are all the sets of size one with coefficients equal to 1. □

Using the lemma above, it follows the main result in this section which can be also found in [38].

Theorem 1 *Let (N, Φ, d, K) be a location situation and let (N, c_{SF}) be its corresponding location game. Denote $l_2 = \min_{S \subset N : |S|=2} L(S)$.*

(a) *Suppose that $2 \le n \le 2 + \frac{l_2}{K}$. If $K(n-1) \ge L(N)$, then c_{SF} has a non empty core.*

(b) *Suppose that $2 + \frac{l_2}{K} < n$. If $K \ge (n-1)L(N) - nl_2$, then c_{SF} has a non empty core.*

Proof In a location game we have for any balanced collection $\mathscr{B}$ with balancing coefficients $\{\gamma_S / S \in \mathscr{B}\}$:

$$\sum_{S \in \mathscr{B}} \gamma_S c_{SF}(S) = K(\sum_{S \in \mathscr{B}} \gamma_S) + \sum_{S \in \mathscr{B}} \gamma_S L(S).$$

Taking into account the monotonicity of L and the fact that $L(S) = 0$ for any coalition S of size one, we have that

$$\sum_{S \in \mathscr{B}} \gamma_S c_{SF}(S) = K(\sum_{S \in \mathscr{B}} \gamma_S) + \sum_{S \in \mathscr{B}: |S| \geq 2} \gamma_S L(S)$$
$$\geq K(\sum_{S \in \mathscr{B}} \gamma_S) + \sum_{S \in \mathscr{B}: |S| \geq 2} \gamma_S l_2.$$

For every minimal balanced collection $\mathscr{B}$ denote

$$m(\mathscr{B}) = K(\sum_{S \in \mathscr{B}} \gamma_S) + l_2 \sum_{S \in \mathscr{B}: |S| \geq 2} \gamma_S$$

(note that, if $\mathscr{B}$ is minimal, the balancing coefficients are uniquely determined). Then, a sufficient condition for the non emptiness of the core is that

$$\min_{\{\mathscr{B}: \mathscr{B} \text{ non trivial and minimal balanced}\}} m(\mathscr{B}) \geq c(N). \tag{3}$$

Suppose that this minimum is achieved in $\hat{\mathscr{B}}$. If $\{a_i\} \notin \hat{\mathscr{B}}$ for every $a_i \in N$, then $\hat{\mathscr{B}} = \{-i \ / \ a_i \in N\}$ (see Lemma 1) and $m(\hat{\mathscr{B}}) = (K + l_2)\frac{n}{n-1}$. If $\hat{\mathscr{B}} = \{\{a_i\} \ / \ a_i \in N\}$, then $m(\hat{\mathscr{B}}) = Kn$. In any other case $\hat{\mathscr{B}}$ can only be a family $\{\{a_i\}, N \setminus a_i\}$ (for an $a_i \in A$) and, then, $m(\hat{\mathscr{B}}) = 2K + l_2$.

Now, since $m(\hat{\mathscr{B}}) = min\{(K + l_2)\frac{n}{n-1}, Kn, 2K + l_2\}$, then it can be easily checked that

$$m(\hat{\mathscr{B}}) = \begin{cases} Kn & \text{if } 2 \leq n \leq 2 + \frac{l_2}{K} \\ (K + l_2)\frac{n}{n-1} & \text{if } 2 + \frac{l_2}{K} < n. \end{cases}$$

This together with (3) completes the proof. $\square$

4.2   *The Continuous Single Allocation Multifacility Case*

Roughly speaking, a multifacility location problem occurs whenever one is going to place several servers to provide some service to a set of users. Assuming that the quality of the service provided decreases with the distance to the user, the most common assumption is that each user will receive its service from the closest server. Ties are solved randomly. In this way, each user is allocated to a unique server.

This situation can be described in the following terms. We are given a set of N users which are modeled by n points in $\mathbb{R}^q$, a feasible region, that for simplification is assumed to be the convex hull of a set of r points, and we wish to place at most m new facilities or service points within the feasible region to provide service to the n users.

Formally, a continuous multifacility location problem is a 5-tuple (N, P, K, Φ, d) where:

- $N = \{a_1, \ldots, a_n\}$ is a set of n different points in $\mathbb{R}^q$ (with $n \geq 2$),
- $P = \{p_1, \ldots, p_r\}$ is the set of extreme points that define the feasible region. p_∞ is a very remote point that models a fictitious placement of those servers that are not actually located.
- $K_j \geq 0, j = 1, \ldots, m$ is the fixed cost to locate the jth facility. We assume that $K_\infty = 0$. In addition, we suppose that there is some economy of scale in those costs so that $K_j \geq K_{j+1}$ for all $j = 1, \ldots, m - 1$.
- $\Phi : \mathbb{R}^n \to \mathbb{R}$ is a lower semicontinuous globalizing function satisfying that: (1) Φ is definite, i.e. $\Phi(x) = 0$ if and only if $x = 0$; (2) Φ is monotone, i.e. $\Phi(x) \leq \Phi(y)$ whenever $x \leq y$, and
- $d : \mathbb{R}^q \times \mathbb{R}^q \to \mathbb{R}$ is a measure of distance, satisfying that, for every $r, s \in \mathbb{R}^q$, $d(r, s) = \| r - s \|$, where $\| \ \|$ is a norm on $\mathbb{R}^q$. For the ease of presentation we assume that this norm is polyhedral, namely its unit ball is a bounded polyhedron. This implies that $\|x\|$ admits a representation as a number of linear inequalities; polynomial in the description of its unit ball.

Next, for any $X \subseteq \mathbb{R}^q$, let $d^N(X) = (d(X, a_1), \ldots, d(X, a_n))$. For any $S \subseteq N$, we denote by $y(S)$ the incidence vector of S. i.e. the vector defined as $y_i(S) = 1$ if $i \in S$ and $y_i(S) = 0$ otherwise. This way $d^S(X) = (d(X, a_1)y_1(S), \ldots, d(X, a_n)y_n(S))$, is the vector whose ith component is equal to $d(X, a_i)$ if $i \in S$ and zero otherwise.

In the following we introduce some instrumental variables to get a mathematical programming representation of this problem. Let x_j be the location of the jth new facility (server). We can write $x_j = \sum_{\ell=1}^{r} \lambda_{\ell j} p_\ell$. The variable $\lambda_{\ell j}$ gives the coefficient of the point p_ℓ in the convex combination of the points in P that describe x_j. Clearly, $\lambda_{\ell j} \in [0, 1]$, for all $\ell = 1, \ldots, r, j = 1, \ldots, m$.

The variable γ_j assumes the value 1 if the jth server is located and zero otherwise. The variable w_{ik} takes the value 1 if the nearest service facility to the demand point a_i is x_j and it is equal to 0 otherwise, for all $i = 1, \ldots, n, j = 1, \ldots, m$. $v_i, i = 1, \ldots, n$, is equal to the distance from the demand point a_i to its nearest service facility. For modeling purposes we introduce a formal point p_∞ with the property that $d(a_i, p_\infty) \gg d(x, a_i)$ for all $i = 1, \ldots, n$ and $\forall x \in conv(P)$. $M \gg 0$ is a big constant to be used in the constraints of the problem. In particular $M \geq \max_{i=1,\ldots,n} d(p_\infty, a_i)$.

Solving the continuous single allocation multifacility location problem (N, P, K, Φ, d) means to find $X \subset \mathbb{R}^q$ with $|X| \leq m$, minimizing the following problem.

$$c_{MF}(N) := \min \qquad \sum_{j=1}^{m} K_j \gamma_j + \Phi(v) \qquad\qquad (4)$$

$$\text{s.t.} \quad x_j = \sum_{\ell=1}^{r} \lambda_{\ell j} p_\ell + (1 - \gamma_j) p_\infty, \quad j = 1, \ldots, m, \tag{5}$$

$$\gamma_j = \sum_{\ell=1}^{r} \lambda_{\ell j}, \qquad j = 1, \ldots, m, \tag{6}$$

$$d(x_j, a_i) \le v_i + (2 - y_i(N) - w_{ij}) M, \quad i = 1, \ldots, n, j = 1, \ldots, m, \tag{7}$$

$$\sum_{j=1}^{m} w_{ij} = y_i(N), \qquad i = 1, \ldots, n, \tag{8}$$

$$v_i \ge 0, \qquad i = 1, \ldots, n, \tag{9}$$

$$\lambda_{\ell j} \ge 0, \qquad \ell = 1, \ldots, r, j = 1, \ldots, m, \tag{10}$$

$$\gamma_j, w_{ij} \in \{0, 1\}, \qquad i = 1, \ldots, n, j = 1, \ldots, m. \tag{11}$$

The objective function (4) accounts for the total fixed cost to locate the servers plus the transportation cost induced by the globalizing function Φ applied to the vector of distances $d^N(X) = (d(X, a_1), \ldots, d(X, a_n))$. The family of constraints (5) defines the location of each new facility as a convex combination of the points in P. By (6), if $\gamma_j = 1$, $x_j \in \text{conv}(P)$ and if $\gamma_j = 0$, x_j equals the point at infinity, p_∞. With (7) and the monotonicity of Φ, it is ensured that v_i represents the distance from a_i to x_j provided that a_i is assigned to x_j, i.e. $w_{ij} = 1$. Otherwise, this inequality is trivially fulfilled since $M \gg 0$. The family of inequalities (8) enforces that one w_{ij} variable assumes the value 1, i.e. each user is allocated to a unique server. Finally, the last three sets of conditions define the domain of the variables. It is worth noting that for any N and suitable choices of Φ, as for instance the ones that we will choose in this chapter, the above problem always has an optimal solution (see, for instance, [2]).

Now, we can extend the location situation to any subset $S \subseteq N$ and it results in finding $X \subset \mathbb{R}^q$ with $|X| \le m$ minimizing the following problem.

$$c_{MF}(S) := \min \qquad \sum_{j=1}^{m} K_j \gamma_j + \Phi(v) \tag{12}$$

$$\text{s.t.} \quad x_j = \sum_{\ell=1}^{r} \lambda_{\ell j} p_\ell + (1 - \gamma_j) p_\infty, \quad j = 1, \ldots, m, \tag{13}$$

$$\gamma_j = \sum_{\ell=1}^{r} \lambda_{\ell j}, \qquad j = 1, \ldots, m, \tag{14}$$

$$d(x_j, a_i) \le v_i + (2 - w_{ij} - y_i(S)) M, i = 1, \ldots, n, j = 1, \ldots, m, \tag{15}$$

$$\sum_{j=1}^{m} w_{ij} = y_i(S), \qquad i = 1, \ldots, n, \tag{16}$$

$$v_i \ge 0, \qquad i = 1, \ldots, n, \tag{17}$$

$$\lambda_{\ell j} \ge 0, \qquad \ell = 1, \ldots, r, j = 1, \ldots, m, \tag{18}$$

$$\gamma_j, w_{ij} \in \{0, 1\}, \qquad i = 1, \ldots, n, j = 1, \ldots, m. \tag{19}$$

As before, it is worth noting that for any $S \subseteq N$ and suitable choices of Φ, as for instance the ordered median function [28], this problem always has an optimal solution (see, for instance, [1, 2]).

The above formulation provides a general version of the continuous multifacility location problem. In this chapter we will restrict ourselves to consider $\Phi(d^N(X))$ as an ordered median function for a given parameter $\mu \in \mathbb{R}_+^n$. Recall that if

$d_{(1)}^N(X) \leq d_{(2)}^N(X) \leq \ldots \leq d_{(n)}^N(X)$, then $Omf_\mu(d^N(X)) = \sum_{k=1}^n \mu_k d_{(k)}^N(X)$ (see [28]), which stands for a utility of the users in N being interested in minimizing the ordered weighted average of their distances to the closest server in X. Formally, the continuous single allocation multifacility location situation is a 5-tuple (N, P, K, Omf_μ, d) where N, P, K and d where defined above; and Omf_μ is an ordered median function. We recall that minimizing Omf_μ can be represented as the following problem [1, 2].

$$\min_X \sum_{k=1}^n \mu_k d_{(k)}^N(X) = \min_X \sum_{i=1}^n \sum_{k=1}^n \mu_k \theta_{ik} \tag{20}$$

$$\text{s.t.} \quad \theta_{ik} \leq d(a_i, X) + M(1 - \psi_{ik}), \ \forall i, k = 1, \ldots, n$$

$$\sum_{i=1}^n \psi_{ik} = 1, \ \forall k = 1, \ldots, n$$

$$\sum_{k=1}^n \psi_{ik} = 1, \ \forall i = 1, \ldots, n$$

$$\sum_{i=1}^n \theta_{ik} \leq \sum_{i=1}^n \theta_{i,k+1}, \ \forall k = 1, \ldots, n-1,$$

$$\psi_{ik} \in \{0, 1\}, \ \theta_{ik} \geq 0, \ \forall i, k = 1, \ldots, n.$$

In this way, the characteristic function $c_{MF}(S)$, for choices of the globalizing function Φ as an ordered median function with parameter μ, can be written in the following form.

$$c_{MF}(S) = \min \ \sum_{j=1}^m K_j \gamma_j + \sum_{i=1}^n \sum_{k=1}^n \mu_k \theta_{ik} \tag{21}$$

$$\text{s.t.} \quad x_j = \sum_{\ell=1}^r \lambda_{\ell j} p_\ell + (1 - \gamma_j) p_\infty, \quad j = 1, \ldots, m, \tag{22}$$

$$\gamma_j = \sum_{\ell=1}^r \lambda_{\ell j}, \qquad j = 1, \ldots, m, \tag{23}$$

$$d(x_j, a_i) \leq v_i + (2 - w_{ij} - y_i(S))M, i = 1, \ldots, n, j = 1, \ldots, m, \tag{24}$$

$$\sum_{j=1}^m w_{ij} = y_i(S), \qquad i = 1, \ldots, n, \tag{25}$$

$$\sum_{i=1}^n \psi_{ik} = 1, \qquad k = 1, \ldots, n \tag{26}$$

$$\sum_{k=1}^n \psi_{ik} = 1, \qquad i = 1, \ldots, n \tag{27}$$

$$\sum_{i=1}^n \theta_{ik} \leq \sum_{i=1}^n \theta_{ik+1}, \qquad \forall k = 1, \ldots, n-1, \tag{28}$$

$$\theta_{ik} \leq v_i + M(1 - \psi_{ik}) \qquad \forall i, k = 1, \ldots, n, \tag{29}$$

$$\lambda_{\ell j} \geq 0, \qquad \ell = 1, \ldots, r, j = 1, \ldots, m,$$

$$v_i \geq 0, \qquad i = 1, \ldots, n, \tag{30}$$

$$\gamma_j, w_{ij} \in \{0, 1\}, \qquad i = 1, \ldots, n, j = 1, \ldots, m,$$

$$\psi_{ik} \in \{0, 1\}, \qquad \forall i, k = 1, \ldots, n.$$

Our interest goes into the direction of analyzing the situation where the costs induced by the above location problem must be shared by the users that are served by the facilities in X. We would like to determine whether there are fair allocation schemes for the overall cost induced by the location problem. We will address this question using a cooperative location game.

Formally, a continuous single allocation multifacility location game is defined by a 6-tuple $(N, P, K, Omf_\mu, d, c_{MF})$ where N is the set of demand points (users), P is the set of extreme points of the convex hull defining the feasible set, K is the set of fixed or installation costs, Omf_μ is the globalizing function that determines the transportation cost, d is the measure of the distance among points and c_{MF} is the characteristic function of the game, namely the set function that for each set $S \subseteq N$ returns the minimal value of a set of servers that solves the location problem for the users in S. We recall that $c_{MF}(S)$ is defined in (21). (Observe that by definition $c_{MF}(\emptyset) = 0$.)

Let F_S be the feasible region of the problem that defines $c_{MF}(S)$ in the space of variables (ξ, y), where for the sake of readability we denote $\xi = (\lambda, v, \theta, \gamma, w, \psi) \in [0, 1]^{r \times m} \times \mathbb{R}^n \times \mathbb{R}^{n \times n} \times \{0, 1\}^m \times \{0, 1\}^{n \times m} \times \{0, 1\}^{n \times n}$. Additionally, for any $(\gamma, w, \psi) \in \{0, 1\}^m \times \{0, 1\}^{n \times m} \times \{0, 1\}^{n \times n}$, let $F_S(\gamma, w, \psi)$ be the feasible domain of the same problem in the space of variables $(\lambda, v, \theta) \in \mathbb{R}^{r \times m} \times \mathbb{R}^n \times \mathbb{R}^{n \times n}$. In the following we denote by $ext(A)$ the set of all the extreme points of the convex set A.

Next, we introduce the following set:

$$F^{\xi y} = \{\gamma \in \{0, 1\}^n, w \in \{0, 1\}^{n \times m}, \psi \in \{0, 1\}^{n \times n}, (\lambda, v, \theta) \in ext(F_S(\gamma, w, \psi)),$$

$$\text{for some } S \subseteq N\}. \tag{31}$$

We can interpret $F^{\xi y}$ as the set of all potential candidate solutions $(\bar{\xi}, \bar{y})$ for the evaluation of $c_{MF}(S)$ whenever $\bar{y}$ is the characteristic vector of S, for all possible $S \subseteq N$.

For the ease of readability, let X_S and $x(i)$ denote the location of the service facilities in the optimal solutions of $c_{MF}(S)$ and $c_{MF}(i)$, respectively.

Proposition 3 If $K_1 - K_2 + Omf_\mu(d^S(x)) + \mu_n d(x(i), a_i) \geq Omf_\mu(d^{S \cup i}(x))$ for any $S \subseteq N$, $i \notin S$ and $x \in conv(P)$ then c_{MF} is subadditive. Namely, for any $S, T \subset N$, $S \cap T = \emptyset$ then $c_{MF}(S \cup T) \leq c_{MF}(S) + c_{MF}(T)$.

Proof Observe that it is enough to prove the claim for S and $\{i\} \notin S$. Under the general assumption of non-negative K_j, there is only one server in the solution for $\{i\}$. If $|X_S| < m$ then clearly X_S is also a feasible solution for $c_{MF}(S \cup i)$. Moreover,

$$c_{MF}(S \cup i) \leq \sum_{j=1}^{|X_S|+1} K_j + Omf_\mu(d^{S \cup i}(X_S))$$

$$\leq \sum_{j=1}^{|X_S|} K_j + Omf_\mu(d^S(X_S)) + K_1 + \mu_n d(x(i), a_i) = c_{MF}(S) + c_{MF}(i).$$

Analogously, if $|X_S| = m$, then

$$c_{MF}(S \cup i) \leq \sum_{j=1}^{m} K_j + Omf_\mu(d^{S \cup i}(X_S)) \leq$$

$$\leq \sum_{j=1}^{m} K_j + Omf_\mu(d^S(X_S)) + K_1 + \mu_n d(x(i), a_i)$$

$$= c_{MF}(S) + c_{MF}(i).$$

$\square$

We note in passing that the condition in Proposition 3 is fulfilled whenever the difference between the set up costs is large enough as compared with the transportation cost. In the case of monotone μ, namely $\mu_1 \leq \ldots \leq \mu_n$, the condition is easier since it always holds that $Omf_\mu(d^{S \cup i}(X_{S \cup i})) \leq Omf_\mu(d^{S \cup i}(X_S)) \leq Omf_\mu(d^S(X_S)) + \mu_n d(X_S, a_i)$. Therefore, to ensure subadditivity, it is simply required that $K_1 \geq \mu_n(d(x, a_i) - d(x(i) - a_i))$ for all $i = 1, \ldots, n$ and any $x \in conv(P)$.

Lemma 2 *The maximum value of $c_{MF}(N)$ so that the core of the game $(N, P, K, Omf_\mu, d, c_{MF})$ is not empty is given as*

$$c_{MF}(N) = \min_{(\bar{\xi}\bar{y}) \in F^{\xi y}} \sum^{m}_{j=1} [\sum_{j=1}^{m} K_j \bar{\gamma}_j + \sum_{i=1}^{n} \sum_{k=1}^{n} \bar{\theta}_{ik}] z_{\bar{\xi}\bar{y}} \qquad (32)$$

$$s.t. \sum_{(\bar{\xi}\bar{y}) \in F^{\xi y}} \bar{y}_i z_{\bar{\xi}\bar{y}} = 1, \ i = 1, \ldots, n,$$

$$z_{\bar{\xi}\bar{y}} \geq 0, \ \forall \ (\bar{\xi}\bar{y}) \in F^{\xi y}.$$

Proof Recall that by Shapley-Bondareva Theorem, the core of the game $(N, P, K, Omf_\mu, d, c_{MF})$ is not empty if and only if

$$c_{MF}(N) \leq \min\{\sum_{S \subseteq N} c_{MF}(S) z_S : \sum_{S \ni i} z_S = 1, \ \forall i = 1, \ldots, n, \ z_S \geq 0, \forall S\}. \qquad (33)$$

We can replace the value of $c_{MF}(S)$ in the above expression by its value as given by minimizing the objective function of (21), namely $\sum_{j=1}^{m} K_j \gamma_j + \sum_{i=1}^{n} \sum_{k=1}^{n} \theta_{ik}$, over the set F_S which is a mixed-integer set of solutions in the space $(\lambda, v, \theta, \gamma, w, \psi) \in [0, 1]^{r \times m} \times \mathbb{R}^n \times \mathbb{R}^{n \times n} \times \{0, 1\}^m \times \{0, 1\}^{n \times m} \times \{0, 1\}^{n \times n}$. Since that objective function is linear, we can also obtain $c_{MF}(S)$ minimizing over the convex hull of the feasible set, i.e. $c_{MF}(S) = \min\{\sum_{j=1}^{m} K_j \gamma_j + \sum_{i=1}^{n} \sum_{k=1}^{n} \theta_{ik} : (\lambda, v, \theta, \gamma, w, \psi) \in conv(F_S)\}$. This representation can be plugged in (33) and we obtain that:

$$c_{MF}(N) \leq \min \sum_{S \subseteq N} (\sum_{j=1}^{m} K_j \gamma_j^S + \sum_{i=1}^{n} \sum_{k=1}^{n} \theta_{ik}^S) z_S \tag{34}$$

$$\text{s.t. } \sum_{S \ni i} z_S = 1, \ \forall i = 1, \ldots, n$$

$$(\lambda^S, v^S, \theta^S, \gamma^S, w^S, \psi^S) \in conv(F_S), \ \forall S \subseteq N.$$

Next, representing each subset S by its characteristic vector $y(S)$, we obtain that the set of all candidate to optimal solutions to (34), for some S, coincides with the set $F^{\xi y}$ defined in (31). Thus,

$$(34) = \min \sum_{(\bar{\xi}, \bar{y}) \in F^{\xi y}} (\sum_{j=1}^{m} K_j \bar{\gamma}_j + \sum_{i=1}^{n} \sum_{k=1}^{n} \bar{\theta}_{ik}) z_{\bar{\xi}, \bar{y}} \tag{35}$$

$$\sum_{(\bar{\xi}, \bar{y}) \in F^{\xi y}} z_{\bar{\xi}, \bar{y}} \bar{y}_i = 1, \ \forall i = 1, \ldots, n, \tag{36}$$

$$z_{\bar{\xi}, \bar{y}} \geq 0, \ \forall \ (\bar{\xi}, \bar{y}) \in F^{\xi y}. \tag{37}$$

$\square$

The construction above, see (36–37), shows that for the evaluation of $c_{MF}(N)$ there is a non-negative variable $z_{\xi,y}$ for each point in $F^{\xi y}$. Thus, finding the optimal values of those variables in (35)–(37) can be interpret as searching on the cone generated by $F^{\xi y}$ intersected by (36).

Let us formally define this set.

$$C^\xi = \text{proj}_\xi (cone(F^{\xi y}) \cap \{(\xi, y) : \xi = (\lambda, v, \theta, \gamma, w, \psi), \lambda \in \mathbb{R}^{r \times m}, v \in \mathbb{R}^n, \theta \in \mathbb{R}^{n \times n},$$

$$\gamma \in \mathbb{R}^m, w \in \mathbb{R}^{n \times n}, \psi \in \mathbb{R}^{n \times n}, y_i = 1, \forall \ i = 1, \ldots, n\}). \tag{38}$$

The discussion above leads to the following alternative condition for the evaluation of the maximal value of $c_{MF}(N)$ that gives rise to a nonempty core for this game.

Proposition 4 *The maximal cost $c_{MF}(N)$ that gives rise to a nonempty core in the location game $(N, P, K, Omf_\mu, d, c_{MF})$ is*

$$\min \{\sum_{j=1}^{m} K_j \gamma_j + \sum_{i=1}^{n} \sum_{k=1}^{n} \theta_{ik} : (\lambda, v, \theta, \gamma, w, \psi) \in C^\xi\}. \tag{39}$$

The result above can be restated equivalently representing C^ξ by means of all its valid inequalities. It is not difficult to prove that C^ξ is the set of points satisfying all inequalities of the form $a' \xi \geq b$ if there exists a homogeneous inequality $a' \xi \geq d' y$ with $\sum_{i=1}^{n} d_i = b$ valid for $conv(F^{\xi y})$.

The results above can be summarized in the following theorem.

Theorem 2 *The core of the continuous multifacility single allocation game* $(N, P, K, \, Omf_\mu, d, c_{MF})$ *is nonempty if and only if any of the following conditions hold:*

1. *The optimal value of Problem (21) for $S = N$ coincides with the optimal value of Problem (39).*
2. *The minimal value of the objective function $\sum_{j=1}^{m} K_j \gamma_j + \sum_{i=1}^{n} \sum_{k=1}^{n} \theta_{ik}$ subject to all the valid inequalities representing C^ξ coincides with the minimal value of Problem (39).*

Moreover, if any of the conditions in 1. or 2. hold then the set of dual optimal solutions to (39) are cost shares in the core $C(N, c_{MF})$.

Proof The proof of items *1.* and *2.* follows from the discussion above. The proof of the last assertion is as follows. From Lemma 2 we obtain that

$$c_{MF}(N) = \min\{\sum_{j=1}^{m} K_j \gamma_j + \sum_{i=1}^{n} \sum_{k=1}^{n} \theta_{ik} : (\lambda, v, \theta, \gamma, w, \psi) \in C^\xi\}$$

$$= \min\{\sum_{S \subseteq N} c_{MF}(S) z_S : \sum_{S \ni i} z_S = 1, \, \forall i, \, z_S \geq 0, \forall S\}. \tag{40}$$

Now, the dual of the second problem in (40) is exact and it is:

$$\max\{\sum_{r=1}^{n} r_i : \sum_{i \in S} r_i \leq c_{MF}(S), \, \forall S \subseteq N\}.$$

Therefore, its optimal solutions are allocations (cost shares) in the core $C(N, c_{MF})$.
□

5 The Continuous Set Covering Location Game

This section considers another location model that gives rise to a different cost sharing problem. This situation is based on the continuous covering location problem, schematically described in Sect. 2. Different versions of covering games applied to discrete location situations have been already studied in [14] in the discrete case and [43] on networks. In this section we study the continuous counterpart of those models. In this case, we are given a set $N = \{a_1, \ldots, a_n\} \subset \mathbb{R}^q$ and each point a_i has associated a radius $r_i \geq 0$. Denote by $R = \{r_1, \ldots, r_n\}$ the set of all radii. We also assume that new facilities (servers) must be located in a bounded domain that is defined by the convex hull of a finite set of points $P = \{p_1, \ldots, p_r\} \subset \mathbb{R}^q$. The goal is to install a set of servers $X \subset conv(P)$, $|X| \leq m$ so that $d(X, a_i) \leq r_i$, for all $i = 1, \ldots, n$ and minimizing the overall installation cost; where $K_j \geq 0$ is the installation cost of the jth server for $j = 1, \ldots, m$ and we assume economy of scale, that is $K_j \geq K_{j+1}$ for $j = 1, \ldots, m - 1$. Let $K = \{K_1, \ldots, K_m\}$. For the ease of presentation we assume that for any $x, y \in \mathbb{R}^q$, $d(x, y) = \|x - y\|$ and the norm is polyhedral, namely its unit ball is a bounded polyhedron. This implies that $\|x\|$ admits a representation as a number of linear inequalities; polynomial in the description of its unit ball.

In order to present a valid formulation for the continuous covering location problem we need to introduce some families of variables. Let x_j be the location of the jth new facility (server). We can write $x_j = \sum_{\ell=1}^{r} \lambda_{\ell j} p_\ell$. The variable $\lambda_{\ell j}$ gives the coefficient of the point p_ℓ in the convex combination of the points in P that describe x_j. Clearly, $\lambda_{\ell j} \in [0, 1]$, for all $\ell = 1, \ldots, r, j = 1, \ldots, m$.

The variable γ_j assumes the value 1 if the jth server is located and zero otherwise. Finally, the variable u_{ij} takes the value 1 if the jth server is at a distance from a_i less than or equal to r_i and 0 otherwise, for all $i = 1, \ldots, n, j = 1, \ldots, m$. For modeling purposes we introduce a formal point p_∞ with the property that $d(a_i, p_\infty) \gg r_i$ for all $i = 1, \ldots, n$ and $K_\infty = 0$. p_∞ is a very remote point that models a fictitious placement of those servers that are not actually located. Recall that $y(S)$ denotes the characteristic vector of $S \subseteq N$, namely $y_i(S) = 1$ if $i \in S$ and 0 otherwise.

Using these variables the continuous covering location problem can be stated as follows.

$$c_{SC}(N) = \min \sum_{j=1}^{m} K_j \gamma_j \tag{41}$$

$$\text{s.t.} \quad x_j = \sum_{\ell=1}^{r} \lambda_{\ell j} p_\ell + (1 - \gamma_j) p_\infty, \ j = 1, \ldots, m, \tag{42}$$

$$\sum_{\ell=1}^{r} \lambda_{\ell j} = \gamma_j, \ j = 1, \ldots, m, \tag{43}$$

$$d(a_i, x_j) \leq r_i u_{ij} + (2 - y_i(N) - u_{ij}) M, \ i = 1, \ldots, n, j = 1, \ldots, m, \tag{44}$$

$$\sum_{j=1}^{m} u_{ij} \geq y_i(N), \ i = 1, \ldots, n, \tag{45}$$

$$u_{ij} \leq \gamma_j, \ i = 1, \ldots, n, j = 1, \ldots, m, \tag{46}$$

$$\lambda_{\ell j} \in [0, 1], \ \ell = 1, \ldots, r, j = 1, \ldots, m,$$

$$u_{ij} \in \{0, 1\}, \ \gamma_j \in \{0, 1\}, \ i = 1, \ldots, n, j = 1, \ldots, m.$$

The objective function (41) minimizes the overall installation cost. The family of constraints (42) describe the location of the different servers as convex combinations of the points in P. By (43), if $\gamma_j = 1$, $x_j \in \text{conv}(P)$ and if $\gamma_j = 0$, x_j equals the point at infinity, p_∞, whose distance to any $a_i \in N$ is greater than r_i for all

$i = 1, \ldots, n$. With (44), it is ensured that the coverage of each demand point is done within the required radius. If $u_{ij} = 0$ then the inequality is satisfied since $2 - y_i(N) - u_{ij} > 0$ and $M \gg \max\limits_{i=1,\ldots,n,x \in \text{conv}(P)} \{d(x, a_i), d(p_\infty, a_i)\}$. If $u_{ij} = 1$ the term $2 - y_i(N) - u_{ij} = 0$ and the server x_j must be within the given radius, r_i, from a_i. The family of inequalities (45) enforces that at least one u_{ij} variable assumes the value 1, i.e. there is at least one server to cover the demand point a_i. The family (46) ensures that coverage is done from open servers. Finally, the last two sets of conditions define the domain of the variables. It is straightforward to check that for suitable choices of radii R the problem above has optimal solutions.

As before, in this section we are interested in the situation where the cost induced by the installation of the servers must be shared by the users or demand points. Formally, a continuous set covering location game is described by the 6-tuple (N, P, K, R, d, c_{SC}) where N is the set of demand points, P is the feasible set, K is the set of installation costs, R is the set of coverage radii, d is the distance and c_{SC} is the characteristic function, namely the function that for each $S \subseteq N$ returns the minimal value of a placement of servers to cover the demand. The characteristic function c_{SC} of the set S is defined as:

$$c_{SC}(S) = \min \sum_{j=1}^{m} K_j \gamma_j \tag{47}$$

$$\text{s.t.} \quad x_j = \sum_{\ell=1}^{r} \lambda_{\ell j} p_\ell + (1 - \gamma_j) p_\infty, \ j = 1, \ldots, m, \tag{48}$$

$$\sum_{\ell=1}^{r} \lambda_{\ell j} = \gamma_j, \ j = 1, \ldots, m, \tag{49}$$

$$d(a_i, x_j) \le r_i u_{ij} + (2 - y_i(S) - u_{ij}) M, \ i = 1, \ldots, n, j = 1, \ldots, m, \tag{50}$$

$$\sum_{j=1}^{m} u_{ij} \ge y_i(S), \ i = 1, \ldots, n, \tag{51}$$

$$u_{ij} \le \gamma_j, \ i = 1, \ldots, n, j = 1, \ldots, m, \tag{52}$$

$$\lambda_{\ell j} \in [0, 1], \ \ell = 1, \ldots, r, j = 1, \ldots, m, \tag{53}$$

$$u_{ij} \in \{0, 1\}, \ \gamma_j \{0, 1\}, \ i = 1, \ldots, n, j = 1, \ldots, m. \tag{54}$$

Let us denote by F_S the feasible set of the problem (47)–(54) that defines $c_{SC}(S)$, $S \subseteq N$. For any $(u, \gamma) \in \{0, 1\}^{n \times m} \times \{0, 1\}^m$ fixed let $F_S(u, v)$ be the feasible region of the above problem in the space of λ variables. Denote by $\zeta = (\lambda, u, \gamma) \in [0, 1]^{r \times m} \times \{0, 1\}^{n \times m} \times \{0, 1\}^m$ and for any convex set A, $\text{ext}(A)$ is the set of extreme points of A. Finally, let $F^{\zeta y}$ be the set of all candidate to optimal solutions of Problem (47)–(54) for all $S \subseteq N$, namely:

$$F^{\zeta y} = \{u \in \{0, 1\}^{n \times m}, \gamma \in \{0, 1\}^m, \lambda \in \text{ext}(F_S(u, \gamma)), \text{ for some } S \subseteq N\}. \tag{55}$$

As usual the question that we wish to answer is under which conditions stable cost allocations exist; or in other words when the core of the game (N, P, K, R, d, c_{SC}) is nonempty. We start by studying some properties related to the characteristic function c_{SC}.

Proposition 5 *The characteristic function c_{SC} of the game (N, P, K, R, d, c_{SC}) is subadditive, that is for any $S, T \subset N$ such that $S \cap T = \emptyset$ then $c_{SC}(S \cup T) \leq c_{SC}(S) + c_{SC}(T)$.*

Proof Observe that it suffices to prove the claim for S and $\{i\} \notin S$. Let X_S, X_i be the optimal location of the solutions of $c_{SC}(S)$ and $c_{SC}(i)$, respectively. It is clear that an optimal solution for $\{i\}$ locates only one server, say $x(i)$ at any point within a distance r_i from a_i and the optimal value is K_1.

Assume that an optimal solution for S requires $|X_S| < m$. Then, clearly $X_S \cup x(i)$ is also a feasible solution for $S \cup \{i\}$ and $c_{SC}(S \cup \{i\}) \leq c_{SC}(S) + c_{SC}(i) = c_{SC}(S) + K_1$. On the other hand, if $|X_S| = m$ then $c_{SC}(S) = \sum_{j=1}^m K_j$. Since we assume that the problem for N is feasible, then it is also feasible for $S \cup \{i\}$ and there must exist a set $X_{S \cup i}$, of at most m servers, that satisfies $d(X_{S \cup i}, a) \leq r_a$, for all $a \in S \cup \{i\}$. This implies that $c_{SC}(S \cup i) = \sum_{j=1}^{|X_{S \cup i}|} K_j$. Hence,

$$c_{SC}(S \cup i) = \sum_{j=1}^{|X_{S \cup i}|} K_j \leq c_{SC}(S) + c_{SC}(i) = \sum_{j=1}^m K_j + K_1.$$

This concludes the proof.   □

Next, we can characterize the nonemptiness of the core of this game.

Theorem 3 *The core of the game (N, P, K, R, d, c_{SC}) is not empty if and only if*

$$c_{SC}(N) \leq \min\{\sum_{j=1}^m K_j \gamma_j : (\lambda, u, \gamma) \in C^\zeta\}, \tag{56}$$

where $C^\zeta = proj_\zeta(cone(F^{\zeta y}) \cap \{(\zeta, y) : \zeta = (\lambda, u, \gamma), \lambda \in [0, 1]^{r \times m}, u \in \mathbb{R}^{n \times m}, \gamma \in \mathbb{R}^m, y_i = 1, \forall i = 1, \ldots, n\})$. Moreover, the set of optimal dual solutions to (56) are cost shares in the core $C(N, c_{SC})$ of the game (N, P, K, R, d, c_{SC}).

Proof Shapley-Bondareva theorem states that the core of the game (N, P, K, R, d, c_{SC}) is not empty if and only if

$$c_{SC}(N) \leq \max\{\sum_{r=1}^n r_i : \sum_{i \in S} r_i \leq c_{SC}(S), \forall S \subseteq N\}$$

$$= \min\{\sum_{S \subseteq N} c_{SC}(S) z_S : \sum_{S \ni i} z_S = 1, \forall i, z_S \geq 0, \forall S\}. \tag{57}$$

Next, for each S we can compute the value $c_{SC}(S)$ minimizing $\sum_{j=1}^{m} K_j \gamma_j$ over the set F_S which is a mixed-integer set of solutions. Equivalently, we can obtain this value as $c_{SC}(S) = \min\{\sum_{j=1}^{m} K_j \gamma_j : (\lambda, u, \gamma) \in conv(F_S)\}$. This representation can be plugged in (57) and we obtain that the right-hand-side of that expression equals the following:

$$\min \qquad \sum_{S \subseteq N} (\sum_{j=1}^{m} K_j \gamma_j^S) z_S \tag{58}$$

$$\text{s.t.} \qquad \sum_{S \ni i} z_S = 1, \ \forall i,$$

$$(\lambda^S, u^S, \gamma^S) \in conv(F_S), \ \forall S \subseteq N.$$

Next, representing each subset S by its characteristic vector $y(S)$, we obtain that the set of all candidate to optimal solutions to (58), for some S, coincides with the set $F^{\zeta y}$. Thus,

$$(58) = \min \sum_{(\bar{\zeta}, \bar{y}) \in F^{\zeta y}} (\sum_{j=1}^{m} K_j \bar{\gamma}_j) z_{\bar{\zeta}, \bar{y}} \tag{59}$$

$$\sum_{(\bar{\zeta}, \bar{y}) \in F^{\zeta y}} z_{\bar{\zeta}, \bar{y}} \bar{y}_i = 1, \ \forall i = 1, \ldots, n, \tag{60}$$

$$z_{\bar{\zeta}, \bar{y}} \geq 0, \ \forall \ (\bar{\zeta}, \bar{y}) \in F^{\zeta y}.$$

The above problem defines one non negative variable $z_{\bar{\zeta}, \bar{y}} \geq 0$ for each point in the set $F^{\zeta y}$. Thus, they can be interpreted as the coefficients of a conic combination of the points in $F^{\zeta y}$. For this reason, minimizing Problem (59) is equivalent to minimize the same objective function over the $cone(F^{\zeta y})$ intersected with the Eq. (60). These equations are nothing but the hyperplanes $y_i = 1, \ \forall \ i = 1, \ldots, n$.

The above discussion implies that

$$(58) = \min \sum_{j=1}^{m} K_j \gamma_j$$

$$(\lambda, u, \gamma) \in proj_\zeta(cone(F^{\zeta y}) \cap (\zeta, 1)) = C^\zeta.$$

Finally, the above chain of equations proves that the dual of (58), which in turns equals (56), is equal to $\max\{\sum_{r=1}^{n} r_i : \sum_{i \in S} r_i \leq c_{SC}(S), \ \forall S \subseteq N\}$. Clearly the optimal solutions of this last problem are allocations in the core $C(N, c_{SC})$. This proves the last assertion of the theorem. □

The result above can be stated equivalently describing C^ζ by means of all its valid inequalities. Since C^ζ is the projection of $cone(F^{\zeta y}) \cap (\zeta, 1)$ on the space of ζ variables, it is well known that all valid inequalities for C^ζ are of the form $a'\zeta \geq b$ such that there exists $a'\zeta \geq d'y$ with $\sum_{i=1}^{n} d_i = b$ which are valid for $cone(F^{\zeta y})$. This implies the following result.

Corollary 1 *The core of the game* (N, P, K, R, d, c_{SC}) *is not empty if and only if* $c_{SC}(N) \leq \min\{\sum_{j=1}^{m} K_j \gamma_j : a'\zeta \geq b,$ *for all inequalities such that* $a'\zeta \geq d'y$ *is valid for cone* $(F^{\zeta y})$ *and* $\sum_{i=1}^{n} d_i = b\}$.

Further, the applicability of this result depends on the availability of obtaining efficient polyhedral descriptions of *cone* $(F^{\zeta y})$ which may not be easy. In any case, any partial description of that set allows to give approximate allocations that differ from actual core allocations an amount given by, at most, the gap between the problem (59) and the approximated version that we are able to solve.

6 Minimum Diameter and Radius Games

In this section we consider two different problems in the field of continuous location, namely finding the Steiner diameter and the radius of sets of points in metric spaces. These two problems are closely related and they have been revisited a number of times in the specialized literature [10, 18, 19, 25, 26]. In our analysis, we are interested in their cost sharing aspects. This perspective has been already studied in [39, 40]. For this reason, our presentation, in this section, follows the material in these two papers.

Let X be a metric space with distance function d and let $N_0 = \{a_0, a_1, \ldots, a_n\}$ be a finite set of points in X. The subset $N = \{a_1, \ldots, a_n\}$ is identified as the set of n players, and we refer to these points as *existing facilities*, or *demand points*. There is also a distinguished point a_0, representing the location of a server that provides service to the players, that can be viewed as an essential element in the system, e.g., each demand point must have access to a_0. Note that a_0 is not a player.

Given a finite subset of points $Y \subseteq X$, its diameter $D(Y)$, is defined by

$$D(Y) = \max_{y_1, y_2 \in Y} d(y_1, y_2).$$

A pair of points $y_1, y_2 \in Y$, satisfying $D(Y) = d(y_1, y_2)$ is called a *diametrical pair*. The radius of Y is defined by

$$R(Y) = \inf_{x \in X} \max_{y \in Y} d(x, y).$$

A point $x \in X$ satisfying $R(Y) = \max_{y \in Y} d(x, y)$ is called a *1-center of Y*. Note that by the triangle inequality

$$R(Y) \leq D(Y) \leq 2R(Y). \tag{61}$$

We now formally define the class of cooperative cost games based on the above facility location problems.

The first game is the Minimum Diameter Location Game (MDLG), (N, c_D), with respect to the metric space X and the set of points N_0 (see [40]). Its characteristic function is defined by

$$c_D(S) = D(S \cup \{a_0\}).$$

The second game is the Minimum Radius Location Game (MRLG), (N, c_R), with respect to the metric space X and the set of points N_0 (see [39]). Its characteristic function is defined by

$$c_R(S) = 2R(S \cup \{a_0\}).$$

(The factor 2 in the above definition is used for convenience and comparison purposes only.)

It directly follows from the definitions that both games are monotone. Also, from (61), for any $S \subseteq N$,

$$c_D(S) \leq c_R(S) \leq 2c_D(S).$$

Next, we recall the concept of network metric space induced by a connected undirected graph and its positive edge lengths. Suppose $G = (V, E)$ is a connected undirected graph with positive edge lengths $\{l_e\}$, $e \in E$, where $V = \{a_0, a_1, \ldots, a_n\}$. When $e = (a_i, a_j)$, we will also use the notation $l(a_i, a_j) = l_e$. Each edge in E is assumed to be rectifiable. We refer to interior points on an edge by their distances (along the edge) from the two nodes of the edge. $A(G)$ is the continuum set of points on the edges of G. For any pair of points $x, y \in A(G)$, we let $d(x, y)$ denote the length of a shortest path in $A(G)$ connecting x and y. We refer to $A(G)$ as the metric space induced by G and the edge lengths.

We first prove, following [40], that $C(N, c_D)$ is nonempty.

Theorem 4 *Given a graph $G = (V, E)$, and a subset $N \subseteq V \setminus \{a_0\}$, let (N, c_D) be the minimum diameter location game, defined over $A(G)$. Then, there is an extreme point of the core $C(N, c_D)$, which has at most two positive components.*

Proof Let $a_i, a_j \in N \cup \{a_0\}$ such that $c_D(N) = d(a_i, a_j)$.

If $a_j = a_0$, define the allocation x' by setting $x'_i = c_D(N) = d(a_i, a_0)$, and $x'_k = 0$, for any $k \neq i$. It is easy to see that x' is in the core since for each coalition S such that $a_i \in S$, we have $x'(S) = x'_i = d(a_i, a_0) \leq c_D(S)$.

Next suppose that $a_i \neq a_0$ and $a_j \neq a_0$. We present two extreme points of $C(N, c_D)$. First, define the allocation x' by setting $x'_i = d(a_i, a_0)$, $x'_j = c_D(N) - d(a_i, a_0)$, and $x'_k = 0$ for any $k \neq i, j$. Note that by the triangle inequality, $x'_j \leq d(a_j, a_0) = c_D(\{a_j\})$.

Then, $x'(S) = c_D(N) = d(a_i, a_j) \leq c_D(S)$, for each coalition S, satisfying $i, j \in S$. Also, $x'(N) = c_D(N)$. If $a_i \in S$ and $a_j \notin S$, then $x'(S) = x'_i = d(a_i, a_0) \leq c_D(S)$. Similarly, if $a_j \in S$ and $a_i \notin S$, then $x'(S) = x'_j \leq d(a_j, a_0) \leq c_D(S)$.

A second extreme point of $C(N, c_D)$, x'', is similarly defined by setting, $x''_j = d(a_j, a_0)$, $x''_i = c_D(N) - d(a_j, a_0)$, and $x''_k = 0$ for any $k \neq i, j$. This concludes the proof. $\qquad\square$

In spite of the facts that $C(N, c_D)$ is nonempty and that $c_D(S)$ can efficiently be computed for any coalition S, it is known that testing membership in the core for a given vector x is NP-hard for general graphs [40]. Note that the latter task amounts to testing whether $\min_{S \subseteq N}(c_D(S) - x(S)) \geq 0$.

Formally, given an MDLG with an underlying graph $G = (V, E)$ with positive edge weights, and an allocation vector x, the *core membership decision problem* is to determine whether x is not in the core $C(N, c_D)$.

Theorem 5 *The core membership decision problem is NP-hard even when $G = (V, E)$ is a complete graph, $N = V \setminus \{a_0\}$, the edge lengths satisfy the triangle inequality, and x distributes the total cost $c_D(N)$ equally.*

Proof We formulate the independent set problem [13] as an instance of the core membership decision problem. An instance of the NP-Complete independent set problem is an undirected graph $G_1 = (V_1, E_1)$ and an integer k, and the decision problem is whether G_1 has an independent set (i.e., a set of nodes such that no pair of them are adjacent) of size greater than k. Without loss of generality we may assume that $|V_1|$ is even and $k = |V_1|/2$. (If $k \leq |V_1|/2$, add $|V_1| - 2k$ isolated nodes to G_1. If $k > |V_1|/2$, add a clique with $2k - |V_1|$ to G_1.)

Let $G_1 = (V_1, E_1)$ be an undirected graph with $V_1 = \{a_1, \ldots, a_n\}$. Let $G_2 = (V_1, E_2)$ be the complete graph with node set V_1. Associate a positive length with each edge of E_2 as follows: If $e \in E_1$ then set the length of e to be equal to n. If $e \notin E_1$ then set the length of e to be equal to $n/2$. Let $G_3 = (V_1 \cup \{a_0\}, E_3)$ be the graph obtained from G_2 by adding the node a_0 and the n edges connecting a_0 to the n nodes in V_1. The length of each one of these n edges is set to be equal to $n/2$. Note that G_3 is a complete graph with $n + 1$ nodes, and its edges satisfy the triangle inequality.

Next, set $N = V_1$ and consider the game (N, c_D), defined on $A(G_3)$. In order to prove our claim, we will show that $x = (1, \ldots, 1)$ is not in $C(N, c_D)$ if and only if the graph G_1 has an independent set of cardinality greater than $n/2$. We assume without loss of generality that E_1 is nonempty, and therefore $c_D(N) = n$.

First note that $c_D(S) \in \{n, n/2\}$ for any $S \subseteq N$. Also, $c_D(N) = n = \sum_{j=1}^{n} x_j$.

Suppose that G_1 has an independent set S with $|S| > n/2$. Then, by definition $c_D(S) = n/2 < |S| = \sum_{a_j \in S} x_j = x(S)$, and therefore $x \notin C(N, c_D)$.

Next suppose that there is a subset $S \subseteq N$ such that $c_D(S) < x(S) = \sum_{a_j \in S} x_j = |S| \leq n$. Therefore, $c_D(S) = n/2$, and $|S| > n/2$. In particular, the subgraph induced by S has its diameter equal to $n/2$. By the definition of the edge lengths, S is an independent set of G_1 (otherwise there would exist a pair $a_i, a_j \in S$ with $d(a_i, a_j) = n$). Since $|S| > n/2$, the result is proven. $\qquad\square$

In view of the above result it is unlikely that there is a formulation of $C(N, c_D)$ involving only a polynomial number of linear constraints. In [40] the authors present

Fig. 1 Graph in Example 2

an efficient representation of $C(N, c_D)$ for the class of minimum diameter games defined on tree graphs.

In contrast, in the following we show that $C(N, c_R)$, the core of the MRLG, can be empty. In view of this result we will show that for several metric spaces the core, which by definition is a polyhedral set in $\mathbb{R}^n$, is nonempty and/or has a polyhedral representation by $O(n^c)$ linear inequalities (c is independent of the number of players n, and depends only on some parameters of the space X.) Such a representation is usually called *efficient* or *compact*.

We have already noted that by definition the characteristic function c_R is monotone. However, when the metric space X is discrete, i.e., $|X|$ is finite, the radius location game, (N, c_R) may not be subadditive. As a result players may have no incentive to cooperate and the core can be empty, as shown in the next example also taken from [39]

Example 2 Consider a 5-node path with edge set $E = \{(a_1, a_2), (a_2, a_0), (a_0, a_3), (a_3, a_4)\}$. The respective edge lengths are 1, 1, 2 and 2, as shown in Fig. 1.

The finite (discrete) space X consists of the 5 nodes (points) with the distance function induced by the edge lengths. X can also be viewed as a set of 5 points on the real line. Consider first the 2-player game on X defined by $N = \{a_1, a_4\}$. It is not subadditive since $c_R(\{a_1, a_4\}) > c_R(\{a_1\}) + c_R(\{a_4\})$.

The above example can easily be modified to show that subadditivity may not hold even for complete discrete games, i.e., when $N = X \setminus \{a_0\}$. Specifically, consider the complete 4-player radius game defined on the above set X, and let $N = X \setminus \{a_0\} = \{a_1, a_2, a_3, a_4\}$.

The smallest discrete neighborhood covering all nodes has radius 4, while the smallest (discrete) neighborhoods covering $\{a_1, a_2, a_0\}$ and $\{a_3, a_4, a_0\}$ have radii 1 and 2, respectively. Hence, $c_R(\{a_1, a_2, a_3, a_4\}) = 8, c_R(\{a_1, a_2\}) = 2, c_R(\{a_3, a_4\}) = 4$, and therefore $c_R(\{a_1, a_2, a_3, a_4\}) > c_R(\{a_1, a_2\}) + c_R(\{a_3, a_4\})$.

It is easy to check that unlike the above 2-player radius game defined on a discrete metric space, every complete 2-player game, defined on a 3 point discrete metric space has a nonempty core.

When the metric space X consists of a continuum set of points $C(N, c_R)$ can also be empty for a 3-player game, as illustrated by the next example of a network metric space $A(G)$. This example corresponds to a very simple geometric planar road network, where the edges are line segments and their lengths are the respective Euclidean distances.

Example 3 Consider the graph $G = (V, E)$ where $V = \{a_0, a_1, \ldots, a_6\}$ and $E = \{(a_0, a_4), (a_0, a_5), (a_0, a_6), (a_1, a_4), (a_1, a_6), (a_2, a_4), (a_2, a_5), (a_3, a_5), (a_3, a_6)\}$. All edges are of unit length, see Fig. 2.

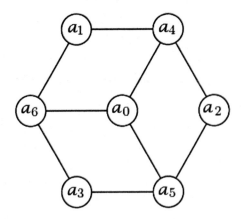

Fig. 2 Graph in Example 3

Set $X = A(G)$. Consider the game (N, c_R), defined on X, with $N_0 = \{a_0, a_1, a_2, a_3\}$ and $N = \{a_1, a_2, a_3\}$. It is easy to check that for each coalition $S \subseteq N$ with $|S| \leq 2$ we have $c_R(S) = 2$, and $c_R(N) = 4$. The following table summarizes these results. By symmetry, if the core was not empty the symmetric

S	Center	Radius
$\{a_1, a_2\}$	a_4	1
$\{a_1\}$	a_4, a_6	1
$\{a_1, a_2, a_3\}$	a_1, a_2, a_3, a_0	2

allocation $x = (4/3, 4/3, 4/3)$ would be in the core contradicting the constraint $x_1 + x_2 \leq c_R(\{a_1, a_2\}) = 2$.

For any metric space X, the definition of c_R ensures the monotonicity of the game (N, c_R), whereas subadditivity is proved in the next proposition, under the following continuity assumption:

Definition 1 Let X be a metric space such that for any pair of points $x, y \in X$, and a real $0 \leq \alpha \leq 1$, there is a point $z \in X$ such that $d(x, z) + d(z, y) = d(x, y)$ and $d(x, z) = \alpha d(x, y)$. Then X is called a "geodesic metric space", [33].

Proposition 6 *If X is a geodesic metric space, then the radius game (N, c_R) over X is subadditive.*

Proof Consider a pair of coalitions, S_1 and S_2. We need to show that

$$c_R(S_1 \cup S_2) \leq c_R(S_1) + c_R(S_2).$$

For $j = 1, 2$, let c_j and r_j be the 1-center and 1-radius of the smallest ball enclosing the points in $S_j \cup \{a_0\}$, respectively.

Let $P(c_1, c_2)$ be a shortest path in X, connecting c_1 and c_2. Let $d(c_1, c_2)$ denote the length of $P(c_1, c_2)$. Then, $d(c_1, c_2) \leq d(c_1, a_0) + d(a_0, c_2) \leq r_1 + r_2$.

Suppose without loss of generality that $r_2 \geq r_1$. If $r_2 \geq r_1 + d(c_1, c_2)$, then a center established at c_2 will ensure a covering radius of r_2 to all nodes in $S_1 \cup S_2 \cup \{a_0\}$. Hence, $c_R(S_1 \cup S_2) \leq 2r_2 = c_R(S_2)$.

If $r_1 \leq r_2 \leq r_1 + d(c_1, c_2)$, then consider a center established at the point c^*, such that $d(c_1, c^*) = (d(c_1, c_2) + r_2 - r_1)/2$, and $d(c_2, c^*) = (d(c_1, c_2) - r_2 + r_1)/2$. It is easy to check that this center will ensure a covering radius of $(d(c_1, c_2) + r_1 + r_2)/2 \leq r_1 + r_2$ to all nodes in $S_1 \cup S_2 \cup \{a_0\}$. (Note that v_0 is in the intersection of the smallest balls enclosing $S_1 \cup \{a_0\}$ and $S_2 \cup \{a_0\}$.) Therefore, $c_R(S_1 \cup S_2) \leq c_R(S_1) + c_R(S_2)$. $\qquad\square$

6.1 ℓ_p Metric Spaces over $\mathbb{R}^q$

In this section we focus on the case in which the MRLG (N, c_R) is defined on the ℓ_p metric space over $\mathbb{R}^q$. Again, we let $N_0 = V = \{a_0, a_1, \ldots, a_n\}$ be a set of points in $\mathbb{R}^q$, and set $N = V \setminus \{a_0\}$.

The following examples, borrowed from [39], show that in general the MRLG is not submodular, and that with the exception of the case $p = \infty$, $c_D(N) = D(V) \neq 2R(V) = c_R(N)$. Hence, the existence of core allocations is not clear in the case where $p \neq \infty$.

Example 4 Consider the planar ℓ_p normed case with $V = \{a_0, a_1, a_2, a_3\}$, where, $a_0 = (0, 0)$, $a_1 = (0, 1)$, $a_2 = (1, 0)$ and $a_3 = (-1, 0)$.

We have $c_R(\{a_1, a_2, a_3\}) = 2$, $c_R(\{a_1\}) = 1$, and $c_R(\{a_1, a_2\}) = c_R(\{a_1, a_3\}) = 2^{1/p}$. Thus, c_R is not submodular in this example for any p such that $2^{1/p} < 3/2$, which in particular applies to $2 \leq p \leq \infty$.

Example 5 Consider the planar ℓ_1 case with $V = \{a_0, a_1, a_2, a_3\}$, where, $a_0 = (0, 0)$, $a_1 = (1, -1)$, $a_2 = (1, 1)$ and $a_3 = (-1, -1)$. We have $c_R(\{a_1, a_2, a_3\}) = 4$, $c_R(\{a_1\}) = 2$, and $c_R(\{a_1, a_2\}) = c_R(\{a_1, a_3\}) = 2$. Thus, c_R is not submodular in this case.

The next two examples show that for any $1 < p < \infty$ in the planar case, and for the rectilinear norm ℓ_1, even in $\mathbb{R}^3$, $c_R(N) = 2R(N \cup \{a_0\})$ can be strictly larger than $c_D(N) = D(N \cup \{a_0\})$. (In $\mathbb{R}^2$ the ℓ_1 norm is equivalent to the ℓ_∞ norm.)

Example 6 Consider the set of points $V = \{a_0, a_1, a_2, a_3\}$ where $a_1 = (a, b)$, $a_2 = (-a, b)$, $a_3 = (0, -1)$, and $a_0 = (0, 0)$. For $1 < p < \infty$, let $a = b = 2^{-1/p}$. Then, the ℓ_p diameter of V is $(a^p + (b + 1)^p)^{1/p}$ whereas the ℓ_1 radius is 1 and the 1-center is $(0, 0)$. Hence, $c_D(N) = D(V) < b + 1 < 2 = 2R(V) = c_R(N)$.

Example 7 Consider the set of points $V = \{a_0, a_1, a_2, a_3\}$ where $a_1 = (1, 1, 1)$, $a_2 = (-1, -1, 1)$, $a_3 = (-1, 1, -1)$, and $a_0 = (1, -1, -1)$. The ℓ_1 diameter of V is 4 whereas the ℓ_1 radius is 3 and the 1-center is $(0, 0, 0)$. Hence, $c_D(N) = D(V) < 2R(V) = c_R(N)$.

Next, we show that for any $p \geq 1$, the core of the game (N, c_R), defined on the ℓ_p metric space over $\mathbb{R}^q$, can be represented as a set described by a polynomial number of linear inequalities, for any fixed q.

Consider first the case where $1 < p < \infty$.

Theorem 6 *Let $1 < p < \infty$, and consider the game (N, c_R), defined on the ℓ_p metric space over $\mathbb{R}^q$. Let $\{S_j\}$, $j \in J$, be the collection of all subsets $S \subseteq N$ with $|S| \leq q + 1$. For each $j \in J$, let $B(S_j)$, be the smallest enclosing ball containing $S_j \cup \{a_0\}$, and let S'_j be the subset of all points in N, contained in $B(S_j)$. Then the core of the game is given by,*

$$C(N, c_R) = \{x \in \mathbb{R}^n_+ : x(S'_j) \leq c_R(S_j), \ \forall j \in J, \ and \ x(N) = c_R(N)\}.$$

Proof For any subset $S \subseteq N$, $c_R(S)$ is the diameter of $B(S)$, a smallest enclosing ball containing $S \cup \{a_0\}$. (Since $1 < p < \infty$, $B(S)$ is unique, [45].)

By the Helly property there is a subset $S_j \subseteq S$, $j \in J$, such that $c_R(S) = c_R(S_j)$. Then, by definition $S \subseteq S'_j$. Moreover, by the monotonicity of the game each vector in the core is nonnegative, and therefore $x(S) \leq x(S'_j)$. Hence, the constraint $x(S) \leq c_R(S)$ is dominated by the constraint $x(S'_j) \leq c_R(S_j)$. This completes the proof. $\square$

Next, consider the case where $p = \infty$. As above, let $\{S_j\}, j \in J$, be the collection of all subsets $S \subseteq N$ with $|S| \leq q + 1$.

Theorem 7 *Consider the game (N, c_R), defined on the ℓ_∞ metric space over $\mathbb{R}^q$. Then there is a collection of subsets of N, $\{S^\infty_j(k)\}, j \in J, k = 1, \ldots, c^\infty_j(n, q)$, such that $c^\infty_j(n, q) = O(2^q n^{(q-1)})$, and the core of the game is given by,*

$$C(N, c_R) = \{x \in \mathbb{R}^n_+ : x(S^\infty_j(k)) \leq c_R(S_j), \ \forall j \in J, k = 1, \ldots, c^\infty_j(n, q)$$

$$and \ x(N) = c_R(N)\}.$$

Proof For each subset S the problem of finding the smallest ℓ_∞ ball enclosing S is reduced to finding a smallest hypercube containing S. Such a hypercube is not unique. The set of centers of all optimal hypercubes is itself a hypercube of dimension less than or equal to $q - 1$. For $j \in J$ consider an optimal hypercube $H(S_j)$ enclosing $S_j \cup \{a_0\}$ and let $P(H(S_j))$ be the maximal subset of N, contained in $H(S_j)$. We can shift $H(S_j)$ along the axes and obtain an optimal hypercube $H'(S_j)$ such that $P(H'(S_j)) = P(H(S_j))$, and for each coordinate $i = 1, \ldots, d$, one of the two faces of $H'(S_j)$ corresponding to the ith coordinate contains a point in N. Thus, there is only $c^\infty_j(n, q) = O(2^q n^{(q-1)})$ such maximal subsets of N, associated with a given subset $S_j, j \in J$. Denote this collection of subsets by $\{S^\infty_j(k)\}, k = 1, \ldots, c^\infty_j(n, q)$.

Using the monotonicity of the game and following the arguments used in the previous proof, we observe that for each subset $S \subseteq N$, there is a subset $S_j, j \in J$, and $k = 1, \ldots, c^\infty_j(n, q)$, such that the constraint $x(S) \leq c_R(S)$, is dominated by the constraint $x(S^\infty_j(k)) \leq c_R(S_j)$. This completes the proof. $\square$

A similar analysis applies to the rectilinear case when $p = 1$ [39].

Theorem 8 *Consider the game* (N, c_R), *defined on the* ℓ_1 *metric space over* $\mathbb{R}^q$. *Then there is a collection of subsets of* N, $\{S_j^1(k)\}$, $j \in J$, $k = 1, \ldots, c_j^1(n, q)$, *such that* $c_j^1(n, q) = O(2^{q^2} n^{q-1})$, *and the core of the game is given by,*

$$C(N, c_R) = \{x \in \mathbb{R}_+^n : x(S_j^1(k)) \le c_R(S_j), \ \forall j \in J, k = 1, \ldots, c_j^1(n, q)$$

$$and \ x(N) = c_R(N)\}.$$

With the exception of the case $p = \infty$, we do not know yet whether $C(N, c_R)$ is nonempty for all ℓ_p metric spaces over $\mathbb{R}^q$. We assume without loss of generality that $a_i \ne a_0$ for all $i = 1, \ldots, n$.

Theorem 9 *The core* $C(N, c_R)$ *of the game* (N, c_R), *defined on the* ℓ_∞ *metric space over* $\mathbb{R}^q$, *is nonempty. Specifically,* $C(N, c_D) = C(N, c_R)$.

Moreover, if $D(N_0) = d(a_0, a_j)$, *for some* $a_j \in N$, *the dimension of* $C(N, c_R)$ *is* $n - 1$, *and there is* $x^* \in C(N, c_R)$ *such that* $x_t^* > 0$, *for any* $a_t \in N$. *Also, if* $D(N_0) = d(a_i, a_j)$, *for some* $a_i, a_j \in N$, *and* $d(a_i, a_j) < d(a_i, a_0) + d(a_j, a_0)$, *then the dimension of* $C(N, c_R)$ *is* $n - 1$, *and there is* $x^* \in C(N, c_R)$ *such that* $x_t^* > 0$, *for any* $a_t \in N$.

Proof When $p = \infty$, it is easy to see that for any set S we have $c_D(S) = D(S \cup \{a_0\}) = 2R(S \cup \{a_0\}) = c_R(S)$. Thus, $C(N, c_D) = C(N, c_R)$, and the nonemptiness of the core follows from the fact that $C(N, c_D) \ne \emptyset$.

Suppose without loss of generality that $D(N_0) = d(a_0, a_1)$. Let $\alpha = (\alpha_1, \alpha_2, \ldots, \alpha_n)$ be an arbitrary real vector satisfying $0 \le \alpha_1 \le \min_{t=1,\ldots,n} d(a_0, a_t)$, $\alpha_1 = \sum_{j=2}^n \alpha_j$ and $\alpha_j \ge 0, j = 2, \ldots, n$.

We show that the allocation $x^\alpha = (d(a_0, a_1) - \alpha_1, \alpha_2, \ldots, \alpha_n)$ is in $C(N, c_R)$. First, by definition $x^\alpha(N) = d(a_0, a_1) = D(N_0) = c_R(N)$. Next consider a coalition $S \subseteq N$. If $a_1 \in S$, then $x^\alpha(S) \le d(a_0, a_1) \le c_R(S)$. If $a_1 \ne S$, then $x^\alpha(S) \le \alpha_1 \le \min_{t=1,\ldots,n} d(a_0, a_t) \le c_R(S)$.

To see that the affine dimension of $C(N, c_R)$ in this case is $n-1$, we show that the core contains n independent vectors. One of them is the vector $x^1 = (x_1^1, \ldots, x_n^1)$, defined by $x_1^1 = d(a_0, a_1)$ and $x_j^1 = 0$ for $j = 2, 3, \ldots, n$. The other $n-1$ vectors are defined as follows:

Let ϵ be a sufficiently small positive real, and consider the $n-1$ independent core allocations $\{x^{\alpha(\ell)}\}$, $\ell = 2, \ldots, n$, where $\alpha(\ell)$ is the vector defined by $\alpha_1(\ell) = \epsilon, \alpha_\ell(\ell) = \epsilon$, and $\alpha_t(\ell) = 0$, for any $t = 2, \ldots, n; t \ne \ell$. The allocation $x^* = \sum_{\ell=2}^n x^{\alpha(\ell)}/(n-1)$ is in the core and has strictly positive components.

Next, suppose without loss of generality that $D(N_0) = d(a_1, a_2)$ and $d(a_1, a_2) < d(a_0, a_1) + d(a_0, a_2)$. Let δ_1, δ_2 be a pair of positive reals satisfying $0 < \delta_1 < d(a_0, a_1), 0 < \delta_2 < d(a_0, a_2)$, and $\delta_1 + \delta_2 = d(a_1, a_0) + d(a_2, a_0) - d(a_1, a_2)$.

Let $\alpha = (\alpha_1, \alpha_2, \ldots, \alpha_n)$ be an arbitrary real vector satisfying $\alpha_1 \le d(a_1, a_0) - \delta_1$, $\alpha_2 \le d(a_2, a_0) - \delta_2, 0 \le \alpha_1 + \alpha_2 \le \min_{t=1,\ldots,n} d(a_0, a_t), 0 \le \alpha_1 + \alpha_2 \le \min\{\delta_1, \delta_2\}$, $\alpha_1 + \alpha_2 = \sum_{j=3}^n \alpha_j$ and $\alpha_j \ge 0, j = 1, \ldots, n$.

We show that the allocation

$$x^\alpha = (d(a_0, a_1) - \delta_1 - \alpha_1, d(a_0, a_2) - \delta_2 - \alpha_2, \alpha_3, \ldots, \alpha_n)$$

is in $C(N, c_R)$. First, by definition $x^\alpha(N) = d(a_1, a_2) = D(N_0) = c_R(N)$. Next consider a coalition $S \subseteq N$. If $a_1, a_2 \in S$, then $x^\alpha(S) \leq d(a_1, a_2) = c_R(S)$. If $a_1 \in S, a_2 \neq S$, then $x^\alpha(S) \leq d(a_1, a_0) - \delta_1 - \alpha_1 + \sum_{\ell=3}^{n} \alpha_\ell \leq d(a_1, a_0) - \alpha_2 - \alpha_1 + \sum_{\ell=3}^{n} \alpha_\ell \leq d(a_1, a_0) \leq c_R(S)$. Similarly, if $a_1 \neq S, a_2 \in S$, we obtain $x^\alpha(S) \leq d(a_2, a_0) \leq c_R(S)$. Finally, suppose that $a_1, a_2 \neq S$. Then, $x^\alpha(S) \leq \alpha_1 + \alpha_2 \leq \min_{t=1,\ldots,n} d(a_0, a_t) \leq c_R(S)$.

To see that the affine dimension of $C(N, c_R)$ in this case is $n - 1$, we show that the core contains n independent vectors. Two of them are the vectors $x^1 = (x_1^1, \ldots, x_n^1)$, and $x^2 = (x_1^2, \ldots, x_n^2)$, defined by $x_1^1 = d(a_1, a_0), x_2^1 = d(a_1, a_2) - d(a_1, a_0), x_j^1 = 0$ for $j = 3, 4, \ldots, n$, $x_1^2 = d(a_1, a_2) - d(a_2, a_0), x_2^2 = d(a_2, a_0)$, and $x_j^2 = 0$ for $j = 3, 4, \ldots, n$. The other $n - 2$ vectors are defined as follows:

Let ϵ be a sufficiently small positive real, and consider the collection of $n - 2$ independent core allocations $\{x^{\alpha(\ell)}\}$, $\ell = 3, \ldots, n$, where $\alpha(\ell)$ is the vector defined by $\alpha_1(\ell) = \epsilon, \alpha_\ell(\ell) = \epsilon$, and $\alpha_t(\ell) = 0$, for any $t = 2, \ldots, n; t \neq \ell$. The allocation $x^* = (x^1 + x^2 + \sum_{\ell=3}^{n} x^{\alpha(\ell)})/n$ is in the core and has strictly positive components. This completes the proof. □

Elaborating on the result in the last theorem, the next example illustrates that when the conditions in the theorem are not satisfied, the dimension of the core can even be zero. Specifically, for any number of players, even in the ℓ_∞ planar case, the core can be a singleton where only two players share the total cost, in spite of the fact that the distance from each player to the server a_0 is positive. (We also note in passing that in the ℓ_∞ case, if $c_R(\{a_i, a_j\}) < c_R(\{a_i\}) + c_R(\{a_j\}) = d(a_i, a_0) + d(a_j, a_0)$, for any pair of distinct players in N, then there is a core allocation where any player which is at a positive distance from a_0 pays a positive amount.)

Example 8 Consider the set of points $N_0 = \{a_0, a_1, \ldots, a_k\}$ where $a_0 = (0,0), a_1 = (0,1), a_2 = (0,-1), a_3 = (1,0)$ and $a_i = (x_i, 0), 0 < x_i < 1$, for $i = 4, 5, \ldots, k$. Since $c_D(S) = 2$, if $\{a_1, a_2\} \subseteq N$, and $c_D(S) \leq 1$, otherwise, it is easy to see that $C(N, c_D) = C(N, c_R) = \{(1, 1, 0, \ldots, 0)\}$.

Corollary 2 *The core of the game (N, c_R), defined on the ℓ_1 metric plane is nonempty. Specifically, $C(N, c_D) = C(N, c_R)$.*

The next example shows that even in $\mathbb{R}^3$, $c_R(N) = 2R(N \cup \{a_0\})$ can be strictly larger than $c_D(N) = D(N \cup \{a_0\})$ for the rectilinear norm.

Example 9 Consider the set of points $V = \{a_0, \ldots, a_3\}$ where $a_1 = (1, 1, 1), a_2 = (-1, -1, 1), a_3 = (-1, 1, -1)$, and $a_0 = (1, -1, -1)$. The diameter of V is 4 whereas the radius is 3 and the 1-center is $(0, 0, 0)$. Hence, $D(V) < 2R(V)$.

It is not known whether the core of the radius location game, $C(N, c_R)$, is nonempty for the rectilinear case on $\mathbb{R}^q$ for $q \geq 3$. Nevertheless, for any fixed q the nonemptiness of the core can be tested in polynomial time since the core can be represented by a polynomial number of constraints as follows. If $S \subseteq N$ then by the Helly property there is a subset $S' \subseteq S$, $|S'| \leq q + 1$ such that $c_R(S) = c_R(S')$. Therefore

$$C(N, c_R) = \{x \in \mathbb{R}_+^n : x(S) \leq c_R(S), \forall S \subseteq N, |S| \leq q + 1, \text{ and } x(N) = c_R(N)\}.$$

The above compact representation also holds for the space $\mathbb{R}^q$ augmented with any metric induced by a gauge with a symmetric unit ball (norm).

6.1.1 Euclidean Spaces

In the particular case of $p = 2$ which is the Euclidean model, in general, the equality $c_D(N) = c_R(N)$ may not hold even in the planar case. From Proposition 6 it follows that the characteristic function $c_R(S)$ is subadditive also for the Euclidean model. However, it does not follow from the general analysis in previous sections that the core of the Euclidean planar game is nonempty.

In spite of that, it is known that $C(N, c_R)$ is nonempty for the Euclidean planar case [39]. More specifically, there is a core allocation where at most 3 players (points) pay positive amounts. These are points defining $C(V)$, the minimal circle in the plane enclosing the set V.

Theorem 10 *The core $C(N, c_R)$ of the minimal radius location game (N, c_R) in the Euclidean planar case is non-empty.*

Proof See [39]. $\blacksquare$

According to [39] the proof of the above theorem is based on a rather long case analysis. It is still unclear whether a shorter and more elegant proof exists and it is applicable to any dimension $q > 2$.

In the minimum radius location game (N, c_R), for each coalition S, c_R is defined as twice the solution value to the 1-center problem for the set of nodes $S \cup \{a_0\}$. Similarly we can consider location games defined by other common optimization criteria often used in facility location models. For example, consider the minimum ordered median location game, (N, c_{OM}), where for each coalition S, c_{OM} is defined as the solution value to the single facility ordered median problem for the set of points $S \cup \{a_0\}$.

We note that from the cooperative point of view the above definition does not even induce the desirable property of subadditivity. Thus, players may not even have the incentive to cooperate. This can be enforced by introducing set up costs as in Sect. 4.

Acknowledgements The research of the author has been partially supported by the Spanish Ministry of Economy and Competitiveness through grants MTM2013-46962-C02-01 and MTM2016-74983-C02-01 (MINECO/FEDER). This support is gratefully acknowledged.

References

1. Blanco, V., Puerto, J., El-Haj Ben-Ali, S.: Revisiting several problems and algorithms in continuous location with l_τ norms. Comput. Optim. Appl. **58**(3), 563–595 (2014)
2. Blanco, V., Puerto, J., El-Haj Ben-Ali, S.: Continuous multifacility ordered median location problems. Eur. J. Oper. Res. **250**(1), 56–64 (2016)
3. Borm, P., Hamers, H., Hendrickx, R.: Operations research games: a survey. TOP **9**(2), 139–216 (2001)

4. Caprara, A., Letchford, A.N.: New techniques for cost sharing in combinatorial optimization games. Math. Program. **124**(1–2), 93–118 (2010)
5. Curiel, I.: Cooperative Game Theory and Applications. Kluwer Academic, Dordrecht (1997)
6. Daskin, M.S.: Network and Discrete Location: Models, Algorithms, and Applications, 2nd edn. Wiley, Hoboken (2013)
7. Drezner, Z.: Facility Location: A Survey of Applications and Method. Springer, New York (1995)
8. Drezner, Z., Hamacher, H.: Facility Location: A Survey of Applications and Theory. Springer, New York (2002)
9. Eiselt H.A., Laporte, G., Thisse, J.-F.: Competitive location models: a framework and bibliography. Transp. Sci. **27**(1), 44–54 (1993)
10. Elzinga, J., Hearn, D.W.: Geometrical solutions for some minimax location problems. Transp. Sci. **6**, 379–394 (1972)
11. Fragnelli, V., Gagliardo, S.: Open problems in cooperative location games. Int. Game Theory Rev. **15**(3), 1–13 (2013)
12. Gabszewicz, J., Thisse, J.: Location. In: Aumann, R.J., Hart, S. (eds.) Handbook of Game Theory with Economic Applications, vol. 1, Chap. 9, pp. 281–304. Elsevier, Amsterdam (1992)
13. Garey, M., Johnson, D.: Computers and Intractability. A Guide to the Theory of NP-Completeness. Freeman, San Francisco (1979)
14. Goemans, M., Skutella, M.: Cooperative facility location games. J. Algorithms **50**, 194–214 (2004)
15. Granot, D.: The role of cost allocation in locational models. Oper. Res. **35**, 234–248 (1987)
16. Hakimi, S.: Optimum locations of switching centers and the absolute centers and medians of a graph. Oper. Res. **12**(3), 450–459 (1964)
17. Hakimi, S.: Optimum distribution of switching centers in a communications network and some related graph theoretic problems. Oper. Res. **13**(3), 462–475 (1965)
18. Handler, G.Y.: Minimax location of a facility in an undirected tree graph. Transp. Sci. **7**, 287–293 (1973)
19. Hassin, R., Tamir, A.: On the minimum diameter spanning tree problem. Inf. Process. Lett. **53**(2), 109–111 (1995)
20. Hotelling, H.: Stability in competition. Econ. J. **39**, 41–57 (1929)
21. Laporte, G., Nickel, S., Saldanha, F.: Location Science. Springer, Cham (2015)
22. Larson, R., Odoni, A.: Urban Operations Research. Prentice-Hall, New York (1981)
23. Love, R.F., Morris, J.G., Wesolowsky, G.O.: Facilities Location: Models and Methods. North-Holland, Amsterdam (1988)
24. Mallozzi, L.: Cooperative games in facility location situations with regional fixed costs. Optim. Lett. **5**, 171–183 (2011)
25. Megiddo, N.: The weighted Euclidean 1-center problem. Math. Oper. Res. **8**, 498–504 (1983)
26. Megiddo, N., Zemel, E.: An $O(n\log n)$ randomizing algorithm for the weighted Euclidean 1-center problem. J. Algorithms **3**, 358–368 (1986)
27. Mirchandani, P., Francis, R.: Discrete Location Theory. Wiley, New York (1990)
28. Nickel, S., Puerto, J.: Location Theory: A Unified Approach. Springer, Berlin (2005)
29. Nickel, S., Puerto, J., Rodríguez-Chía, A.M.: An approach to location models involving sets as existing facilities. Math. Oper. Res. **28**(4), 693–715 (2003)
30. Nissan, N., Roughgargen, Tardos, E., Vazirzni, V.V. (eds.): Algorithmic Game Theory. Cambridge University Press, Cambridge (2007)
31. Owen, G.: Game Theory. Academic, San Diego (1995)
32. Owen, S.H., Daskin, M.S.: Strategic facility location: a review. Eur. J. Oper. Res. **111**(3), 423–447 (1998)
33. Papadopoulos, A.: Metric Spaces, Convexity and Nonpositive Curvature. European Mathematical Society, Zürich (2005)
34. Plastria, F.: Continuous location problems. In: Drezner, Z. (ed.) Facility Location - A Survey of Applications and Methods. Springer Series in Operations Research, pp. 225–260. Springer, New York (1995)

35. Puerto, J.: Lecturas en teoría de localización. Technical Report, Universidad de Sevilla. Secretariado de Publicaciones (1996)
36. Puerto, J., Rodríguez-Chía, A.M.: On the structure of the solution set for the single facility location problem with average distances. Math. Program. **128**, 373–401 (2011)
37. Puerto, J., Rodríguez-Chía, A.: Ordered median location problems. In: Laporte, G., Nickel, S., Saldanha da Gama, F. (eds.) Location Science, pp. 249–288. Springer, Heidelberg (2015)
38. Puerto, J., García-Jurado, I., Fernández, F.: On the core of a class of location games. Math. Meth. Oper. Res. **54**(3), 373–385 (2001)
39. Puerto, J., Tamir, A., Perea, F.: A cooperative location game based on the 1-center location problem. Eur. J. Oper. Res. **214**, 317–330 (2011)
40. Puerto, J., Tamir, A., Perea, F.: Cooperative location games based on the minimum diameter spanning Steiner subgraph problem. Discret. Appl. Math. **160**, 970–979 (2012)
41. Skorin-Kapov, D.: On cost allocation in hub-like networks. Ann. Oper. Res. **106**, 63–78 (2001)
42. Tamir, A.: On the core of network synthesis games. Math. Program. **50**, 123–135 (1991)
43. Tamir, A.: On the core of cost allocation games defined on locational problems. Transp. Sci. **27**, 81–86 (1992)
44. Tamir A., Puerto, J., Pérez-Brito, D.: The centdian subtree of a tree network. Discret. Appl. Math. **118**(3), 263–278 (2002)
45. Zurcher, S.: Smallest enclosing ball for a point set with strictly convex level sets. Masters thesis, Institute of Theoretical Computer Science, ETH Zurich (2007)